RECOVERY FROM DISASTER

Disasters can dominate newspaper headlines and fill our TV screens with relief appeals, but the complex long-term challenge of recovery – providing shelter, rebuilding safe dwellings, restoring livelihoods and shattered lives – generally fails to attract the attention of the public and most agencies. On average, 700 disasters occur each year. They affect more than 200 million people and cause US$166 trillion of damage. Climate change, population growth and urbanisation are likely to intensify the impact of natural disasters and add to reconstruction needs. *Recovery from Disaster* explores the field and provides a concise yet comprehensive source of knowledge for academics, planners, architects, engineers, construction managers, relief and development officials, and reconstruction planners involved with all sectors of recovery, including shelter and rebuilding. With almost 80 years of first-hand experience of disaster recovery between them, Ian Davis (an architect) and David Alexander (a geographer) draw substantially from their work in a variety of recovery situations in China, Haiti, India, Italy, Japan, New Zealand, Pakistan, the Philippines and the USA.

The volume is further enriched by two important and unique features. First, 21 models of disaster recovery are presented, seven of which were specifically developed for the book. The second feature is a survey of expert opinion about the nature of effective disaster recovery – the first of its kind. More than 50 responses are provided in full, along with an analysis that integrates them with their underpinning theories. By providing a framework and models for future study and application, Davis and Alexander seek both to advance the field and to provide a much-needed reference work for decision-makers. With a broad perspective derived from the authors' roles as university professors, researchers, trainers, consultants, NGO directors and advisors to governments and UN agencies, this comprehensive guide will be invaluable for practitioners and students of disaster management.

Ian Davis has specialised in shelter, reconstruction and disaster risk reduction since 1972. He completed a PhD at University College London on *Shelter after Earthquakes*, and in 1982, he edited the first UN guidelines on *Shelter after Disaster*. He has been instrumental in producing guidelines for the United Nations International Strategy for Disaster Reduction (UNISDR), the European Union and the World Bank. He is currently a Visiting Professor in Disaster Management at Lund, Oxford Brookes and Kyoto Universities.

David Alexander is a Professor at the Institute for Risk and Disaster Reduction, University College London. Since 1981, hazards and disasters have been his main focus of professional interest. He is the Editor-in-Chief of the *International Journal of Disaster Risk Reduction*. His previous books include *Natural Disasters, Confronting Catastrophe* and *Principles of Emergency Planning and Management*. He is Visiting Professor of Disaster Management at Lund, Northumbria and Bournemouth Universities.

ROUTLEDGE STUDIES IN HAZARDS, DISASTER RISK AND CLIMATE CHANGE

Series Editor: Ilan Kelman, Reader in Risk, Resilience and Global Health at the Institute for Risk and Disaster Reduction (IRDR) and the Institute for Global Health (IGH), University College London (UCL).

This series provides a forum for original and vibrant research. It offers contributions from each of these communities as well as innovative titles that examine the links between hazards, disasters and climate change to bring these schools of thought closer together. This series promotes interdisciplinary scholarly work that is empirically and theoretically informed, with titles reflecting the wealth of research being undertaken in these diverse and exciting fields.

Published:

Cultures and Disasters
Understanding cultural framings in disaster risk reduction
Edited by Fred Krüger, Greg Bankoff, Terry Cannon, Benedikt Orlowski and E. Lisa F. Schipper

Recovery from Disaster
Ian Davis and David Alexander

RECOVERY
FROM DISASTER

Ian Davis and David Alexander

Routledge
Taylor & Francis Group

LONDON AND NEW YORK

First published 2016
by Routledge
2 Park Square, Milton Park, Abingdon, Oxon OX14 4RN

and by Routledge
711 Third Avenue, New York, NY 10017

Routledge is an imprint of the Taylor & Francis Group, an informa business

© 2016 Ian Davis and David Alexander

British Library Cataloguing-in-Publication Data
A catalogue record for this book is available from the British Library

Library of Congress Cataloging in Publication Data
Davis, Ian, 1937-
 Recovery from disaster / Ian Davis and David Alexander.
 (Routledge studies in hazards, disaster risk and climate change)
 Includes bibliographical references and index.
 I. Alexander, David (David E.) II. Title.
 HV553.D379 2015
 363.34'8—dc23   2015013424

ISBN: 978-0-415-61168-8 (hbk)
ISBN: 978-0-415-61177-0 (pbk)
ISBN: 978-1-315-67980-8 (ebk)

Typeset in Bembo
by Keystroke, Station Road, Codsall, Wolverhampton

DEDICATION

This book is dedicated to all the displaced people and families who struggle against massive obstacles to obtain shelter and safe homes in strong communities after the trauma of disaster.

Ian's dedication: As this is likely to be my final book on disaster risk and recovery management, I would like also to dedicate it to my family who have put up with so many absences and enabled me to work in this demanding field for 43 years. Therefore, sincere gratitude to Judy, who died before the book we often discussed together materialised, and later to a wonderfully encouraging Gill. Gratitude must also be given to ever-supportive children Mandy, Caroline and Simon; and grandchildren Luke, Emily, Ben, Joseph, Ellen, Jude, Sammy and Finn.

I also wish to thank my colleague and co-author David for warm friendship, rich insights and support in our exciting joint venture to write this book. The long process included memorable times together righting the world's wrongs, as well as doing intermittent writing in L'Aquila, Davos, London, Devon and Oxford.

Finally, I give thanks to a loving God for the abundance of rich gifts that have been showered upon me and for the opportunities given to undertake this work.

CONTENTS

Appendix 1 Summary of models   **316**

**Appendix 2 Survey answers to the question: what in your
view are the most important aspects of a successful recovery
operation following a natural disaster?**   **318**

Appendix 3 Key books and websites on disaster recovery   **330**

FIGURES

TABLES

TABLES

PROLOGUE

Winter

Winter strips the persimmon of its leaves,
The bare branches laden with golden fruit
Bend low, like the apples of the garden
Where Hesperides chanted alluring
Songs of immortality.

But starlings peck the persimmons until
The bulbous fruit drops from sagging branches,
Smashes gelatinously on the tiles
Of the courtyard floor, sticky pieces bright
Amid the withered leaves.

At night a freezing wind surrounds our house,
A dull kinetic cage of creaks and groans,
Banging shutters, stealthy shifting objects,
While strings of red peppers dryly rattle
Against the wall.

Through the hills comes a sinuous ripple
That makes our house dance on its foundations.
The stones protest ominously and then
It is over and silence reigns once more
Amid the fear.

One hundred shocks come in a single day,
One by one perturb the frozen landscape,
And terrorise our neighbours in their homes;
Swarms of earthquakes, shoals of tremors, shaking
Us out of complacency.

2009:
In Aquila the night brought sudden death
And a chaos of shouts, dust and rubble,
Occult in the cold, unfathomable
Blackness; javelins of fear, blind running
Away from insidious terror.

2015:
Will this be our fate? Will the main shock come
In the night-time, unfeeling, unannounced,
Turn our home and sanctuary into
A death trap of collapsing masonry
With us inside?

Shocks continue, small and subtle movements,
While the wind blows strong and cold and the house
Flexes yet again. Uneasily we
Listen for vibrations and string our nerves
Taut as aeolian harps.

The persimmons are gone, starlings have left,
No trace is left of the sweet orange fruit.
Stirred by the wind, the branches of the tree
Whip and creak. Meekly we inherit this
Attrition of the winter earth.

May insidious fears not disturb us
In sleep or waking; but if disaster
Should surprise us yet, let us live, let us
Mend our lives, rebuild our shattered homes, and
As equal, not underdog,
Look upon the earthquake.

                                                    David Alexander

PREFACE

At 2.46 p.m. local time on Friday 11 March 2011, the largest tsunami recorded in historical times rolled across the coast of Iwate, Miyagi and Fukushima Prefectures in north-eastern Japan. The waves reached a maximum height of 40.5 metres and in many places crossed the shore more than 10 metres high. As a result, more than 430 square kilometres of coastal land were devastated, 18,500 people were dead or missing and 6,150 were injured, and 400,000 buildings were severely damaged or destroyed. The response by the Japanese authorities was rapid and efficient, yet so many things had been destroyed that it still took more than two years of hard work merely to collect and recycle the debris. The rebuilding of towns and cities effectively razed to the ground by the tsunami posed a remarkable series of challenges in planning, development, architecture, urban design, infrastructure protection, engineering and economics.

On 8 November 2013, Typhoon Haiyan made landfall in the Philippines, where it was given the local name Yolanda. According to some estimates, it was the strongest hurricane to have made landfall over the period for which scientific records are available. The dead and missing amounted to 7,329, and 28,689 people were injured. In Eastern Visayas Province, about 90 per cent of the city of Tacloban (population 220,000) was severely damaged and many people had their homes swept away by a storm surge that, at its maximum, was higher than 5 metres. In addition to encountering many of the same problems as those that resulted from the Japanese tsunami, the Philippines faced the challenge of alleviating extreme poverty among the survivors.

Kesennuma is a city of 73,400 inhabitants in the north-eastern Japanese prefecture of Miyagi. Here, the dead and missing in the tsunami exceeded 2,000 people, and large parts of the urban fabric were destroyed by the waves and pools of burning fuel. To stand in what had been the centre of town two years after the tsunami, when debris clearance was more or less complete, was to experience the rare sensation of disaster as *tabula rasa*, the 'scrubbed table', in which there is much to replace and almost nothing to restore. Standing on the shore of Tacloban months after the typhoon provoked a different feeling. Buildings had been swept away much as they had in Kesennuma. Indeed, in both places it was common to find only the concrete base plates of houses as all superstructure had been swept away by the waves. But while the survivors of the Japanese town had been moved to small but orderly prefabricated homes, many of the Philippine survivors had built shelters amid the rubble of their former homes, such that the shore was populated with a dense 'informal settlement' of small wooden buildings.

In Kesennuma, a large trawler, the *Kyotoku Maru No. 18*, was washed up and deposited well inland amid the ruins of the urban area. Five ships of similar size were beached together in Tacloban. But while the Japanese ship stood alone amid the sites of the razed houses until it was dismantled, a village of 'informal' houses grew up on the seaward side of the Philippine ships. In a sense, both situations drew attention to the impermanence of human settlement, but the contrast in vulnerabilities and resources available for recovery was notable.

The Japanese earthquake and tsunami of 2011 and the Southeast Asian typhoon of 2013 are examples of unusually large events of natural origin, but there is every indication that they will be followed by more such events in future years and decades, especially if extreme meteorological phenomena are further intensified by climate change. Massive destruction and loss of life on a grand scale are at the upper end of a scale that stretches down to events that merely disrupt and damage, not destroy and annihilate. However, in all cases, the process of recovery is likely to be complex and demanding, for that is how modern life sets the parameters.

In recent decades, there has been a gradual accumulation of information on the processes and experiences of recovering from disaster and rebuilding human settlements, communities and environments. Like the processes themselves, much of the information is fragmentary. There have been relatively few attempts to review and summarise it, or to take stock of knowledge about how best to recover from disaster. Even shelter, one of the basic components of recovery, has not been summarised adequately since Ian Davis did so in his 1978 book *Shelter after Disaster*. The basic conclusion of that work, that shelter is a process more than an end product, is as valid now as it was then, but unfortunately many of the mistakes identified in 1978 are still being made today. When it occurs, disaster usually opens a 'window of opportunity' for positive change; but is this being utilised sufficiently? Is the need to recover from disaster being taken as an invitation to recover while simultaneously reducing vulnerability, increasing resilience and creating more functional communities? Or are the same mistakes perpetuated each time such that vulnerabilities are maintained?

We believe that recovery from disaster needs to be taught, organised and practised in ways that are effective, efficient and fair. More notice needs to be taken of good practice, but failures and inefficiencies also need to be understood so that they are not perpetuated. The verification of 'lessons learned' is that there be measurable positive change as a direct result of the experience gained. In our attempt here to summarise and interpret the catalogue of experience and to enrich its theoretical basis, we hope that this book will contribute to the learning of lessons and to changes for the better.

On average around the world, two disasters occur per day, amounting to about 700 per year. Major events such as those described above are thankfully relatively rare, but full recovery from their impacts can take up to a quarter of a century. In some of the poorest and most polarised societies, there are instances in which recovery has simply never been completed. In all societies, disaster represents a setback for most survivors but an opportunity for some. It should be an occasion for seizing the opportunity and ensuring that it is utilised equitably in ways that minimise the setback.

Over several decades, we have observed recovery processes in a wide variety of settings – in rich and poor countries as well as those that fall between the extremes of wealth and poverty. We have sought to identify critical issues and have pondered the solutions to intransigent problems. We have seen recovery and reconstruction from the points of view of the donors, beneficiaries, public administrators and aid agencies and have noted the discrepancies of attitude, approach and expectations.

Advice given to aspiring authors often includes the injunction to 'write what you know'. This has been our primary motivation and intention. Hence, rather than spending most of our time combing through secondary sources among the extensive disaster recovery literature, we have both reflected on the lessons we have learnt from our lengthy direct and continual exposure to the field of disaster recovery. In the case of Ian Davis, this began in 1972 with the Managua earthquake in Nicaragua, while in the case of David Alexander, it started eight years later in 1980 with the Campania–Basilicata earthquake in Southern Italy. Since then, we have accumulated decades of experience derived from analysing recovery after disaster in the field, through the literature and in the academic institutions of which we are a part.

In 2004, the International Recovery Platform and the United Nations Development Programme asked Ian Davis to write a book on reconstruction (Davis 2007). The task was beset with difficulties and Ian soon acquired a hearty dislike of the approach taken by the sponsoring bodies. Two years after the start of the project, at a review session in Kobe, Japan, Ian was asked to remove from the text all material that was anecdotal as it was deemed inappropriate to an 'objective' UN publication. Yet the anecdotes were stories of vital first-hand experience with reconstruction processes. They embodied much wisdom gleaned from the field. One official at the meeting wisely counselled Ian to write a book without UN oversight, in which he could freely include his experiences and opinions. The present work began as a result of this advice. Both of us have enjoyed the freedom of writing without needing to secure the approval of interested parties or paymasters.

Whether they are positive or negative, we have been free to include our experiences of recovery. But to keep us out of the libel law courts or prison cells, we have reluctantly refrained in places from naming individuals or specific agencies or institutions. Where our reflections are sensitive or critical, we have followed a policy of not naming specific UN agencies, using instead the general term 'UN agencies'. Similarly, we have avoided naming specific NGOs or academic bodies. However, we do refer to specific countries and governments as omitting such precise examples and experiences risks the text becoming meaningless.

We have each undertaken assignments in which we gathered information and the insights of survivors and decision-makers on the progress of recovery: in Indonesia, Malaysia and Sri Lanka following the 2004 tsunami; in Pakistan after the 2005 earthquake; in the Philippines following Typhoon Haiyan in 2013; in settlements recovering from the devastating 2011 Tōhoku earthquake and tsunami in Japan; in New Zealand after the Christchurch earthquakes of 2010–11; in New Orleans after Hurricane Katrina in 2005; in Haiti following the earthquake of 2010; in Sichuan Province in the People's Republic of China following the 2008 earthquake; in Italy following the earthquakes in L'Aquila in 2009 and Emilia in 2012; and so on. We have also looked back at events from the more distant past by revisiting disasters in order to examine the long-term recovery of communities; for example, in India following the Latur earthquake of 1993 and in Italy after the Campania–Basilicata earthquake of 1980.

We have dispersed examples of disaster recovery throughout the book and have located them where they seemed to fit chapter themes. Some examples turned into detailed case studies, while others are brief descriptions. The most detailed examples are as follows:

Chapter 1	Earthquake recovery in Malkondji, India; Belice, Sicily.
Chapters 2 and 9	Typhoon recovery in Haiyan, Philippines.
Chapter 3	Earthquake recovery in Wenchuan, China.
Chapter 4	Flood recovery in the UK, the Netherlands and Mozambique; Earthquake recovery in Gujarat, India.
Chapter 6	Earthquake recovery in Christchurch, New Zealand.
Chapter 7	Earthquake recovery in Skopje, Yugoslavia.
Chapter 9	Flood impact to a Frank Lloyd Wright masterpiece dwelling; WWII bombing and evacuation in the UK; Earthquake recovery in Managua, Nicaragua; Guatemala; Southern Italy; Kobe and Tōhoku, Japan; and Bam, Iran; Hurricane recovery after Katrina, USA.
Chapter 10	Earthquake and tsunami recovery in Chile and Japan.
Chapter 12	Cyclone recovery in Andhra Pradesh, India; Tsunami recovery in Indonesia; Earthquake recovery in Pakistan and L'Aquila, Italy.

We have written this book together as we have been close friends and academic colleagues for over 25 years and brought complimentary backgrounds and experiences to the writing enterprise. Ian comes from an architectural background while David was trained as a geographer and geomorphologist. Both of us have worked with research institutes, universities, governments, NGOs, private sector companies, mass media representatives and locally-based community organisations. Our paths crossed at Cranfield University, where we became Professors of Disaster Management.

We believe our separate approaches balance each other – David has written extensively about emergency planning and contingency management while Ian has focused on disaster risk reduction, adaptation to climate change, shelter and the reconstruction of housing. We have both worked extensively on the vulnerability of people to natural disasters and more generally on disaster risk reduction. Moreover, we are interested in disaster recovery in both developing and industrialised countries as we see increasing parallels in disaster impact and recovery patterns regardless of whether countries have severely limited resources or are fortunate enough to be able to draw upon massive capabilities.

As in all our past writing, this book is intended to be of interest to a wide audience of concerned lay persons, researchers, media personnel, academics, professionals, students and recovery management officials. We have made a number of assumptions that underpin our writing. Specifically, we believe the following:

- Disaster survivors need to be closely involved in decisions that affect them. Moreover, the maintenance and recovery of their dignity should be a paramount concern, and efforts should be made by all parties to avoid creating unrealistic expectations and long-term dependencies. In Chapter 13, we conclude our book with a series of principles and a final model (no. 21), linked to principle no. 13, shows in graphic form the strength and weakness of community participation in recovery in the major case studies of recovery cited in the book.
- In most cases, recovery should be a multi-sectoral process that involves closely linked psychosocial, environmental, economic, physical, political and administrative elements. This is represented in Chapter 3 in a hexagonal diagram that forms model 2 – the 'recovery sectors' model.
- National governments need to coordinate recovery actions within their respective countries. They need to harmonise procedures adopted at lower levels of government and ensure that there is strong, positive leadership at the top. See Chapter 4, model 20 for contrasting frameworks of governmental leadership of recovery.
- Whether they are national or international, assisting groups have key roles to play in recovery, but they must avoid duplicating the actions that disaster survivors can take themselves. We believe that aid should empower, not weaken, efforts by beneficiaries to recover and take back control of their

own lives. In Chapter 9, we consider model 17 which reviews multiple options and identifies the providers of shelter and housing. We discuss the roles of external groups as well as spontaneous actions by survivors to provide their own shelter and build their own houses.

- A key factor in all recovery actions should be to improve safety in the light of future disasters. In Chapter 4, through models 9 to 13, we discuss the critical issue of safety during recovery.

- Finally, we have a shared belief that the world of disaster risk reduction and disaster recovery management is rather overpopulated with self-serving publications, produced by UN agencies, NGOs and international financial institutions. The positive side of this plethora of literature has been the wide coverage of topics and issues, but too many of the publications define these to meet their own interests in gaining publicity, securing resources or building agency profiles. Therefore, we have sought to use our independence to write freely. In the over-politicised world of disaster risk reduction, we believe that the critical spirit badly needs to be revived. While not wishing to take his injunction too literally, we remember Bernard Shaw's advice to writers: 'Decide what you want to say and then say it as offensively as possible.'

Having stated the intentions of this book, we need to specify what it is not. For example, it is not a set of technical guidelines – these are already common and examples are listed in our bibliography and in Appendix 3. We have offered practical advice, but it has not been our intention to write a 'cookbook'; rather, we aim to delve into the principles and theory of recovery management in the manner of a 'nutrition guide'. We should also note the professional bias of the book. Neither of us are economists, health professionals, specialists in the recovery of the natural environment, etc. While we have noted the importance of these and other sectors in our second model in Chapter 3, we have concentrated our energies in the sectors where we have most experience, such as shelter, housing, planning, risk management and policy concerns.

We have devoted two chapters (3 and 4) to 20 models that relate to recovery planning and management.[1] Many of these have grown from our teaching in various universities and in leading management training workshops, where we have found that students and practitioners value simple conceptualisations to help explain complex ideas. Throughout the text, we return to these models where they fit given topics (see Appendix 1 for a summary of models).

In order to constrain this book to manageable proportions, we have had to adopt a somewhat restrictive definition of 'disaster'. We do not include conflict, warfare, economic disasters, nutritional emergencies or epidemics and pandemics. Some of our conclusions may apply to such situations, but we recognise that the survivors of disease, spontaneous loss of wealth, wars and displacement may have different predicaments and needs to those that we describe in our focus on natural and anthropogenic disasters, *sensu stricto*. Where appropriate, we have noted the overlapping concerns, but we feel that limiting our scope allows us to ensure a more focused, concentrated coverage of our subject matter. Similarly, we have tended to exclude business continuity management and many examples of recovery from technological disasters as these are less germane to the issues of greatest importance to us, such as shelter management. Finally, we have not considered how humanity might recover from apocalyptic disasters, such as asteroid impact or the collapse of the food chain. Such eventualities are outside our experience and will no doubt be taken up by other authors.

During the course of writing the book, we decided to check whether our convictions concerning the priorities of recovery were widely shared by experts working in the field – 51 generously responded to our survey, in which we asked each to state concisely the essence of effective recovery. Their answers were revealing and we devote Chapter 11 to them, accompanied by our observations. Their responses are recorded in full in Appendix 2. We were tempted to quote from their answers within the book but (with one exception) resisted this impulse as it would have been invidious to quote from one author and not another.

At the end of the book in Chapter 13, we present some general principles of recovery from disaster, which we back up with the models noted above that explain how to apply them; and we cite examples of cases in which they were applied or of situations that inspired us to formulate them. In an ideal world – one that is considerably more prudent than that in which we live – much more effort would be devoted to reducing the risks and impacts of disaster. As it is, such efforts are vastly overshadowed by expenditure on responding to disaster when it occurs. Early relief and intervention dominate the field, in part because the imperative of early response to disaster cannot be ignored and in part because it seems to be much easier to respond than to reduce the reasons why such response is needed by making society less vulnerable. The disaster 'crunch' model (no. 10 in Chapter 4) seeks to identify possible drivers of vulnerability. In any case, the medium- and long-term phases of impact have been studied much less by disaster scholars than have hazards, risks, predictions, warnings and early phases of impact.

The world's news media tend to lose interest in disasters long before they cease to be the main issue in the areas that are affected. Like the international relief community, the global news media move on to the next major catastrophe somewhere else, which can lead the inhabitants of disaster areas to feel that the world has lost interest and is neglecting them. A remarkable example of this was investigated by the sociologist Kai T. Erikson, whose book *Everything in its Path: Destruction of Community in the Buffalo Creek Flood* (1976) chronicled the isolation of a poor, isolated community affected by a devastating flood. Anthony Oliver-Smith's classic work of social anthropology, *The Martyred City: Death and Rebirth in the Andes* (1986), similarly records isolation and neglect in a Peruvian community devastated by a rock avalanche. In Appendix 3, we offer a list of key writings and websites that deal with recovery.

In the present book, we are interested in the long haul, the painstaking process of restoring damaged communities and settlements, activities and environments to a state of functional productivity and hopefully also to a state of resilience against future disasters. In many cases, recovery is a slow process and one that must necessarily proceed on many different planes: social, economic, cultural, psychological, institutional, environmental and physical. The multifaceted nature of the process easily leads to fragmentation. The extension of recovery through various time periods puts it into a dynamic context that forces constant changes upon the guiding parameters. Small wonder that it is a complex and variable process. Nevertheless, we have sought in this book to unravel it and explain the objectives, opportunities, constraints and pressures involved. We hope that this may make a small contribution to enlightening the process of recovery from disaster and suggesting ways in which it can be pursued more vigorously and effectively.

Ian Davis and David Alexander
Oxford, UK, 1 February 2015

Note

1   A further model (no. 21) is introduced in Chapter 3 and described in Chapter 13.

ACKNOWLEDGEMENTS

This book has taken many years to write, and it has benefitted from the contributions of many friends, colleagues and students as well as total strangers who have generously shared their time, information and insights. We have also appreciated the excellent writing in the shelter and housing sector in recent years, as noted in Appendix 3.

Special thanks are due to:

Sheikh Ahsan Ahmed and Babar Tanwir of UN-Habitat for photographs of the Earthquake Reconstruction and Rehabilitation Authority (ERRA) Pakistan earthquake reconstruction;

Yasamin Aysan (Gediz, Turkey earthquake);

Shigeru Ban Voluntary Architects for permission to use the photograph in Figure 9.12;

Mihir Bhatt, Thomas Swaroop and residents in Malkondji (Latur, India earthquake);

Mary Comerio for use of her model (no. 21, Chapter 13) and data on the Chile earthquake and tsunami;

Connor Gallagher for superb graphic design skills in producing the models and diagrams for this book;

Debarati Guha-Sapir and Philippe Hoyois for use of statistics and a chart from the Centre for Research on the Epidemiology of Disasters, Catholic University of Louvain;

Mikio Ishiwatari (Tōhoku, Japan tsunami and earthquake);

Yasamin Izadkhah and Mahmood Hosseini (Bam, Iran earthquake);

Titus Kuuyour (Mozambique floods);

Steve Platt (Chile earthquake and tsunami);

Olga Popovic Larsen (Skopje earthquake);

Paul Sherlock, Ian Bray and Oxfam for permission to use the photograph in Figure 9.18;

Anil Sinha (Gujarat earthquake);

Maggie Stephenson (Pakistan and Haiti earthquakes);

Thomas Swaroop and John Noble for photography and interviews in Chinthayapalem, India related to the Andhra Pradesh cyclone;

Dr Joern Birkmann and members of the EC FP7 project MOVE, Improvement of Methods for Vulnerability Assessment in Europe, for use in Chapter 6 of the diagram of the project's theoretical basis;

the following for their encouragement and general advice – Bill Flinn, Mo Hamza, Terry Jeggle, Rumana Kabir, Fred Krimgold, Melissa Lindros, Everett Ressler, Graham Saunders and Paul Thompson;

the 51 international experts who gave their insights in the field of disaster recovery, as listed in Appendix 2.

The authors have made every effort to contact the owners of the copyright of visual material and each figure has been credited appropriately. We would be grateful if anyone who believes we have used their copyrighted material without permission would contact the publishers so that any omissions can be corrected in future editions.

1

THE DYNAMICS OF RECOVERY

Two examples

Experts talk of 'building back better', of concepts like 'resilience' and 'sustainability', of crisis being opportunity in the way that it was for the devastated cities of Germany and Japan in 1945.

The practice ... can be very different; piecemeal, dilatory, bureaucratic, venal even. Urban planners, it seems, never miss an opportunity to miss an opportunity. But occasionally, just occasionally, they surprise on the upside too, and reimagine the city in ways that might have been impossible had disaster not struck.

(Rice-Oxley 2014)

Examples of positive and negative recovery

A rich collection of metaphors is used to describe the complicated process of disaster recovery: melting pots, moving targets, spiders' webs, roots and branches, jigsaw puzzles, and so on. Writers have used these pictures as they attempt to capture the right image of a recovery operation that involves people, institutions, structures, issues, policies, plans, pressures, progress or stagnation, and hopes. We consider varied models of recovery in Chapters 3 and 4 (see Appendix 1 for a summary of models), but we have found that these complex realities and relationships are best conveyed in the narrative of observed experience. Therefore we launch straight into our book with a pair of contrasting examples of disaster recovery after earthquakes in India and Sicily.

One of these was highly successful while the other was less successful and gave rise to some opportunities that were missed. We selected them because we have first-hand experience of both situations over different stages of the recovery process; and we make no apology for including the broad 'macroscopic' elements in each story as well some of the minor details conveyed in our drawings and photographs, for the 'magic is in the details'. These examples relate to the first and second models (the 'progress with recovery' model and the 'recovery sectors' model) that appear at the beginning of Chapter 3.

Successful recovery of Malkondji village following the 1993 Latur earthquake in Maharashtra State, India[1]

In a remote rural town, Latur, and surrounding villages in the Indian state of Maharashtra, a magnitude 6.4 earthquake occurred at 3.30 in the morning on 30 September 1993 while people were sleeping in heavy masonry dwellings. In a manner typical of earthquakes that occur at night, casualties were high. It is estimated that 7,928 people were killed and 16,000 were injured. Some 37 villages were catastrophically damaged and about 30,000 houses collapsed. The earthquake did not occur in one of the well-known seismic zones of India; this lack of history of recent disasters partly explains the absence of building codes designed to guarantee seismic resistance in Latur, as well as the consequent high levels of damage to dwellings (Figure 1.1).

The Government of Maharashtra managed the earthquake recovery process with considerable skill, setting standards of cost and space for dwellings and specifying the minimum earthquake resistance of reconstructed buildings. They then assigned the reconstruction of various villages to those agencies that were

FIGURE 1.1 Earthquake damage to stone masonry dwellings in Malkondji (sketch by Ian Davis).

able to produce evidence that they had adequate resources and experience to complete the job well. One village, Malkondji, had a population of 1,562 distributed among 281 households; in the earthquake, seven people lost their lives and five were injured. The village was made up of houses constructed with unreinforced stone masonry walls, locally called *malwad*, with heavy wooden roofs covered with a thick layer of mud – a form of construction that is highly vulnerable to seismic damage. The reconstruction of this agrarian village was entrusted to a Christian NGO called the Evangelical Fellowship of India Commission of Relief (EFICOR). This decision was taken despite the fact that disaster reconstruction was not EFICOR's normal area of expertise – its particular strength lay in community-based projects. Meanwhile, the UK Government's Department for International Development (DFID) had £750,000 (about US$1.5 million) to spend on reconstruction and decided to allocate it to a single community rather than to disperse it more widely. In recognition of its accountability to British taxpayers, DFID also decided to work through Tearfund – a British NGO which had built a good working partnership with EFICOR.

At that time, Ian Davis was leading a disaster management consultancy organisation, and one of the organisation's staff was sent out to India by Tearfund in 1994 to discuss how to begin the design of the village. We supported the idea that the design should adopt a cluster approach that had been proposed by renowned British architect Laurie Baker, who had worked in South India from 1945 until his death in 2007 (Bhatia 1991):

> I learn my architecture by watching what ordinary people do; in any case, it's always the cheapest and the simplest. They didn't even employ builders but the families did it themselves.
>
> [M]y feeling as an architect is that you're not after all trying to put up a monument which will be remembered as 'a Laurie Baker Building' but Mohan Singh's house where he can live happily with his family.
>
> *(Bhatia 1986: 218)*

Baker's 'architecture as service' approach was radically different from the way most architects were operating, and his architectural layout was a refreshing alternative to the monotonous rows of regimented 'barrack-style' dwellings that were the default design used by the Government of India and most agencies. Tearfund also encouraged the appointment of a progressive Indian firm of consultants called Development Alternatives to provide the detailed design and technical services. Baker's planning approach included the allocation of communal spaces, public buildings and extensive tree planting to provide shade, fruit and medicine. An important aspect was to retain a traditional feature in villages in the region by building compound walls to surround properties so as to provide privacy as well as protection for livestock that were brought in from the fields every night to avoid risk of theft (Figure 1.2).

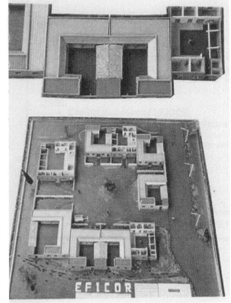

FIGURE 1.2   Model of housing cluster layout. In 1996, the site engineer, Mr Mandappa, is explaining the cluster layout to Ian. After many failed attempts to explain the house and cluster layout to the residents using architects plans, a model was prepared, and immediately everyone understood the design concept (photographs by Ian Davis).

The settlement was located about 600 metres from the destroyed village (illustrated in Figure 1.1). A strong participatory process in decision-making was characteristic of EFICOR's approach, and the agency and their planning consultants decided to build 336 identical houses, each being 34.5 square meters, using a 'core house' principle, carefully planned with sufficient space for later expansion. Houses were constructed in cement blocks, strengthened with ring beams made of reinforced concrete that were positioned at plinth and lintel levels. The roofs were constructed in reinforced concrete. The layout was for a single-family unit and was composed of two rooms, one for use as a kitchen and the other room for sleeping. A separate unit was provided with a bathroom and pit latrine.

In 1996 as the reconstruction project was nearing completion, Ian was asked to lead an evaluation team to assess the recovery of the village, including the reconstruction of houses. The evaluation included examination of the following sectors: health, community development, engineering and planning. It benefitted from the expertise of Mihir Bhatt, the founder and leader of the All India Disaster Mitigation Institute (AIDMI) (Davis and Bhatt 1996)

The results of the evaluation were generally positive, indicating a good level of involvement of the community in the design of the settlement and the

FIGURE 1.3 Completed houses in the main street of Malkondji. An avenue of trees has been planted by the side of the road (photograph by Ian Davis).

dwellings (Figure 1.2). The sensible cluster layout enabled farming families to bring their cattle into communal courtyards at night to provide protection against animal theft. The team also recognised the value of training masons to build safe, robust houses. It was, however, divided over the wisdom of the decision to incorporate toilets in each dwelling as only a small fraction were being used for their intended purpose, most of the others being used for grain storage. Some of the team believed that it would take a generation for these rural communities to become accustomed to toilets instead of their traditional practice of going to the fields.

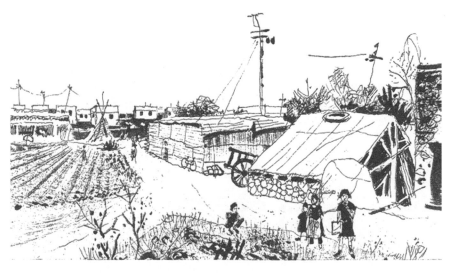

FIGURE 1.4 Sketch of Malkondji in 1996, three years after the earthquake. In the foreground, there are varied forms of improvised shelters that were built by survivors. Others have been replaced with the permanent new dwellings that appear in the background (sketch by Ian Davis).

In the disaster recovery field, long-term 'longitudinal' evaluations are exceedingly rare, but, as Ian remembers: in January 2011, our curiosity about progress in Malkondji during the intervening 15 years led us to return to the village. We asked three members of the original evaluation team to revisit the village in order to gauge long-term progress. The team included Mihir Bhatt, with his AIDMI colleagues, and the Director of EFICOR's Reconstruction Programme, Thomas Swaroop, together with his colleagues. None of the team had visited the village since 1996.

We were warmly welcomed back and the results were beyond our expectations in virtually all areas. Social progress had been significant. The most unexpected result was that 98 per cent of all dwellings had toilets in use following action by the headmistress of the local school, who insisted that on health grounds all children used the school toilets and also those at home. The parents and other family members were therefore 'educated' by their own children who, in current development jargon, became the 'agents of change'. This reminds one of the social reformers in nineteenth-century Britain who argued that working-class people should have running water and baths in their homes, at which the critique was levelled that the baths would be used to store coal. Of course, that was not the case.

Generally, social organisation in Malkondji was excellent. Various public buildings and facilities had been added since our 1996 evaluation: the school, a health centre, a temple and a sports ground. Local leaders took great pride when, in 2008, Malkondji was awarded a prize for being the best village in Maharashtra.

We reviewed progress in the provision of dwellings and established that the distribution of houses had been equitable and had taken place according to

FIGURE 1.5 Headmistress of Malkondji School and a class of schoolgirls. Together they were the 'change agents' of Malkondji School in 2011 who introduced the use of toilets into the village (photographs by Ian Davis).

FIGURE 1.6   Many house occupants added covered spaces to their dwellings for various household and occupational tasks (sketch by Ian Davis).

family needs. The cluster layouts of houses were working well, with full use of courtyards for the protection of cattle as envisaged by the architect. The compound walls that enclosed individual dwellings were greatly appreciated by residents (Figure 1.2). Throughout the town, dwellings had been enlarged – in a few cases, with the addition of a second storey.

Support from the World Bank had enabled the infrastructure to be improved, including the arrival of electricity and piped water as well as surfaced roads that allowed a bus route to be added. Trees that had been planted beside roads by the Indian Department of Forestry were now fully grown and provided excellent shade. The medicinal Neem trees, wisely planted in each reconstructed village, were greatly valued by villagers.

The village economy was in good shape, with some useful diversification of employment beyond farming. We discovered that the original training course in safe building construction had enabled some local farm labourers to secure occasional building work and thus diversify their sources of income.

A minor regret experienced by our team was the decision the village had taken to demolish historic buildings possessing rather beautiful masonry and carved timber that had been ruined in the disaster. Their retention, in their ruined form, could have provided a valuable link with the past; but for the residents, these ruins were merely a painful memory of the disaster they had suffered and were therefore not to be retained. It was also clear that more public education was needed regarding community preparedness against future earthquake disasters and other hazard threats.

As we left Malkondji after its second assessment, members pondered on the possible reasons why this community had recovered successfully. One underlying factor may be that only seven villagers died in the earthquake. In contrast to other villages where there was a high toll of death and injury, this may be a significant influence upon the psychosocial recovery of the community. Where family losses were extensive, it was always likely that recovery would be incomplete.

We concluded that a decisive factor had been the culturally and environmentally sensitive design of the settlement and its dwellings. Qualities that contributed to success included the clustering of buildings instead of grouping them in tedious rows, the space allotted to expansion, streets that are curved rather than straight, extensive planting of trees, ample provision of community buildings and vital new infrastructure. We noted that there had been a high level of participation by the beneficiaries from the outset of the project, in both formal and informal ways. This occurred in the design and allocation of dwellings, the management of projects and the construction of dwellings. All of these were vital components of success.

A persistent theme throughout this book is that *process* is often more important than *product*, and indeed the manner in which the village had been reconstructed seems to have been particularly important. When building work began, EFICOR and Development Alternatives set up basic camp-style living accommodation in the village for their construction teams, rather than allowing them to return each night to more comfortable hotels in Latur. Thus, for the three years it took to rebuild, the construction teams shared in village life and interacted closely with the residents. As a result, the process grew far beyond the establishment of a mere contractual relationship, with the creation of genuine warm friendships between agency construction staff and village residents of all ages. There were joint sports activities and children's clubs, and a village delegation was even taken to view development projects in India, including a memorable sightseeing excursion to Delhi. We were shown many prized photographs of the trip, which decorated the walls of their homes. Thus, the EFICOR team let their Christian values and extensive development experience guide them to invest in people, who had become trusted friends, and a strong community was formed as they shared in the creation of homes and not merely anonymous houses for unknown families. (See Chapter 12, in the section 'Red hat – emotions', for a contrasting example of an Indian housing cyclone reconstruction project that had been built 18 years earlier by EFICOR.)

The training of masons had proved to be effective, but there were regrets that subsequent years did not bring more opportunities to apply the building skills acquired during the reconstruction. However, the direct involvement of members of the Malkondji community in building houses was clearly one of the main reasons for the ubiquitous pride in their homes and the village. Though the team's visit took place 18 years after the earthquake disaster, references were still made to the reconstruction process, and the statement 'we built it' (or parts of it) was repeated often. This close identification between residents and their built environment would not have occurred without their own direct and active involvement in building. This must be one of the strongest arguments in favour of active 'user-build' construction practice as compared with passive 'contractor-build' approaches (see Chapter 9, 'Option 7a: user-build permanent dwellings').

In writing a book on disaster recovery, it is all too easy to devote disproportionate attention to definitions, concepts, models and sectors and thus to lose sight of the needs of tens of thousands of families and individuals who are at the very heart of the recovery process. (This is discussed further in Chapter 8 in 'Voices of survivors'.) Ian was reminded of this reality in 1996 as he climbed into a Land Rover with his colleagues, about to leave the village following the first evaluation of Malkondji. He remembers: a man came up to our vehicle and asked if he could speak to me. I climbed out and had a conversation with a Mr Rangappa.

He was a farm labourer and the earthquake had destroyed his house, although no one in his family (mother, wife and two small children) had been injured. Soon afterwards, he had been asked whether he would like to attend a masonry training course to prepare for the reconstruction of his village. He did take part, and his work included building the stone plinths of the dwellings using stones from ruined houses. The plinths supported cement-rendered concrete block walls. He told me with obvious pride that he had developed the skills of a stonemason and bricklayer in just over two years, and that his previous income as a farm labourer had been multiplied by ten in the process. Then he came to the point of his query: 'What should I do now that the reconstruction work is complete in Malkondji and in the surrounding villages?' He had found that there were no opportunities for *normal* building work in the Latur region, other than those created by the reconstruction. He asked for my opinion about whether he should return to farm labouring, with its low wages, or leave the village and take his family to Mumbai or Hyderabad where he understood there was plenty of building work.

I found it hard to advise him, not knowing sufficient facts to make a useful comment. However, I suggested that if he took the major step of moving to a large city, he would probably end up living in a slum with an uncertain future, far inferior to the newly reconstructed house his family currently occupied. He understood this but said that if they moved, unlike in Malkondji, there would be schools, doctors, electricity and public transport, as well as employment opportunities. I tried to reassure him that these vital services would come to Malkondji within the coming two years – I had seen the World Bank plans and timetable. I recall saying that if I were in his shoes, I would not take the risk of moving to the city.

So when we returned to Malkondji 15 years later, I was curious to see what Mr Rangappa had done. I brought a photograph of him with me and asked whether he was still living in the village. It turned out that he lived very close to our base, and a few hours later, he arrived on his brand new motorbike. He told me his story, of how his family had decided not to leave (although he was quick to point out that this was *not* on account of my advice!). Over the years he had developed dual occupations with some building work as well as farming. He told me that farm incomes had increased as India had prospered, bringing higher prices for farm products. He was also happy that the village had acquired a good school for his children, as well as water and electricity, a health clinic, a bus service – even broadband Internet services.

This encounter was a reminder that reconstruction is an effective generator of livelihoods, but this can be a purely temporary 'boom' unless thought and action follow to ensure continuity of work when the reconstruction is completed in order to capitalise on all the newly acquired skills.

Organisational support also proved to be a key element in the success of this recovery. The roles of the Government of Maharashtra, DFID and Tearfund were all important. However, the success also stemmed from the vision and work of some inspired individuals. This included support from the Government of Maharashtra by Krishna Vatsa, who implemented the World Bank-funded Maharashtra Emergency Earthquake Rehabilitation Programme during 1995–9, and the wise decisions of the District Collector[2] Praveen Pardeshi (both of whom became key international leaders in disaster risk management). Laurie Baker, the architect with a vision, and Eddie May and Thomas Swaroop of EFICOR also contributed much. Moreover, the local village leader and the village school headmistress both had a positive influence, as well as Mr Rangappa and his mason colleagues who rebuilt the village. Finally, perhaps Malkondji has benefitted in particular from being located in India's richest state and for being able to take advantage of India's rapid economic expansion.

This case study is a particularly good example of an effective planning and implementation sequence as outlined in model 14 (Chapter 4). In this example, all the stages of this 'project planning and implementation' model were well covered, including short- and long-term evaluations.

Long-term longitudinal learning

Members of the team discovered that this form of long-term evaluation is a crucial learning experience that must cease to be a rarity and become the normal pattern for agencies involved in disaster recovery. If we are to learn from experience and decode the complex elements that contribute to success, funding will have to be found for such studies. In 2014, a groundbreaking longitudinal study of housing reconstruction by agencies was published. The research in Maharashtra (including Malkondji), Gujarat and Tamil Nadu undertaken by Jennifer Duyne Barenstein and her Indian colleagues is particularly significant in outlining the long-term lessons from housing reconstruction (Duyne Barensein *et al.* 2014).

A further good sign is that the British-based Earthquake Engineering Field Investigation Team (EEFIT), established in 1984, has now started returning to the sites of previous earthquake disasters in order to evaluate progress in recovery (e.g. EEFIT 2013). However, longitudinal studies remain rare in this field, and that is unfortunate. The Malkondji case also reminds us that evaluations of recovery from large disasters need to avoid the trap of overgeneralisation as one community or village may be successfully rebuilt while another, perhaps next to it, may fail to be adequately reconstructed. In order to identify recovery patterns correctly, it is necessary to conduct studies at the local scale of the village, community or neighbourhood, if possible with comparative evaluation of other locations.

Failure to recover following the Belice Valley earthquakes, western Sicily, 1968

Twenty-five years before the Latur earthquake, a medium-power earthquake sequence occurred in the Belice Valley of western Sicily on 14–15 January 1968. Five of the tremors had magnitudes greater than 5. The cumulative effect was to cause destruction to ten villages. At least 231 people were killed and 623 were injured, a sign of the very high seismic vulnerability of local building stock. About 100,000 people were left homeless (Haas and Ayre 1969). The predominant local construction was random rubble masonry with thick timber roof beams supporting heavy roofs covered with terracotta tiles. Many such buildings collapsed entirely.

In Italy, disaster management was in its infancy at the time of the Belice earthquakes. Moreover, this rural extremity of an island region was absolutely marginal to the mainstream of national political and economic life and priorities. In short, it could easily be forgotten without major political consequences.

Wooden prefabricated dwellings with 35 to 40 square metres of floor space were procured, but a year after the earthquake, many residents were still living in tents. A mixture of planned and *ad hoc* transitional housing was eventually created in the area, and some of it survived, occupied by families, for a further 20 years. Not all of it was functionally justifiable. At Gibellina, for example, the barracks housing provided by the Italian state was quickly abandoned because it was set up in an area far from the fields that most farmers tended and the sources of employment of people with other occupations. It was later taken over by North African migrant squatters.

During the 1970s and 1980s the economy of rural western Sicily showed no signs of growth. Locally, a start was made on a fast, limited access highway, the 'Belice axis', but for many years only the junctions had been built, accessible by cart track. However, it was eventually completed and opened using money derived from other public works initiatives directed towards regional development.

By the early 1980s, there had been some limited reconstruction, much of it with very little reference to functional needs or planning constraints. For example, in Santa Margherita di Belice (population 6,678), a multistorey apartment block had been constructed amidst the ruins of the town. Many years later, a partial reconstruction was achieved with the rebuilding and adaptation of the noble Palazzo Filangeri-Cutò, immortalised in Giuseppe Tomasi di Lampedusa's classic novel *The Leopard*. The ruins of the parish church of Santa Margherita had been incorporated into a museum rather than being rebuilt.

Down the road at Gibellina (population 4,298), the rebuilt settlement had been moved 20 kilometres from its original site, in part to be near an exit of the proposed high-speed road and in part because the location was determined by two mafia bosses and local landowners (Parrinello 2013). The ruins of the original settlement of Gibellina were transformed into a gigantic environmental sculpture, the *Grande Cretto* of Alberto Burri, which could be seen from 50 kilometres away. The sculpture incorporates the formalised street pattern of the destroyed town (Figure 1.7).[3] The mayor of the new town, Ludovico Corrao, decided to 'humanise' it by making it a mecca for avant garde architects, sculptors, artists and literary folk. Unfortunately, the futuristic parish church designed by the eminent architect Ludovico Quaroni, a cube and a sphere, collapsed before it was completed, although it was successfully rebuilt. Modern Gibellina is a curious mixture of postmodern architecture, open-air sculpture and learned references to the town's 2,770-year history (see Figure 1.8). Its inhabitants bear this situation with stoicism and seem to be more interested in modern consumer values than the Ancient Greek tragedies recited in their original language in the town's theatre. A certain amount of vandalism has occurred to the sculptures and

FIGURE 1.7 The ruins of Gibellina, destroyed by the 1968 earthquakes, in the process of being supplanted by a gigantic environmental sculpture, a monument to the disaster (photograph by David Alexander).

FIGURE 1.8   The monumental tower (Torre Civica) at the centre of new Gibellina, an artistic symbol of the reconstruction. The architect was Alessandro Mendini (photograph by David Alexander).

ornamentation of streets, and the triumphal arch at the entrance to the town bears a depressing resemblance to similar work commissioned by Saddam Hussein for Baghdad. New Gibellina remains 'off the beaten track', a small rural town 95 kilometres from the regional capital of Palermo and 65 kilometres from the provincial capital Trapani. The population was firmly opposed to moving the town from its original location until that decision was foisted upon them as a condition of achieving any reconstruction at all (Parrinello 2013).

Santa Ninfa (population 5,125) was completely devastated by the tremors. According to estimates (which are at variance with official government figures), 337 people died there and 560 were injured. For more than 15 years, only the street plan remained at the centre of town as the rubble of buildings had been removed, and a motley collection of *ad hoc* prefabricated buildings lined the approach roads. Santa Ninfa was reconstructed to anti-seismic standards in the mid 1980s, but with more respect to the pattern of landownership than the need to institute rational planning of urban form and function. This was probably an effect of the ways in which rebuilding was financed, which promoted individual initiative over collective needs.

The Belice Valley earthquakes were followed by more than a decade and a half of stagnation in recovery. This was the result of several factors, principally laissez faire attitudes in central government, bureaucracy and corruption with links to organised crime (Angotti 1977; Chubb 2002). During this period, Sambuca di Sicilia (population 6,179), which was the only town to have a progressive administration allied with the main national opposition party, struggled to institute rational, well-planned recovery and reconstruction in the near absence of funding by the central state. The Belice Valley, with its largely agricultural economy and weak tourism, has never had a strong tax base from which to draw such funds.

In the end, the 14 to 16 towns that were damaged by the earthquakes of February 1968 were reconstructed, but the process took somewhere between 20 and 40 years to complete, and very little happened during the first 15 years. The process was jump-started by the arrival of funds connected with other disasters in Italy, in which the financing of reconstruction in Belice was 'piggybacked' onto acts of parliament concerned with other events – what in America is called 'pork-barrel legislation'. Meanwhile, the population of the affected towns has continued to decline and tourism income remains a weak substitute for agricultural employment.

When it did finally begin to gather momentum, progress was insufficient for years. Developmental aspects included the creation of new towns and growth poles (pre-planned industrial areas), and the improvement of the road network (but not the restoration of rail links devastated by the earthquake, which were permanently abandoned – the line had in any case been a loss-making rural branch). Resources remained scarce and were not provided systematically. Although anti-seismic construction did, thankfully, proliferate, the planning strategies employed tended to be haphazard, despite being heavily centralised. Bureaucracy and organised crime damaged and slowed down the recovery. There was a propensity to commission grandiose projects (such as Gibellina, the 'open-air museum town') with no underlying evaluation of whether this was the best way to allocate scarce resources. In synthesis, leadership, democratic participation, availability of resources, and planning foresight were at best haphazard, experimental phenomena (Parrinello 2013). As the population of the affected area continued to decline in the decades after the earthquake, the measures taken did not amount to a fully fledged economic recovery, even though the damaged towns were rebuilt.

Comparison

Table 1.1 compares certain characteristics of these two case studies. Of course, this comparison is unscientific as it does not match like with like: Malkondji is merely a small village while Belice is an entire valley full of settlements. However, it is clear that good leadership, participatory approaches, transparency and careful, responsive planning all lead to outcomes that are better and achieved faster than those which result from less well conceived approaches.

Underpinning this prescription for good recovery are concepts of human rights, democracy, participation, trust in government and the rule of law. If any

TABLE 1.1 Comparison of reconstruction in Malkondji and Belice

	Duration of reconstruction	Vision	Planning	Local involvement	Government involvement	Continuity of funding
Malkondji, India	4–8 years	Strong at every level	Well controlled	Strong in design, construction, management	Government cooperation, zero corruption	Good
Belice, Italy	20–30 years	Largely absent	Haphazard	Weak public participation	Bureaucracy, corruption	Lacking

of these are weak, the conditions may exist to slow down, stultify or pervert reconstruction from a course that embodies the greatest benefit for disaster survivors in need. These broad, moral and political issues are developed later in the book as they are of fundamental importance.

In Chapter 13, model 21 (see Figure 13.2), we contrast the strength of government in recovery operations with the strength of community participation. The Malkondji recovery is represented as very high in both government strength and community participation, while the Belice case is represented as below average in strength of government and community participation.

Some factors operate at the scale of micro decisions, such as whether to incorporate toilets into dwellings that previously would not have had them, while other aspects relate to social concerns, such as how to engage with survivors. The preservation or regeneration of livelihoods is vital to recovery, and so are political dynamics, such as ensuring the continuity of funding or coping with the curse of corruption. Likewise, technological and administrative concerns need to be addressed, and finally 'process' is vitally important, meaning that attention must be given to *how* reconstruction occurs. This is a critical issue of equal importance to the 'products' that are generated. These form the questions, constraints, challenges, dilemmas and achievements that recur throughout this book.

'Golden rules' may seem elusive and impossibly demanding, but success can be achieved. Positive evidence of this is provided by the Indian case study from Malkondji, with which this chapter opens. This was a project in which the richness of the qualities that are important to recovery became apparent, and where there was a shared vision for recovery. It all began with an inspired architect, who challenged the prevailing official orthodoxy with his advice and designs. His vision was supported by both the government and an international donor, and the settlement was built by a deeply committed contractor that worked hand in hand with community 'owner-builders' who shared a commitment to build a much better and safer village than the one that had been destroyed. The success of the project was monitored and evaluated over time. It is shared with the reader in order to encourage and inspire others to seek excellence rather than to be content with mediocrity.

Chapters 3 and 4 of this book present some models of recovery from disaster. Malkondji was chosen to open this book since it aptly fulfils the four scenarios in the first of these models 'progress with recovery', as well as the four strands of recovery: vision and leadership, resources, participation and ownership, and organisation (see Chapter 3, models 1 and 2). In relation to model 1, the recovery at Malkondji followed a holistic approach that integrated the five sectors: physical, environmental, institutional and governmental, economic and livelihoods, and psychosocial recovery.

In the case of the Belice earthquakes, leadership was seriously diluted by the ambiguous relationship that prevailed between the central state and organised crime in Sicily at the end of the 1960s. Vision was diluted by uncertainty in the compact between the state and citizens regarding how much welfare could, or should, be provided to survivors after natural disasters. This is a dilemma that has not abated in Italy or, indeed, in many other countries. Western Sicily emerged slowly and with difficulty from a quasi-feudal system of society to one with the modern trappings of democracy and governance. Public administration was by no means uniformly bad or corrupt: indeed over the years, as Sicily gained more and more autonomy from the central state, it achieved a degree of far-sightedness and made some serious innovations. However, that was not particularly the case in 1968 when many more decisions were made in Rome than in Palermo or more locally. As the Belice Valley was an economic backwater, there was little incentive to divert substantial public resources to its problems. Many other areas of the country also needed regional aid.

As a result of these problems, there was little sense of local participation and ownership of the reconstruction process in the Belice Valley. In any case, it was largely stalled, and what advances did occur had a somewhat unpredictable motivation. David recalls a conversation with the mayor of Sambuca di Sicilia, one of the damaged towns, in the searing midday heat of August 1983. In 15 years he had managed to get some housing fully reconstructed, but it had been a hard battle against official indifference and shortage of funds. He felt that the central state had been unreasonably stingy. Yet Sambuca was then a local example of progressive attitudes: other towns were afflicted by more profound stasis. Part of the problem related to the tremendous difficulty in stimulating employment in an area that offered no natural advantages to employers. Its only economic safety valve was outmigration.

One positive sign was that the artistic community rallied around the people of the Belice Valley. Through their efforts, coordinated in part by the mayor of Gibellina, they managed to bring dignity, recognition and a modicum of tourism to the area. There was a certain clash of cultural values between the remnants of the pre-existing peasant society and the new veneer of artistic sophistication; but at least much of the art called attention to, and celebrated, the distinguished intellectual, cultural and artistic traditions of western Sicily, which stretch back over millennia and across continents.

Conclusion

This chapter opens with a quotation from Rice-Oxley (2014) which suggests that on rare occasions there is evidence that one can 'reimagine the city in ways that might have been impossible had disaster not struck'. Malkondji provides us with a particularly encouraging example of 'developmental recovery' that moved beyond the pre-disaster status quo (see models 1 and 2 in Chapter 3). Gibellina provides us with an intriguing example with many subtleties and indications of what might have been if circumstances had been only moderately different.

Having looked at these contrasting examples of disaster recovery, we now consider in Chapter 2 how the context of recovery changes continuously. We examine a series of themes that lay down the foundation of this book. Our introduction includes the need to adapt to climate change and associated meteorological extremes, and then we consider differing points of view on the recovery process. This leads to a discussion about the language of recovery followed by a discussion concerning the likelihood of being able to define a 'vision for recovery'.

Notes

1   A description of the relocation policy following the Latur earthquake is given in Chapter 7 (see 'Third dilemma: reconstructing existing unsafe settlements vs relocation to safer sites') and an example of permanent housing reconstruction appears in Chapter 12 – see Figure 12.14.
2   A District Collector, known simply as a 'Collector', is the chief administrative and revenue officer of an Indian district. He is also referred to as the District Magistrate, the Deputy Commissioner and, in some districts, the Deputy Development Commissioner. District Collectors are appointed by a state government and are members of the Indian Administrative Service (IAS).
3   A similar monument was created in Armero, Colombia where the volcano Nevado del Ruiz erupted in November 1985 causing a series of lahars (volcanically-induced flows of mud and debris). The lahars buried Armero and surrounding villages, causing 23,000 deaths. The authorities sought to prevent reconstruction on the original site, which is now regarded as a cemetery. To create the memorial they planted trees in a pattern that delineates the original street plan of the town (see Chapter 9, 'Option 2: evacuation to safe shelter to escape impending hazards').

2

THE CONTEXT OF RECOVERY

Context – from the Latin *contextus*: a joining together, a scheme or a structure; *contexere*: to join by weaving, to plait

Recovery – from the Latin *recuperatio*: recovery or recuperation

Joining together

The pair of examples considered in Chapter 1 clearly demonstrate the inherent complexity and collective nature of effective recovery. We saw that dynamic forces are an ever-present reality to be found in good or bad relationships, in the positive or negative exercise of power, in the presence or absence of creative vision and, crucially, in the understanding or ignorance of local cultures.

In this chapter, we seek to dwell on the wider context in which recovery may fail to materialise or to flourish. Thus, we consider a series of foundation blocks that support recovery, weaving together varied strands, concepts and challenges, including 'dwarfs standing on giants', to gain better viewpoints or perceive visions. But our understanding of 'context' is not just about understanding past and present realities – as Jean-Luc Godard perceptively observed: 'It's not where you take things from – it's where you take them to.' Therefore, this chapter forms a bridge between the physical and cultural environment where recovery is firmly anchored and all the subsequent chapters where we explore some positive directions for this important subject.

The context of disasters and recovery: changing hazards, increased vulnerability and the need to adapt to climate change

Sudden-impact disasters, such as earthquakes, flash floods, hurricanes and torna- does, can cause dramatic increases in homelessness and almost instantaneous mass demand for temporary as well as permanent shelter. Moreover, 700 catastrophes occur each year on average, some 220 million people are affected and US$150 billion of damage is caused (IFRC 2013). Frequently, major crises occur as a result of the sudden creation of situations of mass homelessness in the wake of catastrophe. Moreover, in the future, climate change is likely to intensify the physical impact of meteorological and hydrological disasters, while at the same time many of the areas at risk (including floodable coasts and river valleys, unstable slopes, tectonic fault lines) are experiencing rapid rates of population growth. Throughout the world, there has been a tendency to migrate to coastal areas, which offer superior economic opportunities. Rise of sea level and intensifying tropical storms are bound to increase the level and frequency of destruction in coastal disasters.

The loss of an estimated 138,000 lives in Cyclone Nargis in Myanmar (Burma) in 2008 is indicative of what a major storm surge can do in a country that lacks basic preparations. Tsunami damage to the countries of the Indian Ocean basin in 2004 and to Japan in 2011, and the devastation caused in Southeast Asia by Typhoon Haiyan in 2013 highlight the perils of living on coasts in areas subject to seismic sea waves or tropical storms – or both. In regulating coastal development, governments face a considerable dilemma between encouraging residency and economic activity in areas where it has potential for growth, which is how many coasts can be characterised, and protecting the population against the effects of disasters. Typhoon Haiyan, called Yolanda in the Philippines, not only washed out the homes of 90 per cent of residents of the city of Tacloban (population 220,000) but also devastated basic infrastructure, beginning with the local airport which is located almost on the beach.

In natural disasters, it is common for the relationship between cause and effect to be non-linear, to the extent that increases in the energy input to the causal phenomenon (wind, wave, strength of currents, etc.) produce disproportionately large increases in damage to the built environment or human activities. There may also be thresholds; for example, when small waves do no significant damage but those that exceed a certain amplitude abruptly bring damage into play. The result of this is that any intensification of extreme meteorological phenomena that occurs as a result of climate change may cause disproportionately large increases in their destructive effects and, hence, in the demand for shelter and other aspects of reconstruction. There may also be disproportionate damage to the infrastructure (roads, seaports, airports, etc.). Finally, there may be health implications of climate change that interact with disaster; for example, in increasing tropical disease incidence rates and making the challenge of controlling disease after disaster more difficult.

Figure 2.1 illustrates the possible connections between the imperatives of climate change adaptation (CCA) and disaster risk management (DRM). This diagram indicates that both DRM and CCA share the common goal of increasing resilience to climatic/hydrological/geophysical threats. The main overlap between the DRM and CCA agendas is the management of hydro-meteorological hazards where DRM needs to take account of changes in weather hazards and both DRM and CCA aim to reduce their impacts. CCA considers long-term adjustment to changes in mean climatic conditions, including the opportunities that this can provide as well as how government organisations can develop capacities to stimulate and respond to a much longer-term process that has been a traditional focus of practical applications of DRM. Hence, scientific policy and practice on CCA needs to be better integrated with DRM in order to create a solid foundation for action.

Figure 2.1 indicates that there are high levels of certainty in many aspects of DRM. For example, we broadly know which rivers will flood, when and to what extent. However, that level of certainty is not the case with climate change. Another contrast is in the political profile, where climate change and adaptation measures attract a far higher level of political support and consequent funding – though also opposition from 'climate change deniers' – than measures to reduce disaster risks. As we discuss later in this chapter, DRM – often built into disaster recovery – has a very long history while CCA has a recent history, only becoming an international concern in the mid 1980s.

Over the past decade, more attention has been given to converging DRM and CCA agendas, conceptually and in practice at international, sub-national and local levels. Despite the linking of DRM and CCA concerns and projects, the current institutional context discourages collaboration between and within levels of government. Governments have traditionally divided their responsibilities into discrete areas, such as emergency services, housing, infrastructure, agriculture, etc. This strict demarcation has led to a silo mentality within organisations that encourages narrow views of the issues and tends to overlook the broader cross-agency implications. These kinds of rivalries are exacerbated by issues such as CCA and DRM that cut across defined areas of responsibility (Burton 1998). As perceptions grow through the enhanced attention that both CCA and DRM are now receiving, there will be growing opportunities for effective response in the integration of climate change and disaster risk management (Handmer and Dovers 2013).

There is no doubt that future disasters will generate massive and widespread demand for temporary shelter and permanent reconstruction. Experience of both good and bad practice has accrued over the years, but there has been no comprehensive survey of the problems of shelter and reconstruction after disaster since the 1970s. It is time to summarise the state of the art and provide a concise, comprehensive source of knowledge for academics, planners, architects, engineers, construction managers, relief and development officials, reconstruction planners and the commercial firms involved in recovery projects who are all involved with shelter and the built environment after disasters.

Since the mid 2000s, the planning and implementing of reconstruction, previously a seriously neglected humanitarian concern, has been thrust onto the priority agendas of donor governments, UN agencies, NGOs and the private sector. This is the result of catastrophic damage and high human losses from eight very large earthquakes, a hurricane and a tsunami: the Bam earthquake in

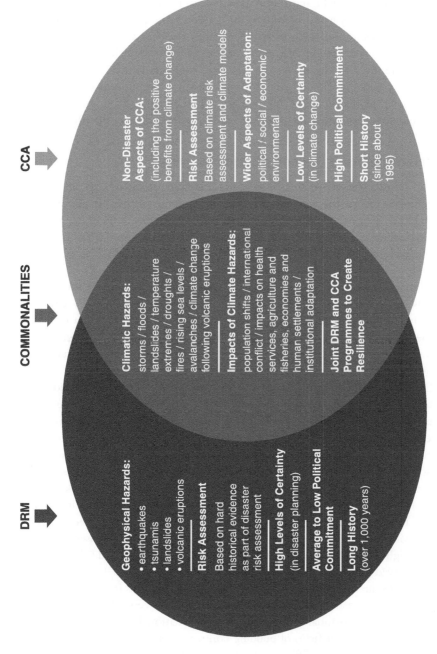

DRM ➡ **COMMONALITIES** ➡ **CCA** ➡

Geophysical Hazards:
• earthquakes
• tsunamis
• landslides
• volcanic eruptions

Risk Assessment
Based on hard historical evidence as part of disaster risk assessment

High Levels of Certainty
(in disaster planning)

Average to Low Political Commitment

Long History
(over 1,000 years)

Climatic Hazards:
storms / floods / landslides / temperature extremes / droughts / fires / rising sea levels / avalanches / climate change following volcanic eruptions

Impacts of Climate Hazards:
population shifts / international conflict / impacts on health services, agriculture and fisheries, economies and human settlements / institutional adaptation

Joint DRM and CCA Programmes to Create Resilience

Non-Disaster Aspects of CCA:
(including the positive benefits from climate change)

Risk Assessment
Based on climate risk assessment and climate models

Wider Aspects of Adaptation:
political / social / economic / environmental

Low Levels of Certainty
(in climate change)

High Political Commitment

Short History
(since about 1985)

FIGURE 2.1 Differences and commonalities in disaster risk management and climate change adaptation.

Iran in 2003, the South Asian earthquake and tsunami of 2004, the Pakistan earthquake of 2005, Hurricane Katrina of 2005, the Sichuan earthquake in China of 2008, the earthquakes in Haiti and Chile in 2010, the earthquake and tsunami in Tōhoku, Japan in 2011 and Typhoon Haiyan in the Philippines in 2013. While each of these events had a unique form of impact, the patterns of damage and human losses were similar. So were the recovery needs and associated responses. There is a pervasive sense that many of the people responsible for managing recovery are not learning the lessons of past events sufficiently. Later chapters of this book consider where and why that might be the case, but now it is important to take note of the fact that different protagonists, and people with different forms of experience, view the process of recovery from disaster in different ways.

Different points of view on disaster recovery

The size, form of impact and recovery needs of disaster can differ very considerably from one event to the next. Given that no country or area of the world is immune to calamity, it is not surprising that the organisers of recovery planning and management are equally varied in their competencies, approaches and vision of the work to be done. Therefore, it is important to discuss the culture of individuals and organisations in relation to their roles, approaches and performance. Relevant groups include politicians, public administrators in government, national NGOs, international donors, commercial companies and, in every case, the surviving community and its leaders or representatives. This topic is a recurring theme throughout our book, and we devote Chapter 12 to a consideration of Edward de Bono's 'thinking hats' where we look in detail at varied attitudes that shape perceptions of recovery: optimism, discernment, creativity, etc.

Each group has its own general approach that is inevitably influenced by its operating mandate as well as the self-interest that embodies its values and determines how it goes about its tasks. However, for any given organisation, this 'internal logic' may not always seem particularly logical to the external observer, especially when applied to the complex demands of disaster recovery. In fact, some approaches can appear to be dictated by the vested self-interests of institutions rather than the needs of the affected communities. Kates and Pijawka (1977) noted this when they observed that banks have the best access to credit and may be first to rebuild after disaster, but there is no rule that they must necessarily be as generous to other survivors who need credit as they are to themselves.

Throughout the book, we discuss the differing approaches and values that we have observed at first hand and compare them to what we call a 'logical approach' or 'good practice' in recovery. We have attempted to state this from the viewpoint of the survivors needs as this priority concern is not necessarily the same as the internal logic of national political leaders, international financial institutions, international NGOs or firms of building contractors. Thus, definitions of disaster recovery can be radically different when seen from the standpoints of those who direct recovery operations and those who are on the receiving end. For example, a report on the progress of recovery from the earthquake in Port-au-Prince, Haiti highlighted the issue of differing standpoints: 'In a post-earthquake city that has seen coups, invasions and disease, redevelopment is either going strong or standing still depending on who you speak to and what their agenda is' (Lall 2014).

To illustrate this point, here are two different perspectives on disaster recovery.

Disaster recovery – a governmental view: This refers to decisions and actions taken after a disaster in order to restore and improve pre-disaster living conditions and the environment of the affected community. Disaster recovery provides a unique opportunity to introduce improved safety measures in all sectors as well as to strengthen, rebuild or restore weak or damaged social, physical, economic and environmental elements and conditions. A key task is to rebuild any damage sustained to vital local institutions, with special emphasis on local government. Many governments will need external assistance to support the recovery, but neither at the price of interfering with their authority nor that of accepting inappropriate aid that reduces local capacity. In summary, governments need adequate resources and support, an effective plan, good management and strong leadership, but most of all a clear vision of the intended future. The population – i.e. the governed – need to have a sense of trust in government, which is best fostered by democratic participation, transparency and a good record on human rights.

Disaster recovery – a survivor's view: The recovery of family life is the key aim, as far as this is feasible in the light of deaths, injuries, disability and the loss of homes, jobs and identity. Survivors need to be guaranteed the opportunity to participate in making decisions that affect their recovery and in the work opportunities created by the reconstruction process. In Chapter 1, we describe how Mr Rangappa, a farmer and mason in Malkondji, seized the opportunity presented by the reconstruction of his village to acquire new skills that are helping to pull him and his family out of poverty. The confidence of survivors in government depends on fairness and lack of corruption in the distribution of assistance. The rapid restoration of livelihoods is an important means of enabling them partly to fund their own recovery. Survivors need 'homes', not merely 'houses', future safety from hazards, and the repair or rebuilding of precious cultural heritage. Families need safe schools for their children and adequate health-care facilities, as well as the infrastructure of essential roads and services on which their lives depend. Normally, unless there are pressing safety concerns, they need to rebuild as close as possible to the original locations of their homes. In summary, survivors need dignity, respect, employment, safety, practical help and ability to recover their identity. From all those who assist them, they have no need of dependency, largesse, paternalism, political posturing or false promises. In Chapter 8, we discuss ways to hear the 'voices of survivors'.

Ian vividly recalls an example of the tension over differing standpoints in a heated exchange during a workshop that he organised in Oxford in 1993. An eminent epidemiologist, Chief Medical Advisor to an international NGO, was making a presentation to senior officials concerning health aspects of disaster management. He outlined a definition of disaster that immediately resulted in a challenge from the floor by a leading geophysicist and former director of a United Nations disaster relief agency. The delegate protested that the definition proposed by the speaker did not conform to the recently agreed set of definitions published by the UN after a period of consultation. He suggested that the speaker replace his 'incorrect' definition with the more orthodox one. The speaker's response was something of a classic, as he made a significant point that still eludes many in authority: 'Well, you, as a geophysicist and senior UN official, have a definition that suits you, and that is fine, but we in [my organisation] have our own definition and this suits our needs well and reflects our standpoint and we will continue to use it.'

The accuser responded with some irritation that there could only be one definition for a given topic, to which the speaker retorted: 'On the contrary, there are different perceptions and definitions of any given issue, depending on your specific needs and where you stand.' While some of the participants in the workshop were enlightened, others were thoroughly confused at this representation of two versions of reality: the fixed, finite perception of the questioner and the variable, pragmatic viewpoint of the speaker.

Before considering definitions, let us offer some thoughts on the more general question of the importance of choosing one's words.

The language of recovery

One advantage of the rich vocabulary that characterises the English language is the repertoire of words that describe the nuances of complex meanings. Thus there is a cluster of words that describes 'recovery', and all of them begin with the prefix 're' which derives from the Latin for 'again' or 'again and again' or 'back to the starting point', as in '*recedere*' or '*revocare*'. The words are recovery, redesign, recreate, renew, reconstruct, rebuild, repair, rectify, relieve, recompense, rehabilitate, recuperate, regroup, re-establish, restitution, restore, replicate, reform, revive, resurrect, rebirth, renaissance and resurgence. The words from this list that we use most frequently in this book are recovery, reconstruction, rehabilitation, restoration and resilience. All of them have a decidedly optimistic flavour, expressing confidence that matters will improve and normality will eventually return (Wisner *et al.* 2004). Various cases throughout the book indicate that this is not always the case. However, for obvious public relations reasons, all parties involved in recovery, whether governments, international bodies or NGOs, adopt optimistic terminology.

At the other end of the scale, negative words that begin with the prefix 'de' can also be traced back to the Latin words for removal or reversal. Thus we have decay, degeneration, despair and deterioration. However, even if everyone knows that decline or stagnation is the reality on the ground, such words are never the favoured vocabulary of official post-disaster reports. In this context, as authors with the freedom to write without deferring to any official body, one advantage we have is to be able to look with realistic and sceptical eyes at the confident 'phoenix from the ashes' proposals of colourful brochures and upbeat progress reports. We can identify where inconvenient realities have been glossed over and the truth has been dealt with too economically. Where there have been missed opportunities, we can try to explain why.

As it can shape our attitudes, the terminology we use is of considerable importance. Thus throughout the book, we use the term 'survivors' rather than 'victims' as the latter term carries with it a notion of passive helplessness which is in contrast to the normal behaviour of those who have survived disaster. We will also frequently use the term 'sheltering' in preference to 'shelter' as the *process* of finding or providing accommodation is more significant and accurate than the *product* in terms of walls and roof.

Having shown how terms can be used in different ways, it is imperative to tackle the thorny problem of how to establish definitions of them.

Defining recovery and other terms

In reality, many phenomena associated with disaster have been given a whole range of definitions. A leading sociologist working in disaster research for over half a century, Henry Quarantelli (1998a) edited a book of reflections on the meaning of the term disaster; and Perry and Quarantelli (2005) subsequently produced a sequel. In the progenitor to the first book, Quarantelli offered the following justification:

> to be concerned about what is meant by the term 'disaster' is not to engage in some useless or pointless academic exercise. It is instead to focus in a fundamental way on what should be considered important and significant.
>
> *(Quarantelli 1995: 225)*

But he later all but admitted that definitional problems are undermining the field of disaster studies:

> If workers in the area do not even agree on whether a 'disaster' is fundamentally a social construction or a physical happening, clearly the field has intellectual problems.
>
> *(Quarantelli 1998a: 3)*

The dilemma of what definition to accept is not limited to the term 'disaster'. For example, Weichselgartner (2001) offered 22 definitions of 'vulnerability', and since the publication of his paper at least another 22 have appeared in the literature. O'Brien and O'Keefe (2013) listed 28 definitions of 'resilience' – and so the story goes on.

What can we conclude for disaster recovery from this tale of arguments over definitions? First, in the field of disaster risk reduction, and affiliated topics, the phenomena that must be characterised are complex, subtle and multifaceted. Second, as more than 40 disciplines and professions are involved in this field, it is hardly surprising that the interpretations of terms differ with the interpreter. Third, definitions can – and should – be used in appropriate contexts according to the requirements of the work with which the terms are associated. Fourth, definitions can be pluralistic, but it is important to obtain a consensus among the users that each chosen definition is acceptable for the use to which it is put. This is the concept of the 'working definition'.

The process of applying a definition to a practical problem underlines the importance of establishing the context in order to select the correct one. Quarantelli noted the richness, as well as the confusion, in the way a number of words have been used to describe recovery in the English language. In commenting on the complexity and implications of these definitions, he asked 'what's in a name?' His answer was: 'A great deal, and in this instance there is far more involved than semantic quibbling. There are, for example, policy and legal implications linked to different labels. What something is called does make a difference' (Quarantelli 1998b: 3). In the opening section of Chapter 4, we provide an example of this process, with a discussion concerning the desire of governments in Algeria and London to deceive by changing the names of places suffering disasters in order to reassure residents that 'all was well' when it clearly was not.

The sets of definitions of disasters, disaster management and disaster recovery that the United Nations and other official bodies regularly produce inevitably reflect their own perceptions as coordinators and donors of assistance (UNISDR 2009). These are valid and useful, but they may be at variance with the humble viewpoint of survivors and their communities, or of the needs and mandates of other interested parties. The reason for this is that there may be a substantial difference between the exigencies of those who supply aid and those who receive it.

Let us enquire more deeply into the definitions of disaster recovery used by the various agencies that practice it, noting their inherent strengths and weaknesses. We then offer our own.

In 2009, the United Nations International Strategy for Disaster Reduction defined recovery as follows:

> The restoration, and improvement where appropriate, of facilities, livelihoods and living conditions of disaster-affected communities, including efforts to reduce disaster risk factors.
>
> *(UNISDR 2009: 10)*

The International Recovery Platform expands this definition:

> The restoration, and improvement where appropriate, of facilities, livelihoods and living conditions of disaster-affected communities, including efforts to reduce disaster risk factors.

The recovery task of rehabilitation and reconstruction begins soon after the emergency phase has ended, and should be based on pre-existing strategies and policies that facilitate clear institutional responsibilities for recovery action and enable public participation. Recovery programmes, coupled with the heightened public awareness and engagement after a disaster, afford a valuable opportunity to develop and implement disaster risk reduction measures and to apply the 'build back better' principle.

(IRP 2014)

The Global Facility for Disaster Reduction and Recovery (GFDRR) of the World Bank provide the following definition:

Decisions and actions taken after a disaster to restore or improve the pre-disaster living conditions of the affected communities while encouraging and facilitating necessary adjustments to reduce disaster risk. Focused not only on physical reconstruction, but also on the revitalization of the economy, and restoration of social and cultural life.

(Jha et al. 2010: 365)

Definitions can also refer to particular sectors of the general problem of recovering from disaster. For example, the International Labour Office describe disaster recovery through the lens of work generation:

Decent work matters in crisis. It is a powerful, tested rope that pulls people and societies out of crises and sets them on a sustainable development path. Decent and stable jobs offer crisis-affected people not only income, but also freedom, security, dignity, self-esteem, hope, and a stake in the reconciliation and reconstruction of their communities.

(ILO 2014)

Likewise, the International Federation of Red Cross and Red Crescent Societies emphasise the human recovery and resilience needed in order to withstand future disaster impacts:

Recovery refers to those [IFRC] programmes which go beyond the provision of immediate relief to assist those who have suffered the full impact of a disaster to rebuild their homes, lives and services and to strengthen their capacity to cope with future disasters.

(IFRC n.d.)

These five definitions provide useful descriptions of recovery that relate to the roles and priorities that characterise each organisation. Together, they cover the varied *sectors* of recovery, missing, however, the recovery of the natural environment. They also discuss the *functions* of recovery, namely, to 'build back better', restore, reduce risks, improve the situation, restore dignity, and so on. However, they make no reference to the deeply rooted need to preserve *genius loci*, or 'sense of place', in any local recovery process (we discuss this in Chapter 6, '*Genius loci* and preservation of the identity of places and human settlements'). They also include the *roles* of stakeholders with emphasis on public participation and institutional responsibilities, but without explicit reference to the primary role of national and local governments.

There is an implicit concern in these definitions to ensure safety during the process of recovery, which derives from an ethical recognition of the sanctity of life. Other than that, ethical issues are conspicuous by their absence. Perceptively, Lawrence Vale and Thomas Campanella (2005) note that the process of post-disaster recovery: 'is a window into the power structure of the society that has been stricken. . . . What we call "recovery" is also driven by value-laden questions about equity.' They follow this with a list of recurring issues that go to the heart of the values that underpin recovery operations (Vale and Campanella 2005: 12):

*   Who sets the recovery priorities of the recovering communities? (See the section on 'Shelter preferences and functions' in Chapter 8.)
*   How are the needs of low-income residents valued in relation to disrupted business interests?
*   Who decides what will be rebuilt where? (See Chapter 9.)
*   Which voices carry forth the dominant narratives that interpret what transpires? (See the section on 'Voices of survivors' in Chapter 8.)
*   Who gets displaced in the recovery process? (See 'Option 8: relocated dwellings in relocated settlement' in Chapter 9.)
*   What role do non-local agencies and bodies have in setting down the guidelines for reconstruction?
*   How can urban leaders overcome the lingering stigma inflicted by their city's susceptibility to disaster?

We return to ethical concerns in Chapter 4, model 14 ('project planning and implementation'), which gives decision-making a solid ethical base. However, we recognise a more general need to build ethics into any definition. Therefore, we propose a clutch of new descriptions of disaster recovery that extract good material from existing definitions, and try to rectify the omissions noted above.

1   In contrast to the vulnerable status quo that resulted in failure during the disaster, recovery requires significant and sustained development in the form of improved, safe and sustainable environments. All sectors – physical, economic, psychosocial, environmental and governmental – need to recover in an integrated manner. (See Chapter 3, 'Model 2: recovery sectors'.)
2   The foundation of genuine recovery requires an ethical concern for equity, honesty, transparency and the protection of lives, livelihoods, culture, property and the environment. Effective recovery is most securely founded where the

government plays the lead role, citizens are actively involved at all stages and their future safety is of paramount concern to all. (See Chapter 4, model 14 and Chapter 6, 'Accountability'.)

3   All recovery assistance has to be provided in a manner that builds and strengthens the capacity of the surviving community and does not leave a deadly legacy of dependency. (See Chapters 8 and 9.)

4   In essence, recovery from disaster is a process of restoring the physical, socio-economic and mental conditions of society to a state of functionality that at least corresponds with what was present before the calamity and hopefully improves upon it by providing greater safety, more secure economic prospects and a more stable, healthy, happy society. This definition is by no means all-embracing and will not be adequate for some uses, but perhaps it is more important in this field to appreciate the plurality of views rather than to seek to unify them. (See Chapter 3, 'Model 1: progress with recovery'.)

Evolution of recovery studies: 'dwarfs on the shoulders of giants'

In a saying, to be quoted later by Isaac Newton, the twelfth-century philosopher Bernard of Chartres stated that we are like

> dwarfs perched on the shoulders of giants. . . . [W]e see more and farther than our predecessors, not because we have keener vision or better height, but because we are lifted up and borne aloft on their gigantic structure.
> *(as recorded in John of Salisbury's* Metalogicon *1955 [1159]: 167)*

This captures an underlying principle of the entire book: *while all disasters are different in scale, nature, impact and recovery actions, critical lessons can nonetheless be deduced from past experience and applied in order to create patterns of effective recovery.* Thus, all who manage the recovery process are indebted to those who have worked in this field before them whether or not they recognise their influence, whether these predecessors were giants or dwarfs in their fields, and whether the emerging lessons are positive or negative. Thus, there should be a continual learning process of interconnectedness and interdependency.

Illustrating this process of continual learning, we consider in this section some of the pioneers whose work has been built on over time. Any account that concentrates on the efforts of particular individuals is likely to miss out some of those whose influence has been significant, but this is an occupational hazard of the historian. We begin our list of pioneers with two brilliant Greek innovators working on the Hagia Sophia in Istanbul. This was the first great Christian Church, built between AD 532 and AD 537 after its predecessor was destroyed in an arson attack that was part of the Nika Riot – civil strife that accompanied the succession of Emperor Justinian in AD 532, causing the deaths of 30,000. The new basilica, built to celebrate the emperor's reign, was designed by the partnership

FIGURE 2.2 Hagia Sophia, Istanbul, built between AD 532 and AD 537. Following the collapse of the dome in an earthquake in AD 586, an iron ring beam was inserted at the base of the dome and the large buttresses were inserted on both sides of the dome to support the structure (sketch by Ian Davis).

of mathematician *Anthemius of Thralles* and physicist *Isidore of Miletus*. Later in around AD 586, another disaster – this time an earthquake – caused the collapse of the domes. This resulted in reconstruction designed by Isidore's nephew, completed in AD 588, using the brilliant innovative technology of an iron ring beam to resist the loading pressures surrounding the central dome as well as four massive buttresses. The ring beam technique was to be used later in cathedral architecture in Florence, Rome and London. However, another earthquake in AD 989 caused the dome of Hagia Sophia to collapse again, to be rebuilt in its present form (Davis 1983).

Over a thousand years later, we come to *Christopher Wren* (1632–1723), the English architect who devised urban recovery plans after the 1666 Great Fire of London and who superintended the reconstruction of public buildings including 52 churches and his masterpiece, St Paul's Cathedral – the product of 35 years' architectural work. Wren was also influential in the development of the London Building Acts that were introduced after the fire, which required fire-resistant plans, building techniques and materials. The Acts are still in force 350 years later. When Wren died, he was buried in the crypt of St Paul's Cathedral. His son placed an inscription on his tomb: *'Si monumentum requires circumspice'* (If you seek his monument, look around you). The same can be said for the other

FIGURE 2.3   Christopher Wren's St Paul's Cathedral dome and the type of fire-resistant brick architecture for urban buildings that was required in the 1667 London Building Acts (photograph by Jon Sullivan, public domain).

pioneers of this subject listed in this chapter, since their impact on the history and development of disaster recovery has been massive; all we have to do is to 'look around' to observe their impact on specific places that have recovered successfully – some described in the book – as well as their role in innovative recovery approaches, improved safety legislation, technological innovation and hazard-resistant urban planning.

The next example is the *First Marquis of Pombal, Sebastião José de Carvalho e Melo* (1699–1782) who, along with *General Manuel de Maia* (1672–1768), *Captain Eugénio dos Santos* (1711–60) and *Colonel Carlos Mardel* (1695–1763), planned, designed and executed the reconstruction of Lisbon after the devastating earthquake, fire and tsunami of 1755. Under the leadership of Pombal, these military engineers and planners created the first urban reconstruction plan in the world that placed emphasis on comprehensive safety against fires, earthquakes and tsunamis. They also used timber-framed construction to design earthquake-resistant buildings. In a similar vein, but much more recently in time, the Chicago Fire of 1871 prompted the American architect and engineer *William Le Baron Jenney* (1832–1907) to design fire-resistant steel skyscrapers.

The Canadian clergyman *Samuel Henry Prince* (1886–1960) ministered to the survivors of the Halifax, Nova Scotia munitions ship explosion of 1917.

At Columbia University, New York, he wrote a doctoral thesis on his experience and turned it into a book, which is the first significant work in the social sciences on recovery from disaster (Prince 1920). His work was narrowly preceded by an essay on the psychological impact of earthquakes written by *William James* (1842–1910), brother of the novelist Henry, who derived his knowledge from participant observation during the 1906 San Francisco earthquake when he was able to witness the destruction of part of Stanford University (James 1906). Coeval with Prince, the Chicago geographer *Harlan Barrows* (1877–1960) developed the concept of human ecology, in which disaster was seen as making people adapt to extreme conditions (Barrows 1923).

Gilbert Fowler White (1911–2006) was an American geographer who trained under Harlan Barrows at Chicago University. His doctoral thesis on flood risk management was published in 1945 as a research paper entitled *Human Adjustment to Floods*. From here he exerted a profound influence on American, and worldwide, hazard management policies by advocating a pluralistic mixture of structural and non-structural approaches, rather than one that relied heavily or exclusively on building structural defences (Hinshaw 2006).

Directly after the destruction of 230 schools in the 1933 Long Beach earthquake, Assemblyman *Charles Field* of the California State Legislature propounded legislation on building codes for such structures. Although the schools were not occupied when the earthquake struck, had it occurred at a different time of day, the death toll among children and teachers could have been very high. The 1934 Field Act is a pioneer model of anti-seismic building codes and one that provided the inspiration for similar measures throughout the world (Olson 2003).

Otto Koenigsberger (1908–99) was a German architect and planner who founded the Development Planning Unit (DPU) at University College London. He taught a vital lesson to recovery planners: in the developing world, they should be prepared to adapt their plans dynamically and involve local communities and techniques – an approach he called 'action planning' – as opposed to imposing a static master plan based on Western ideas. Koenigsberger always regarded disasters wearing the 'yellow hat' of optimism that we describe in Chapter 12. Ian recalls: in 1978 while supervising my PhD research, Koenigsberger (cited in Davis 1978: 66) sent me a note to remind me that:

> Relief is the enemy of reconstruction. Therefore minimise relief. Even the minimal relief operation stretches the public sector executive capacity to the utmost. Therefore avoid paternalism. The public sector must not touch any jobs the people can do themselves. The last thing the public sector should do is the construction of houses of any kind.

Then Koenigsberger turned his attention to the opportunities presented after disasters: 'Under the immediate impact of a disaster, people are ready to change long-standing methods and customs. Therefore, act quickly to introduce improved construction methods and bye-laws' (Koenigsberger, cited in Davis 1978: 66).

Adolf Ciborowski (1919–87) was a Polish architect and planner who took charge of the post-war rebuilding of Warsaw for eight years starting in 1956. He then had a major role in rebuilding Skopje after the 1963 earthquake, which was the first reconstruction of a major town or city to use seismic microzonation as a means of identifying the local hazardousness of place (UNDP 1970). The reconstruction of Skopje is described in Chapter 7 ('Reform vs continuity in the reconstruction of Skopje, Yugoslavia (1963–90) following the earthquake of 1963').

The American sociologists *Henry Quarantelli* (1924–) and *Russell Dynes* (1924–) founded the Disaster Research Centre at Ohio State University (they subsequently brought it to the University of Delaware). They have conducted many systematic studies of recovery and reconstruction after disaster, and their findings are reviewed in various places in this book. Dynes has had a particular interest in the reinterpretation of the Lisbon earthquake of 1755 while Quarantelli has produced social models of the processes of community and organisational recovery. In the beginning of Chapter 9, we discuss Quarantelli's important research finding that the progress of recovery rarely follows the tidy linear sequence so frequently assumed by agencies.

John F. C. Turner (1927–) is a British architect whose central thesis is that housing is best provided and managed by those who are destined to live in it, rather than being centrally administered by the state (Turner and Fitcher 1972). Since the Guatemala earthquake of 1976, this focus on user-build housing has been highly influential in disaster reconstruction and has now become the approach favoured by UN-Habitat and the World Bank (Kreimer *et al.* 1998). There are descriptions of user-build housing reconstruction in Guatemala in Chapter 9 ('Option 7a: user-build permanent dwellings') and with respect to rural Pakistan in Chapter 12 ('Yellow hat – optimism').

The Greek-born scientist *Nicholas N. Ambraseys* (1929–2012) can be considered to be the founding father of engineering seismology (Bilham 2013). He was actively involved in providing seismic design and planning expertise for the reconstruction of Skopje and other sites in Europe and the Middle East.

The American scholars *Robert W. Kates* (1929–), *John Eugene Haas* and *Martyn J. Bowden* (1935–) were the authors of the first book to propose an overarching theory of reconstruction (Haas *et al.* 1977). This was articulated in their first chapter, written by Kates and his student David Pijawka, and consolidated in the examples included in the rest of their book (see Chapter 3, 'Model 6: the Kates and Pijawka recovery model').

Last, this brief round-up of academic studies of recovery would not be complete without mention of the redoubtable Texan architect and urban planner *Fred Cuny* (1944, disappeared 1995). He had a decisive impact on disaster recovery and safe low-income housing recovery in developing countries (Cuny 1983; PBS 2013). Following the Guatemala earthquake of 1976, he developed the theory of user-build housing reconstruction (see Chapter 9, 'Option 7a: user-build permanent dwellings'). He was active in refugee camp planning and water supply in various contexts, including the siege of Sarajevo, 1992–5. Tragically, he disappeared (assumed murdered) in 1995 while on a humanitarian assignment to Chechnya (Anderson 1999).

Regrettably, the history of academic studies of recovery has so far been dominated by men, and it is all the poorer for the absence of women. However, given the high proportion of women currently studying disaster risk and recovery in universities, that situation may soon change for the better. Nonetheless, in a cumulative manner, these pioneers have defined the subject of recovery from disaster. They, and others, have mapped some of the most important dimensions of the field, including: the achievement of safety by planning settlements and designing buildings appropriately; the promulgation of appropriate regulations and standards; the search for new and adaptable ways to design and build structures; the roles, perceptions and behaviour of communities, their members and the groups that assist them; the essential phases of recovery; and the critical link between disaster recovery and development processes.

Defining a vision for recovery

> A visionary – regards difficult situations, not just as problems to solve, but as opportunities for creation and collaboration. To present a challenge that calls forth the best in people and brings them together around a shared sense of purpose, leading to a community united around an inspiring goal.
>
> *(Anon.)*

This chapter has shown that effective disaster recovery relies on cumulative and collective action by a wide range of participants. They inhabit the varied departments and levels of government, the UN and international agencies, national and international NGOs and local associations, as well as the private sector and many elements of civil society. All these individuals and groups need to be united around a common endeavour, which is the search for a clear and realistic vision of what recovery should entail, how each actor can play a part in it and where the task is heading.

This requires continual encouragement from visionary people who are capable of challenging political apathy or infighting and are able to maintain the sense of direction despite bureaucratic, financial, political and logistical obstacles. Inspired leadership is needed from the top to the bottom and from all the agencies that are responsible for recovery, but most of all from the source of political authority.

The vision is most likely to be created and nurtured by inspired individuals, such as the pioneers discussed in the previous section. But they are only one of the three principal elements in recovery. The others are structures and policies. It is possible to have a visionary leader who has to operate within a hidebound structure and is tied to a string of lacklustre policies. On the other hand, there are also examples in which there has been a remarkable sense of vision in creating innovative structures and policies. Moreover, there are rare examples in which a strong vision can be recognised in leaders, structures and policies. The rural reconstruction of Pakistan following the 2005 earthquake is a particularly good example of this creative fusion (see Chapter 12, 'Yellow hat – optimism').

The next section will articulate vision in terms of the 'rules of the game' and the prerequisites for intelligent and effective recovery from disaster.

Vision and reality

By their very nature, visions may be transient and may easily vaporise. They are real and relevant when they are expressed but subject to dynamic change as their context develops. Furthermore, there are all manner of reasons why hindsight can prove visions to have been mistaken. Models 1 and 2 in Chapter 3 highlight a progression of visions of recovery within recovery scenarios. The scenarios are: (1) no vision, (2) limited vision, (3) wrong vision, and (4) inspired vision.

When, on 8 November 2013, Typhoon Haiyan (Yolanda) devastated the eastern-central regions of the Philippines, an urgent request was sent by local NGOs and their international partners working in relief operations for guidance on recovery actions. This was received by Dr Anshu Sharma, an Indian architect and planner, who is the Director of the NGO Safer World Communications. He contacted Ian with a request to develop a brief statement constituting an initial 'vision for recovery'. Based on their shared experience, the two architects developed seven 'golden rules of disaster recovery', though at that stage neither had visited the disaster site (Sharma and Davis 2013). David visited the region four months after the typhoon and his case study is reported in Chapter 9. This provided the opportunity for David to reflect on whether Ian and Anshu's 'golden rules' are fully valid, limited or simply wrong. Therefore, the following set of original statements by Ian and Anshu is complemented by David's field experience. We include this dialogue to indicate the dynamic nature of perceptions of the initial phase of disaster recovery.

Seven golden rules for disaster recovery

Suggestions for officials supporting recovery actions following Typhoon Haiyan (Yolanda) in the Philippines.[1]

IAN AND ANSHU: *1 Trust survivors and avoid paternalism.* The recovery process is totally demanding and all resources – international, national and local – need to be mobilised for the task. However, assisting groups must at all costs avoid paternalism, and must not treat survivors as passive 'victims' instead of active participants in the recovery process. Here, there is a principle to respect: the 'survivors' are the primary actors in their own recovery, and while assisting groups can play vital roles, they must avoid creating dependency. Thus, they must *never* undertake any task that duplicates what the survivors can accomplish themselves. Their active role in the process is a vital part of their own psychosocial recovery. The recovery process needs to be 'demand driven' by the community of survivors rather than 'supply driven' in such a way that agency and commercial self-interest dominates it.

DAVID: *Philippines perspective at four months.* In the Tacloban area of the Philippines, the 'epicentre' of destruction caused by the typhoon, many of the survivors enjoyed a high degree of autonomy, but this may have been because they were still waiting for assistance that they were unable to provide themselves. Others were sent to 'bunkhouses' – areas of cramped transitional housing. The Government's decision to institute a 'no-build', or setback, zone along the coast left many shoreline residents in limbo four months after the disaster, with no idea of what the future would hold in store. While many families showed considerable resourcefulness in finding shelter and minor forms of employment, in the devastated shoreline neighbourhoods, there was a palpable sense of abandonment and detachment from the processes of government.

IAN AND ANSHU: *2 Enable survivors to assess their own needs.* Where possible, encourage communities and individuals to assess their own post-disaster requirements and thus avoid inadequate or duplicated needs assessments by external agencies.

DAVID: *Philippines perspective at four months.* In Tacloban and neighbouring towns, survivors did assess their own needs. Seldom did any agency do it for them. However, the scarcity of assistance was probably more important in this than any deference to their judgement. Mere assessment of needs did not guarantee that there was any means of supplying them. Areas that were poor and deprived before the disaster seemed destined to remain so after it.

IAN AND ANSHU: *3 Provide survivors with cash rather than kind.* Trust survivors by providing them with cash grants and trust them to determine whether they need shelter materials, blankets, boats, and so on. If families are not yet in the banking system, help them to get into personal banking, and distribute aid directly through this mechanism, thus avoiding middlemen, which is where corruption often starts.

DAVID: *Philippines perspective at four months.* One international NGO autonomously followed this precept and the cash it supplied to survivors saved them from destitution. The money was used to buy food, building materials of the means of following a trade (a fishing boat, rickshaw, goods for a small shop, etc.). Problems such as gambling, corruption, mafias, theft, and so on do not seem to have been augmented by the distribution of money. The main problem with this approach was that injections of cash into the local economy drove up the price of basic supplies, such as building materials. The Government did not control prices or increase supply enough to drive them back down.

IAN AND ANSHU: *4 Think locally.* Localise all interventions. For example, inject cash into local markets and use locally available materials and resources.

Through this approach, it is possible to restore livelihoods and resurrect the devastated local economy. People with work can help finance their own recovery. Instead of the common approach of allocating contracts to external groups that compete with each other (and with local concerns) and take over the recovery process, support local agencies and suppliers. Support the recovery of local governments, which should play a vital role in local coordination processes. International NGOs must avoid hiring local officials, at inflated salaries, which may take them away from their work in government – an approach that can seriously weaken local administration.

DAVID: *Philippines perspective at four months.* Efforts to use local building materials were hampered by problems of supply. For example, lumber supply was limited by inadequate sawmill capacity. Roofing material was supplied from other parts of the Philippines, but this did little for the local economy. Cash-for-work programmes were instituted by the Government and some international NGOs, but many of these were of short duration. This is a pity as there was much work to do, a great shortage of employment and a sense of stagnation in the local economy.

IAN AND ANSHU: *5 Give priority attention to vulnerable groups.* Pay attention to vulnerable groups such as minorities, the sick, people with special needs, people with disabilities, elderly citizens and pregnant women. Also pay attention to communities in remote areas.

DAVID: *Philippines perspective at four months.* Field investigation found little evidence that this was occurring. Some provisions were made to improve obstetrics in the local health centres, but it seemed that vulnerable people were largely left to their own devices or cared for by families, not the NGOs or Government.

IAN AND ANSHU: *6 Think 'process' not 'product', 'sheltering' not 'shelters', 'housing' not 'houses'.* Regard sheltering as a dynamic process, rather than merely a collection of tangible products. For example, encourage sheltering of displaced families with host families and provide support to those who have opened their houses to survivors. Where possible, avoid a three-stage sheltering process that extends from emergency shelters (such as tents) to transitional housing to permanent reconstruction. Seek to extend the first stage and accelerate the construction of safe permanent dwelling, thus avoiding the waste of resources in building transitional housing, which can delay reconstruction [see Chapter 4, models 16 and 17].

DAVID: *Philippines perspective at four months.* Many families did shelter with relatives, but this does not appear to have been an organised or subsidised process. It seems that the main reason why shelter was a dynamic process in the Tacloban area was because the final outcome was unclear to those in need of shelter. Almost none of the survivors that we interviewed had had any help in constructing shelter. Given the lack of clarity in long-term processes, that which they had built was necessarily to be considered temporary or transitional. This was especially true for the shelters to be

FIGURE 2.4   Entrance to an informal settlement on the seaward side of beached trawlers, north Tacloban, Eastern Visayas, Philippines (photograph by David Alexander).

found on the seaward side of a collection of beached ships, locally and informally designated 'Yolanda Village' (Figure 2.4). After four months, the long-term strategy was unclear and, hence, so were the prospects for recovery of the people living in bunkhouses and informal shelters.

IAN AND ANSHU: 7 While addressing short-term needs, *adopt a long-term perspective.* Start focusing on recovery from the initial stages. Early on, identify possible exit strategies and follow-up measures. Prepare for the next major disaster as soon as possible: building resilience will save lives and livelihoods.

DAVID: *Philippines perspective at four months.* Eastern Visayas is not an economically buoyant part of the Philippines. Instead, it is a marginalised and relatively poor area. As many assets that provided work were destroyed, Typhoon Yolanda was marked by a fall in employment. Heavy emphasis on reconstructing infrastructure was probably justified, in that this (roads, bridges, electricity distribution network, and so on) is fundamental to many other activities, but the picture for survivors was opaque regarding the long-term prognosis. Hence, the survivors continued to live in precarious conditions, involving extreme vulnerability to further natural hazard impacts. Opportunities to lift people and families out of a state of destitution were not being taken, yet this could have been done with relatively modest resources.

Summary

In this chapter we have set the scene for the rest of the book. We have tried to weave together some essential elements as we have described examples of recovery and emphasised their dynamic and varied contexts. We then explained how differing standpoints result in diverse understandings of the processes of recovery and introduced various definitions that contribute to the 'language of recovery'. We have acknowledged our debt to some of the giants of this subject and have concluded the chapter with a quest for a vision of recovery, though we showed how this can change substantially in a very short period. We trust that we have succeeded in revealing the scope and complexity of the subject. While many deeply rooted problems have been described, there have also been examples of positive approaches and successes, and these will be elaborated in subsequent chapters.

In the next two chapters we enter the world of models that seek to capture and clarify disaster recovery using memorable visual images. As these chapters lay out ways of regarding the nature, the phases, the safety factors and the organisation of disaster recovery, they underpin the entire book.

Note

1 For a further description of early recovery following Typhoon Haiyan, see also Chapter 9, 'Option 6a: transitional shelter (temporary)' and Figure 9.11.

3

MODELS OF RECOVERY

Development and phases

Management models are designed to resolve common problems and challenges. . . . At best they will provide a new way of seeing a situation that will result in positive change. The models may be applied strategically, tactically or operationally; some are problem-solving tools, designed to improve efficiency and effectiveness; most are designed to solve specific problems arising out of a specific situation.

(Van Essen et al. *2003: x)*

Introducing models

Having laid down a foundation for the book in the first chapter with positive and negative examples of full and limited recovery, and having established the context for the book in Chapter 2, we now turn to a description of the models that we use throughout the book as visual references or representations for the varied themes and ideas that fill our remaining chapters. We have grouped the models into four categories:

- development and recovery;
- phases of recovery;
- safe recovery;
- organisation of recovery.

By definition, a model is a simplification of reality; and a good model simplifies the complexity of real situations elegantly. Busy readers may wish to focus their attention on this chapter, in which we have selected those models that we regard as the best at simplifying complexity down to its essential elements, helping explain it and showing the most important connections between related factors.

Some of these models have helped us as we have struggled over many years to make sense of the complexities of recovery, and we trust that this will also be the experience of readers.

According to Krogerus and Tschäppeler (2011), to be effective, models of decision-making must fulfil some or all of the following criteria:

- simplify a given situation or issue to its essential components;
- provide new ways of viewing common situations;
- expand our perceptions and understanding;
- contain strong visual images that clearly express ideas which are not easy to convey in words;
- express complex interrelationships and processes;
- provide simple ways to communicate;
- act as aids to memory, or tools to analyse situations and explore a range of solutions.

We echo the quotation from Van Essen *et al.* (2003) that heads this chapter: 'At best they [management models] will provide a new way of seeing'. That is the purpose of this chapter and the reason why we use models in our work. But we have a rather more jaundiced view of management models than that of Krogerus and Tschäppeler. Many models seem highly artificial and some are downright misleading when applied to disaster situations. Our concern in using or applying any model is that there is an inherent risk of oversimplifying or distorting a situation. Representations of the complexities of reality are inevitably rather crude two-dimensional, static images that seek to convey multidimensional dynamic processes. In describing various models in this chapter, we note some of their inherent problems and limitations. However, we sincerely hope that Til Schuermann's comment does not apply in the following two chapters: 'when models turn on, brains turn off' (cited in Strachnyi 2012: 6).

Moreover, users should not endow models with excessive authority: they are always approximations of reality. As we describe, some models have serious defects, and problems have arisen when they have been used repeatedly with unquestioning acceptance rather than being discarded or revised. Therefore, in this chapter, we have tried to avoid models that have accumulated too many weaknesses. Model formulation is a dynamic process for, as new insights occur, each model needs to be adapted. In addition, *all* the models presented in this chapter need to be modified by their users to suit specific situations.

Dr Gustavo Wilches-Chaux, a highly experienced Colombian expert with extensive experience in managing disasters in his country, commented as follows on the difficulties of using models:

> The idea of developing 'models' that can be copy pasted with equal results in every disaster situation is a big mistake that should be avoided. One strategy that has had successful results in a given historical moment and/

or in a given environmental, social or cultural context could be a total failure in a different context. And the opposite: one strategy that fails in one situation could be successful in a different context. The most that one can expect is to have a toolbox, with concepts, principles and strategies that should be tested and adapted (or discarded) in front of each unique challenge.[1]

We offer a total of 21 models, classified according to the four categories listed above. Each category is introduced with a summary chart that describes the models, their application and their source, as well as indicating the main places where they are referred to in the book. See Appendix 1 for a summary of models introduced in Chapters 3 and 4.

Development and recovery models

TABLE 3.1 Summary of development and recovery models (1 to 4)

Model	Graphic representation	Application	Source	Location in book
1 Progress with recovery		Describes four strands of effective recovery with the characteristics of four recovery scenarios; can be used as a monitoring device	New model developed for this book by the authors; tested in Davis (2012)	Chapter 1: case studies of Malkondji, India and Belice, Italy; Chapter 3: case study of Wenchuan, China
2 Recovery sectors		The twin of model 1; here, the recovery scenarios are applied to five recovery sectors; can also be used as a monitoring device	New model developed for this book by the authors; tested in Davis (2012)	Chapter 1: case studies of Malkondji, India and Belice, Italy; Chapter 3: case study of Wenchuan, China
3 Development recovery and elapsed time		This 'timeline' model shows the relationship between development quality and elapsed recovery time	This model appears in various publications; original source is unknown	Chapter 10: example of effective recovery following the Chile earthquake

Model	Graphic representation	Application	Source	Location in book
4 Relationship between disaster and development		Positive and negative realms of disaster and development are represented here; usefully draws the linked, but often divided, disaster/development spheres together	Stephenson (1991)	Examples of development occurring in recovery in Chapter 1 in Malkondji and Chapter 12 ('Yellow hat – optimism') in rural Pakistan

Model 1: progress with recovery

The value of this model (Figure 3.1) is in its description of the stages or scenarios of recovery.

This model owes part of its origin to the work of Kates and Pijawka (1977), who distinguished between recovery designed to replace what was lost and that which goes beyond mere replacement and is intended to show that the area affected has overcome the problems and setbacks caused by the disaster (see models 6 and 7). In many respects, this diagram is the key to the entire book, as it conveys the message that effective recovery management has to develop well beyond the vulnerable status quo that gives rise to any disaster.

A challenge from the floor in a disaster management workshop in 1983: Ian Davis was leading an international workshop on disaster management in Oxford in 1983 when a revealing exchange took place between a speaker and a participant. A speaker from the International Federation of Red Cross and Red Crescent Societies (IFRC) was making a presentation on disaster recovery. He projected a diagram that stated that the aim of disaster recovery is to restore normality. This resulted in an outburst from the back row, where a senior nurse from Jamaica with extensive disaster experience was sitting:

> May I protest; please do not use that slide ever again! You are stating that the aim of disaster recovery is to return to the status quo, but that must be fundamentally wrong! In the city where I live, Kingston, Jamaica, we have people living in cardboard boxes; so are you seriously suggesting that after a future hurricane or earthquake in Kingston, we need to rehouse them back in these boxes? No Sir! Any recovery must move forward to a better and safer future since 'normality' equals vulnerability.

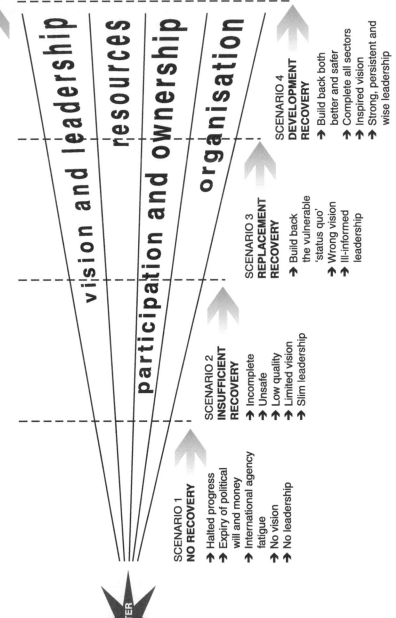

FIGURE 3.1 Model 1: progress with recovery.

The speaker was reduced to stunned silence by the force of her interruption, but it is not known whether these passionate words had any effect on his subsequent lectures.

This model has a double message. The first concerns an expanding vista that grows from a point that indicates the start of disaster recovery into four strands, each of which is necessary to its success. Without 'vision and leadership', recovery is likely to drift into gross inefficiency or stagnation. In a frequently quoted saying from the Book of Proverbs, the wise King Solomon declared: 'Where there is no vision, the people …' – evidently the next word is inaccurately translated from the original Hebrew as 'perish', but a more correct rendering would be: 'the people run wild' or 'the people act without restraint' (Proverbs, Chapter 29, Verse 18). We are aware of situations after disaster in which Solomon's words would accurately describe some exceedingly unwise mismanagement that lacked vision and failed to promote recovery. Some of these cautionary tales are described in this book.

Without political, financial, material or human 'resources', nothing substantial can be created. Without 'participation', affected residents and stakeholders will certainly have no sense of 'ownership' of the reconstructed environment. Finally, without 'organisation', recovery will be erratic, piecemeal and wasteful.

In synthesis, this means that in most cases, recovery from disaster opens a 'window of opportunity' for the improvement of safety, the reduction of vulnerability and the creation of environments that are more efficient at satisfying their users' needs. It represents a situation in which change is inevitable and it is imperative to ensure that this is positive. A 'negative window of opportunity' would give free reign to criminal elements, corruption and vested interests and is, of course, to be prevented. Transparency and a participatory approach are the means by which recovery can be kept on a positive track.

Throughout the book, we provide a series of positive and negative examples that substantiate our conviction that these four strands are essential. The entire history of disaster recovery is a procession of frequent failures and rare success stories.

In relation to the second message derived from this model, we relate the unfolding recovery process to four scenarios, the first of which is the worst solution and the last of which is clearly the best target.

- *In Scenario 1, there is no recovery:* Typically, for years after the disaster, ruined buildings abound, infrastructure is damaged and inefficient, half-finished projects are littered about the landscape and there is clear evidence of failure to deliver on promises made early in the recovery process (e.g. Managua, Nicaragua after the 1972 earthquake).
- *In Scenario 2, recovery is insufficient or erratic:* Some buildings have been reconstructed, but there is widespread unemployment, the economy is limping along, society remains fragmented and the environment remains degraded – for example, trees have not been replanted (e.g. Port-au-Prince, Haiti following the 2010 earthquake).

- *In Scenario 3, replacement recovery restores the status quo ante:* Significant attempts have been made to recover the pre-disaster situation, but alas, this so-called 'normal' situation is characterised by high levels of the same vulnerability that gave rise to the disaster. (For centuries, this was the case in Europe and the Middle East when floods or earthquakes occurred — only recently has there been any attempt to reduce vulnerability during rebuilding).
- *In Scenario 4, developmental recovery occurs:* There are rare examples of recovery in which positive development has taken place to produce a better and safer environment. In Figure 3.1, four key elements are listed: build back better and safer, complete all sectors, make use of an inspired vision, and ensure that leadership is strong, inspired and wise. (The example of the reconstruction of the Indian village of Malkondji described in Chapter 1 is a fitting example of recovery that fulfilled this scenario.)

This model can be used as a monitoring device to enable government and donor officials to check on progress as they strive to reach the final scenario.

Having proposed this model, we can immediately see some of its limitations. For example, the four radiating strands simplify reality to its barest essentials, but they contain words that raise problems. While the word 'vision' is certainly uplifting, and few would deny its importance, the questions remain as to *whose* vision should be represented and vision of *what*? The vision for recovery of the chief executive or shareholders of a firm of building contractors will inevitably differ from that of the leader of an NGO concerned to make recovery sustainable for the most vulnerable citizens. For example, Ian Davis gave advice to the manufacturing association of prefabricated housing following the 1980 Campania–Basilicata earthquake in Italy. An exhibition organised in order to present what was on offer had a banner in letters about two metres high, that stretched across the entire exhibition hall with the slogan: 'One man's disaster is another man's marketing opportunity.' This could be rephrased as 'one man's tragedy leads to another man's commercial gain, or vision'. Morally speaking, leadership vision should not be confused with exploitation. The issue of 'provided shelter and housing' is discussed in Chapter 9, options 5b ('provided shelter') and 7b ('contractor-build permanent dwellings').

Another problematic word is 'participation'. It describes an essential process and has become the mantra of every fundraising application or project evaluation. However, the first reality is that in most countries, and almost certainly in countries with authoritarian regimes, the very notion of participation is alien. Moreover, participation is often used in conjunction with the word 'community'. There is a prevalent feeling that communities are the solution to all problems of disaster vulnerability, risk and recovery (Berkes and Ross 2013). We would like to point out that, useful and fundamental as they are, some communities are heterogeneous, at odds with themselves, dominated by vested interests (including the process of so-called 'elite capture' — see Platteau 2004) or lacking a sense of direction or purpose. Moreover, there is no inherent geographical scale at which community exists: it can vary from a single street to a global phenomenon.

Nevertheless, it may be perilous, or at least unfair, to ignore the community. An example occurred in the late 1990s when one of Ian's close friends, David Oakley, an architect and planner, conducted an evaluation of a UN-sponsored project that involved disaster risk reduction activities in rural Pakistan.[2] Oakley concluded his evaluation report with the criticism that there had been no consultation whatsoever with the residents of settlements in the floodplain before measures had been put in place that significantly affected their houses and livelihoods. He proposed that in future, participation should be built into all projects. He completed the study and it was duly sent to the relevant ministries of the Government of Pakistan. Just before leaving to return to the UK, Oakley received an unexpected invitation to meet one of the ministers in Islamabad to discuss the evaluation report. He arrived and was met by the minister who had been a military general before his political career. This man's explanation of why he had sent for him 'spoke volumes' in Oakley's words:

> Mr Oakley, I have read your report from cover to cover and I have invited you here to ask you just one question. You say that there should have been full participation of the local communities before the flood risk reduction project was undertaken. Well, I want to ask you this, since we do not consult these local communities on any other issues, why should we single out flood protection for consultation?

David Oakley told Ian that for the first time in his life he was reduced to total silence by this extraordinary question.

The second reality to recognise is that the price of the extensive participation of those involved will almost inevitably be long delays. It is also possible that conflicts between rival approaches to recovery will be intensified through the participatory process. 'Participation' is often used in conjunction with the word 'community', and much faith is put in the ability of communities to be the source and location of reconstruction (Duyne Barenstein and Leemann 2012). However, communities are not necessarily homogeneous, receptive to good practice or capable of working in harmony (Davidson *et al.* 2007). Moreover, in any society, rich or poor, community agendas can be subject to 'elite capture' by the most powerful members (Kundu 2011). Hence, 'community' and 'participation' are words that should be used with caution. The strong links between 'participation' and 'accountability' are discussed in Chapter 6 ('Accountability').

Perhaps the main reason why recovery from the Sichuan earthquake in China was completed in an astonishing four years (2008–12) may be due to authoritarian leadership and minimal participation of the surviving population in decision-making (Bernal and Procee 2012). The same could be said of the exceedingly rapid rebuilding of the areas of north-eastern Japan affected by the earthquake and tsunami of March 2011, in which many key decisions were made remotely by national leaders in Tokyo, regardless of local concerns.

Model 21 (Chapter 13) compares the strength of participation with the strength of government.

Recovery from the Wenchuan earthquake, Sichuan Province, People's Republic of China, 12 May 2008

The recovery following the 2008 earthquake in Wenchuan is discussed here as it relates to model 1, 'progress with recovery', and model 2, 'recovery sectors'.

According to Han *et al.* (2014), the Sichuan disaster was the most destructive earthquake with the largest devastated area and the most difficult emergency relief challenges to have occurred in China since the founding of the People's Republic. This magnitude 7.9 seismic event occurred at 2.30 p.m. local time. It struck an area of about 130,000 square kilometres (roughly the size of South Korea) in which 46.3 million people lived (more than the population of Canada), 43 per cent of whom (19.9 million) were directly affected by the tremors. In all, 69,227 people were killed, but almost 18,000 others were unaccounted for and so the final death toll may have been as high as 87,150. Some 4.3 million people were injured. An estimated 20 million houses were destroyed or damaged in rural and urban districts, which left 15 million people homeless. Temporary housing had to be provided for 12 million people (comprising about 4 million families), and livelihood support was given to almost 9 million. More than 11,000 medical centres, 7,444 schools and 47,000 kilometres of roads were damaged. Floods, rockfalls and landslides damaged 60 per cent of farmland in the affected area. Damage was valued at US$116 billion, of which US$31.3 billion (27 per cent) represented the cost of damaged housing.

Like the earthquakes and tsunamis in the Indian Ocean in 2004 and the Tōhoku area of Japan in 2011, the Wenchuan earthquake remains one of the most complex and challenging disasters of modern times. The scale and nature of the disaster and its unique characteristics make it an essential case to be taken into consideration in any study of recovery. When observing the effects of a disaster, emotions tend to swing back and forth from the negative to the positive, from anger to awe. One may experience acute sadness at the appalling waste of human lives, particularly the loss of children, which is so often an effect of human weakness in the form of corruption or official negligence; but this can change into admiration at the rich ability of diverse societies, at all levels, to cope with their respective trials and tribulations, or even triumph over them.

In June 2014, Ian visited Sichuan to observe progress in reconstruction six years after the earthquake. The following are some of his observations on the situation.

Human rights and the vulnerability of schools

While this book is about disaster recovery, not disaster vulnerability, we still need to consider how the failure of buildings gives rise to disaster since vulnerability

TABLE 3.2 Comparison between the impacts of the Kashmir and Wenchuan earthquakes

	Kashmir, Pakistan 2005	Sichuan, P.R. China 2008
Deaths	74, 676	69,227 deaths
		17,923 missing
		(totalling 87,150)
No. of children killed	18,091	5,333
	(24% of total)	(6% of total)
No. of teachers killed	853	Unknown
No. of schools destroyed	3,424	7,444

and risk can be perpetuated by reconstructing buildings that are as shoddy and unsafe as those they replace (see the 'crunch' model – no. 10 in Chapter 4). As data are not available on the vulnerability of rural and urban dwellings, the following discussion concentrates on school buildings in Wenchuan. In total, 7,444 schools were damaged in the earthquake, and this led to a disproportionate number of deaths among children. It is worth making a comparison here with the losses of schools and pupils in the 2005 Kashmir earthquake in Pakistan, which were deemed to be very large (Table 3.2).

It is probable that the much higher ratio of deaths of children to total deaths in Pakistan relates to the much larger number of children in the average Pakistani family, where they constitute a higher percentage of the population compared to that of Sichuan Province where the 'one child per family' policy prevails.

Sophie Richardson, Advocacy Director of the Asia Division of Human Rights Watch, wrote the following about government actions in China:

> Rather than conduct impartial investigations into allegations of shoddy construction, or provide a full accounting of those who died in the schools, national and local officials opted instead to persecute those who were asking the questions. Grieving parents were told not to try to take cases to court, or bought off in order to drop their complaints.
>
> *(Richardson 2010)*

Richardson observed that within hours of the Sichuan earthquake, the Government's Central Publicity Department (which had previously been identified in English as the Propaganda Department) began to dictate to the Chinese press precisely how the disaster was to be covered. The coverage would 'uphold unity and encourage stability' and stress 'positive propaganda'. By the end of the month in which the earthquake occurred, Chinese journalists had been instructed to minimise coverage of the school collapses.

Throughout the region, grieving parents mounted protests against the authorities over the collapse of the school buildings and the lack of information they were receiving. The deaths of children were particularly tragic in China on account of the 'one child per family' law, and this factor added to the suffering of the parents. As a response to the protests, the government offered each family that had lost a child in a school collapse a lump sum, equivalent to around US$8,800, and a guarantee of a pension in return for silence (*The New York Times* 2009).

Three Chinese activists – campaigning journalist Huang Qi, literary editor and environmental activist Tan Zuoren and the artist Ai Weiwei – took up the parents' cause by seeking to find out the names of all the children who died, in which schools their deaths occurred, and exactly how shoddy construction had led to the school collapses. The Chinese authorities persecuted the three men violently for taking this initiative. In 2009, Huang Qi and Tan Zuoren were imprisoned for 'revealing state secrets' and 'subversion'. In reality, what they had done was criticise the Chinese authorities (US Government 2009). Tan was released in March 2014 after five years of confinement in prison, while Huang spent three years in prison. On Huang's release, he continued to suffer from violent headaches as a result of the beatings he had received from the police in Chengdu.

Ai Weiwei became internationally famous when his concept of the 'bird's nest' inspired the design of the Olympic Stadium in Beijing. He played a key role in helping the parents who lost their children to find out how the tragedy happened. With the help of 50 volunteers, he attempted to collect data about the deaths in schools; he then placed the information on his blog and used it in art exhibitions in Germany in 2009 and Washington in 2013. His intention was to draw international attention to the denial of human rights in Sichuan (Elegant 2009). His museum exhibits included the display of school satchels, which symbolised the missing children who died in the earthquake. In the 2009 exhibition, the satchels covered the entire façade of the Haus der Kunst art gallery in Munich. A series of Chinese characters formed by the backpacks spelt out the words of a parent whose daughter had died in the disaster: 'She lived happily for seven years in this world.' Ai explained that his idea to use the backpacks:

> came from my visit to Sichuan after the earthquake in May 2008. During the earthquake many schools collapsed. Thousands of young students lost their lives, and you could see bags and study material everywhere. Then you realize individual life, media, and the lives of the students are serving very different purposes. The lives of the students disappeared within the state propaganda, and very soon everybody will forget everything.
>
> *(Ai Weiwei, cited in McMahon 2009)*

A poignant exhibit was included in the exhibition in the Hirshorn Museum in Washington in October 2013, which spoke volumes about the reluctance of the authorities in China to provide information to disaster survivors. Ai Weiwei included a scan of his head to show the cerebral haemorrhage he sustained on 12 August 2009, two weeks before he attempted to give evidence at the trial of Tan Zuoren, when he was beaten by the Sichuan Police. The circumstances that led up to his beating by the police as well as the MRI brain scans showing his cerebral haemorrhage have been fully documented (Mason 2014).

The persistence of these brave activists in gathering statistics eventually compelled the Chinese Government to reveal that 5,333 children died in the earthquake. The activists may also have pushed the authorities to develop China's first national human rights action plan, issued on 13 April 2009 (Government of China 2010). Ai Weiwei continued to receive information from survivors about deaths in their families, and he estimates that the official figure of child deaths represents 80 per cent of the actual total. Significantly, he suggests that 3,500 of these deaths occurred in only 18 of the 14,000 schools that were damaged or destroyed (Grube 2009).

It is important to note that the imprisonment and beatings of the activists violate Chinese citizens' rights to freedom of expression and information, as guaranteed under both international law and China's constitution. Such action against critics also contravenes the aforementioned human rights action plan, which contains clauses of direct relevance to the Wenchuan earthquake; for example, by committing the government to 'respect earthquake victims (and) register the names of people who died or disappeared in the earthquake and make them known to the public' (Government of China 2010).

When interviewed by *The Economist* in 2009, Ai Weiwei suggests a reason why the schools had been so shoddy and unsafe. He believes that these structures:

> were erected across China because of the government's drive to provide enough classrooms for all children to undergo nine years of compulsory education. Building costs were supposed to be shared by central and local authorities, but the latter often failed to chip in. This led to quality problems.
>
> (The Economist *2009)*

In February 2009, a government report noted that many school buildings were poorly constructed, with the disturbing information that 20 per cent of the primary schools in one south-western province 'may be unsafe' (*The New York Times* 2009). Further confirmation that catastrophic failures among schools and other buildings may have been strongly related to corruption and criminal negligence came from another direction: although there has never been official confirmation, Ian has been reliably informed from an internal source that 15 officials in charge of enforcing the quality of building work, who approved substandard, unsafe school buildings and thus failed in their inspectorate roles, were executed by the Chinese authorities.

Shelter provision

The government had stockpiles of 300,000 tents, but within a month, this was expanded to 900,000 due to the demand. In addition, 300,000 quilts were allocated.

FIGURE 3.2 Mass production of temporary 'cabins' (photograph by Ian Davis, taken in Wenchuan Earthquake Museum).

Some 677,131 wooden-plank, so-called 'makeshift', cabins or houses and some larger prefabricated school units were stockpiled in eight localities distributed around China, including Chengdu, Sichuan. The temporary house units (Figures 3.2 and 3.3) were 5.4 metres long and 3.6 metres wide, making a total of 19.4 square metres. These units were occupied by displaced families for up to three years from around July 2008 until permanent dwellings were available; the latter began to reach completion in the summer of 2010 (Jing 2014; Han *et al.* 2014).

In June 2008, it was reported that media had been instructed not to refer to miscarriages that were reported to have occurred among women living in temporary houses due to the presence of formaldehyde, a material that had been used in the construction (*The New York Times* 2009). This building material also caused a problem following Hurricane Katrina, when toxic air quality was found inside post-disaster shelters (Hagerman and Doherty 2009).

Permanent reconstruction

After the earthquake, it was estimated that about ten million dwelling units were required. They were provided with astonishing speed as required by the National Emergency Relief Plan on Natural Disaster Reconstruction, which states that houses need to be reconstructed within one year of a disaster. This may be

FIGURE 3.3   Lifelike display of a family of waxwork models occupying a temporary cabin. They are wearing overcoats since there was no heating in the shelters (photograph by Ian Davis, taken in Wenchuan Earthquake Museum).

a reasonable goal when the massive resources of the Chinese state are pitted against the effects of a small disaster, but it is hardly realistic when dealing with something the size of the Wenchuan earthquake. Initially, the government stated that two years would be required, but this was later changed to three years. Reconstruction of dwellings began in October 2008, five months after the earthquake, and was completed by December 2012, 55 months after the disaster, representing a total construction period of four years and three months.

Given the vast area affected by the disaster, varied approaches to reconstruction were adopted. In the rural area of Mianzhu County, Sichuan, a cash distribution approach was adopted to support 63,000 families in the reconstruction of homes, this money being placed directly into each individual's bank account. These dwellings ranged in size from 50 to 150 square metres. The cost of materials for the dwellings was US$9,000–US$18,000, and the construction cost per dwelling unit was US$440–US$1,500. Precise selection criteria were adopted in the allocation of rural housing to families who:

- had lost a family member in the earthquake;
- had a family member who sustained a permanent disability from the earthquake;
- lived with an elderly family member, over 60 years old;
- lived with an elderly family member who was seriously ill prior to the earthquake;
- lived in one particularly vulnerable township where total relocation was necessary (Ashmore 2010).

Government relocation of communities and settlements

In the immediate aftermath of the earthquake, Chinese scientists advised the government to move many communities away from unsafe sites, subject to landslides resulting from earthquake aftershocks. On 17 June 2008, 110,000 residents of Aba Prefecture were moved, and further relocations occurred in Maoxian, Lixian, Heishui and Jinzhai. A total of 25 townships were relocated. Ian has been reliably advised that these were 'forced relocations' in which families were given no choice to remain in their original settlements and accept the risks in so doing.

A decision was made to relocate the surviving residents of Wenchuan to new towns at Beichuan (Figures 3.4 and 3.5) and Yinxu. These settlements were built extremely rapidly and completed by 2013, just five years after the earthquake. Special consideration was given to the rebuilding of minority communities in a traditional architectural style. However, it is possible that some of the new residents of Beichuan were not offered the choice of whether to remain in their original settlements or be relocated.

In the past, relocation of settlements after disasters has been a source of persistent failure. Reasons include social objections from relocated communities who wish to remain in their existing site for cultural motives, the fact that the abandoned settlement often continues to exist in parallel to the new relocated settlement, the massive costs involved, and the pressing issues of families not securing a house or plot comparable to their abandoned home. However, from what Ian could gather when visiting the area in June 2014, there appeared to be

FIGURE 3.4   Entrance to the relocated town of Beichuan (photograph by Ian Davis).

FIGURE 3.5   Beichuan town centre (photograph by Ian Davis).

widespread public support for the decision to relocate. This may in part be on account of the bodies of family and friends that remain buried under earthquake or landslide debris. Thus, Wenchuan was no longer a town; it had become a cemetery as well as a tourist venue.

Significance of the recovery from the Wenchuan earthquake

There are several important reasons why the Wenchuan earthquake and its aftermath are particularly significant to a book on recovery from disaster.

The fundamental role of the 'risk drivers' of vulnerability: Inadequate or non-existent certification of construction played a significant role in the collapse of schools and apartment blocks. As noted above, after these failures, the authorities engaged in a violent crackdown on demands for justice by civil rights protestors and the parents of children who died. This case study records the struggle of activists to gather information on dead children to assist grieving families and to help establish reasons why so many schools collapsed. The gathering of information to inform policy and share with involved parties and the forensic identification of causes of failure to guide future policies and practice are regarded in any progressive society as essential processes and part of an established international procedural convention. All the undoubted successes of the recovery process in Sichuan are diminished by the actions undertaken by national and local governments to stamp out enquiries that are both essential and legitimate (see model 10, Chapter 4).

Speed of recovery: The fact that dwellings were reconstructed throughout the region for more than 15 million inhabitants in less than five years is a remarkable achievement by any standard. The work required selection of the site, planning the new settlements, construction of buildings and creation of associated infrastructure. It employed a workforce of 100,000 people, which involved devising opportunities to create livelihoods. It counts as a major feat that in less than five years the town of Beichuan was occupied by 40,000 earthquake survivors (see Chapter 7, 'Fourth dilemma: speed of reconstruction vs vital requirements in reconstruction planning').

A similar case is that of Monterusciello, a new town for 67,000 people displaced by volcanic activity from Pozzuoli in Southern Italy, which was built in 1985 by the Italian Government in only six months. However, in the Italian case, that was the sum total of relocation needs whereas the Chinese authorities were dealing with millions of displaced survivors. In Sichuan, the relocation of survivors into two major towns and countless smaller settlements appears to have been a popular decision – a rare example of public acceptance of complete relocation. Such rapid progress depended on the vast human, financial and material resources of the People's Republic of China, with its centralised political system and government ownership of land.

Partnerships to assist recovery: The World Bank report that about 41,130 projects for reconstruction and rehabilitation were undertaken, 99 per cent of which were completed within a two-year period. This astonishing speed resulted from innovative measures, such as a partnership scheme in which the central government paired up each affected province with an unaffected province that worked in close partnership to provide financial and technical assistance for reconstruction and restoration. It is estimated that US$146 billion were invested for reconstruction (Bernal and Procee 2012). A similar approach has been used in recent Italian earthquakes. For example, transitional accommodation provided in the L'Aquila area of Central Italy after the 2009 earthquake was partly provided by the Province of Trento and the Autonomous Region of Friuli–Venezia Giulia, both of which are located in the Alps where there is a tradition of prefabricating small buildings in wood.

Unlike disasters in such locations as Indonesia, India, Sri Lanka and Pakistan, the Chinese survivors played only a minimal role in creating their own provisional or temporary shelters or in building their own houses. Instead, they relied on government agencies to provide for their varied needs, and on high levels of cooperation with other regions of China. This provides an example of how a centralised communist state manages disaster recovery as an expression of its political ideology. This is similar to the values that underpinned the reconstruction of Skopje in communist Yugoslavia in the years after the 1963 earthquake. For commentary on Skopje by a resident, see Chapter 7, 'First dilemma: reform vs continuity'.

Disaster tourism or risk education? The decision was taken to conserve the ruins of Wenchuan as a cemetery and memorial of those whose bodies lie under the ruins or under a massive landslide. This appears to have had the approval of the survivors, many of whom had relatives whose bodies were buried by collapsed buildings, particularly the school where 400 children and teachers were buried when the earthquake triggered a debris flow that engulfed the building.

This was not the first such instance of ruins being retained. After the devastating lahar caused by the eruption of Nevado del Ruiz in Colombia in 1985, the town of Armero and 22,000 of its citizens were buried under ten metres of mud and volcanic rock debris. The Colombian Government decided not to rebuild on this site but to conserve the entire area as a memorial to the people who died in the eruption (see Figure 9.4 in Chapter 9). Similar considerations were raised after the destruction of Yungay by the Mount Huascarán earthquake and debris avalanche of 1970 (Oliver-Smith 1986).

When the authorities decided to abandon Wenchuan, they put in place stabilisation measures to prop up damaged structures, which were fenced off to prevent public access. Large hoardings were erected with 'before and after' photographs of the town.

In May 2009, the Wenchuan Earthquake Museum was opened. The focus is to show the triumphant way the survivors responded and the vast scale of the relief operation and to set on record the contribution of national and provincial officials as well as the army and international agencies. Unlike the excellently conceived Kobe Earthquake Museum in Japan where there is a clear desire to explain the nature of the earthquake and inform the public of all ages what they can do to reduce risks, this museum does not address such matters.

The likely positive intentions of both the 'living museum' of the ruined town and the displays in the Earthquake Museum may involve paying tribute to those who suffered, remembering a massive event in the history of the region, celebrating the recovery achievements and raising public awareness of earthquake risk. However, the captions in various languages that were set next to the ruined structures missed a golden opportunity to explain why this or that structure failed and why some appeared to be totally unscathed.

But is there a negative consequence to the decision to retain the ruins with public access? This is a site where a massive tragedy occurred where thousands died, many after acute suffering and many who were children denied a future by the disaster; and this is also a vast cemetery where they are still buried. Thus, to regard this as a major tourist attraction, with revenue-earning potential, seems inappropriate and questionable on ethical grounds. The issue of tourist fascination with disaster reconstruction is also discussed in Chapter 9, 'Option 7b: contractor-build permanent dwellings', in relation to reconstruction of dwellings in New Orleans following Hurricane Katrina.

Relevance to models

As mentioned at the start of this case study, the Wenchuan example relates to model 1, 'progress with recovery', and model 2, 'recovery sectors'.

The 'progress with recovery' model contains a progressive set of scenarios, or stages, that lead towards Scenario 4, 'developmental recovery'. There can be little doubt that all these stages have been reached in this ambitious recovery programme. However, the other aspect of this model concerns four strands of recovery: vision and leadership, resources, participation and ownership, and organisation.

The shortcoming in the Wenchuan recovery operation relates to the absence of the third strand, participation. The values expressed in the Sichuan recovery are different and in some cases virtually the opposite of the philosophy and components of 'development recovery', with its emphasis on the participation of disaster survivors in decision-making in such a way as to create local 'ownership', self-help, the development of skills, sustainability, the mobilisation of civil society, transparency and accountability. From a Western perspective, the Sichuan recovery would be regarded as decidedly 'paternalistic'; but when seen in the light of Chinese traditions (or possibly from an Asian viewpoint as similar dynamics seem also to apply to Japan), a highly active government was expected to deliver, at great speed, an abundant supply of goods and services to what they perceived to be a passive population, one generally regarded as composed of 'victims' rather than 'survivors' with little ability to help themselves.

In essence, the Government of China regarded itself as 'provider' rather than 'enabler'. Model 21 (Chapter 13) contrasts the strength of government in recovery operations with the strength of community participation. The Wenchuan recovery is represented as very high in government strength but extremely weak in community participation.

Model 2: recovery sectors

The value of this model (Figure 3.6) is in its description of the sectors of recovery linked to the scenarios of recovery set out in model 1.

The scenarios in model 1 are broken down according to the five sectors shown in model 2:

- physical recovery;
- recovery of the economy and livelihoods;
- psychosocial recovery.
- environmental recovery;
- institutional and governmental recovery;

Model 18 (Chapter 4) considers typical strengths, weaknesses, opportunities and threats pertaining to these recovery sectors. These are set out in Table 4.3 'The SWOT model applied to recovery sectors'.

As each of these sectors depends in some manner on the others and close integration is therefore needed, people in charge of recovery must aim to move

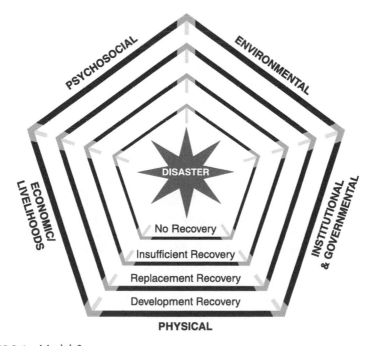

FIGURE 3.6   Model 2: recovery sectors.

each sector, in a balanced manner, towards Scenario 4 of model 1, 'developmental recovery'.

An encouraging example of holistic recovery planning – involving economic, social, natural environment and built environment recovery – is the approach taken by the Government of New Zealand after the four Christchurch earthquakes: 4 September 2010, 22 February 2011, 13 June 2011 and 23 December 2011.

Effective integration can occur in the following manner. After a major disaster, one of the first requirements is for *recovery of government*. In many disasters, damage or destruction of government buildings, records and equipment, and deaths and injuries to government staff result in the absence or weakening of authority. Yet this is precisely where it is most needed in order to manage the recovery process. Despite this, international donors rarely regard the fundamental need to re-establish effective local government as one of their sectors for intervention or funding, and there are even examples of central governments that have neglected this local need. A restored, functioning and enlightened local government is vital to the integrated management of the sectors in this model.

In Banda Aceh, Indonesia following the 2004 tsunami, an estimated 20 per cent of public officials working for the local government were killed. Thus the functioning of local government at a critical time was severely affected. Yet despite the obvious need to strengthen the weakened local administration, several international NGOs (some being leading players who should have known better) hired some of the remaining government officials at inflated salaries, way above government levels, to work for them as they urgently needed staff with local knowledge and language skills. A UN team, including Ian, met the Program Director of a leading agency who had the nerve to moan about the inadequacies of local government, without recognising that his own agency had further contributed to that weakness by hiring key local government staff to work in its own recovery programme.

By providing contracts to local firms and by providing them with incentives to employ disaster survivors, *physical* and *economic* recovery are closely linked. If a 'user-build' approach is followed, this will generate vital work for survivors in rebuilding their own homes and the surrounding environment. Moreover, providing work for disaster survivors who may be grieving the loss of family members can be a useful therapy for *psychosocial recovery*.

The selection of materials for rebuilding can be based in part on considerations of *environmental* recovery by encouraging the use of building materials that will support environmental regeneration. Another example is to reuse damaged material such as timber and stone in building and to use crushed disaster debris as ballast for roads in order to avoid the need to find capacious landfills.

The integration of all key sectors into a balanced approach requires an agreed policy framework and decisive coordination backed by the ultimate authority of the prime minister or president of the affected country.

Model 2 can be a useful monitoring device to enable government and donor officials to check on relative progress across the five sectors of recovery. Therefore, the hexagon can be filled in at regular intervals to indicate whether recovery is evenly balanced or confined to a single sector.

In the case study of the Wenchuan earthquake recovery, discussed above, progress is assessed particularly in relation to the physical recovery of buildings and settlements. While noting the achievements of rapid reconstruction, the absence of participatory involvement is a major criticism of action by the authorities. This omission represents a major lost opportunity that is highlighted by this model of recovery sectors and the need for integrated planning and action. If the traumatised, grieving population had been able to play a significant role in their own recovery, this would have contributed to their *psychosocial* recovery.

While the need for close multi-sectoral integration seems self-evident and essential for the well being of a society that has suffered damage and destruction across all aspects of life, the mutual isolation of government departments is frequently a major obstacle to holistic, integrated actions. Individual departments of government, at both central and local levels, often compete for resources, staff and status with the consequent neglect of one sector at the expense of others. As already noted, this is a blinkered approach in which self-interest prevails at the expense of close working collaborations between related sectors. It has to be dealt with through education, strong direction and close coordination of the kind that can only come from the apex of political power (see model 20, 'organisational frameworks of government for recovery management', in Chapter 4).

Model 3: development recovery and elapsed time

This model deals with resilience and needs to be considered together with model 11, 'resilient communities and settlements' (see Chapter 4). Its value is to depict the trajectory of resilience over time in relation to a theoretical concept of progress in development and safety.

The model (Figure 3.7) has the following elements: the horizontal axis is drawn as a timeline from the pre-disaster context through the disaster event to the period of recovery. The vertical axis relates to the state or quality of development at a given location, which may be a city, a town or a region, and is certainly a place that has suffered the impact of disaster and is attempting to recover. Quality is defined in terms of the functionality, efficiency and safety of the elements of life, including the urban fabric, infrastructure, economy and social services. If a location has a high level of development or quality, the starting point of the bold line would be much nearer the top; however, it would be unlikely to attain the 100 per cent target. Conversely, in a country such as Haiti with a low tenor of development and high levels of vulnerability, the starting point would inevitably be near the base of the diagram, as a sign of low quality. However, despite these reflections, the term 'quality' is difficult to define operationally as it involves a relative judgement of what is good and bad. The same is true of the term 'safety'. As the risk expert Chauncey Starr observed decades ago, 'a thing is safe if its risks are seen as acceptable by society' (Starr 1969).

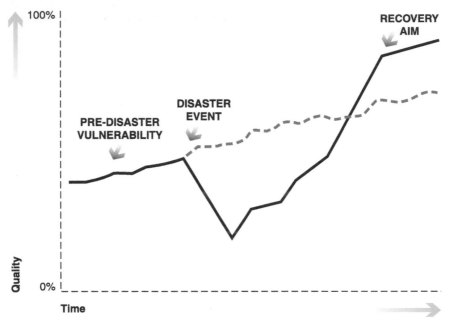

FIGURE 3.7 Model 3: development recovery and elapsed time.

Perception of risk, experience with hazards, and expectations of safety and security mould the public's expectations of acceptable risk in widely varying ways.

The gradually rising line that traverses the model indicates a trajectory of steadily improving development over a defined span of years. As all societies seek gradual improvements in quality, safety and economic growth over time, even if they often fail to realise their high expectations, the trajectory moves upward. The starting point of the model is roughly at the halfway stage, indicating that the country in question is one of relatively low development – unlike Sweden or Switzerland, for example, where the starting point would be near the top of the quality scale.

The first part of the rising line represents a period of pre-disaster vulnerability during which the aim of resilience should be to prepare society to absorb the disruptive and damaging effects of disaster. Hence, the first quality of a resilience framework should be to create a planned programme of risk reduction measures (see model 12, Chapter 4). The bold line represents the real situation. The disaster causes a sudden drop in quality, which is not vertical as the losses are not felt instantly but include indirect consequences that follow in the period after the disaster.

The resilience framework described in model 11 indicates a second aim, namely to 'bounce back rapidly' through the application of detailed recovery plans and efficient measures. However, we noted above that the aim of recovery is not merely to restore pre-existing conditions, and hence it is better to think in terms of 'bounce forward' conditions (Manyena *et al.* 2011) in which recovery goes beyond mere restoration.

At the point of the disaster, the rising line of development becomes a dotted line that represents the projected future for the given locality that has been interrupted by the disaster. Following the impact, recovery has to begin, and from this point onwards, the process starts on its rather variable progress. Spurts of growth may be followed by stagnation. In resilience terms, the entire recovery phase is one of adaptation and change as both processes are needed for effective recovery. The graph makes an important point about the aim of recovery. It is not merely a question of reconnecting with the rising trajectory line; there is also a need to overtake it. To reconnect could be to return to the vulnerable status quo ante that gave rise to the disaster; therefore, an adapted and changed recovery has to rise above this to the development recovery stage, as described in Scenario 4 in models 1 and 2.

This model graphically highlights one of the great challenges of disaster recovery: not merely to restore all the recovery sectors but also to improve on what used to exist in order to create a safer and better environment and one that more closely responds to the basic needs of citizens. To accomplish this with limited resources is an immense challenge, and in the rare cases when this happens, it represents a massive achievement (see model 7 in this chapter, which focuses on the cost-effectiveness of recovery).

Model 4: relationship between disaster and development

This model (Figure 3.8), developed by Rob Stephenson (1991), endeavours to depict succinctly the relationships that exist between the setback of disaster and the forward process of development.

The four quadrants of this model need little explanation. They describe the linked realms of 'disaster' and 'development' and indicate the positive and negative possibilities of each. There are numerous instances of how each quadrant may exist in disaster recovery situations. Here are some typical examples.

Development can increase vulnerability: In disaster recovery, plans are sometimes implemented in great haste before risks have been assessed, building regulations have been revised and builders have been trained in safe construction techniques. The persistent pressure from the survivors, mass media and political leaders to reconstruct rapidly can result in shortcuts that compromise future safety. After the 1906 San Francisco earthquake, debris was dumped on the shore, thus creating reclaimed land from the sea. Eventually, houses were built on this land, and in the Loma Prieta earthquake of 1989, extensive damage was suffered due to the irregularity of ground motions through unconsolidated terrain.

Development can reduce vulnerability: There are many examples in which radical changes in patterns of land use have resulted in safe disaster recovery. In Skopje after the 1963 earthquake, the recovery planners wisely decided to designate a floodplain beside the Vardar River as public parkland in order to prevent new

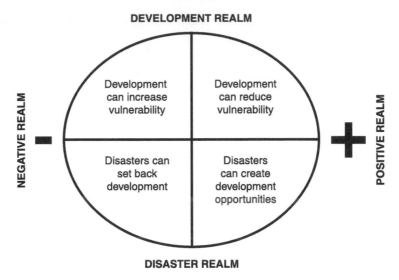

FIGURE 3.8 Model 4: relationship between disaster and development (Stephenson 1991).

buildings from being constructed in areas that had been subject to repeated flooding, and also to avoid an area in which alluvial sediments accentuated seismic shaking. Likewise in the Christchurch, New Zealand earthquakes of 2010–11, 65 square kilometres of land suffered liquefaction and much of it was occupied by suburban development. The areas of greatest liquefaction risk will in future be given non-urban uses.

Disasters can set back development: All major disasters result in widespread failures in which the tangible results of development, such as school buildings, roads, bridges and economic investments, are destroyed. In many cases, these assets will be of recent origin, which may indicate faulty construction, corrupt building practice or failure to implement building codes. The cumulative effect of these failures is to seriously set back development. However, loss of life is far more serious than the destruction of the built environment. The death of children, who had all their lives ahead of them, is the greatest tragedy in any community. Unlike a bridge that can be replaced, such losses are irreparable and will have lasting negative impacts on development hopes. See model 3 for a graphic timeline representation of the drop in quality and development caused by a disaster.

Disasters can create development opportunities: This topic is discussed at various points in our book. For example in the Malkondji case study in Chapter 1, we noted how the community were introduced to using toilets that were incorporated into their new dwellings. A further example of an 'opportunity grasped' is found in Chapter 12 in the section 'Yellow hat – optimism'. Here we describe how cash grants were made available to rural families in Pakistan when they secured new dwellings and how this cash was distributed with the dual advantage of cutting out 'middlemen' – a frequent source of corruption – and enabling families to enter the formal banking system – a vital prerequisite to future development opportunities. In Chapter 2, we identify pioneers of the subject – all of whom regarded disasters as unique opportunities for development with numerous examples of how 'form followed failure'. Such opportunities include new building bye-laws, the development of land use planning controls to prevent building in floodplains or on precipitous slopes, safer construction techniques, the adoption of community preparedness plans, and fiscal and insurance incentives designed to encourage prudent building practices.

Phases of recovery models

TABLE 3.3 Summary of phases of recovery models (5 to 8)

Model	Graphic representation	Application	Source	Location in book
5 Disaster cycle		The 'original' disaster management model, indicating five progressive phases of activity	Original source unknown	Discussed in this chapter
6 The Kates and Pijawka recovery model		Four phases of recovery are related to varied levels of activity	Kates and Pijawka (1977)	Chapter 5, 'Economic recovery and the question of livelihoods'; 'Cultural recovery'; and 'Environment'
7 Cost-effectiveness (unit cost)		Phases of recovery are related to escalating unit costs	Alexander (2000)	Chapter 5, 'Economic recovery and the question of livelihoods'
8 Disaster timeline		Expresses the 'ebb and flow' of the strands of disaster management and recovery management	Original source unknown	Discussed in this chapter

Model 5: disaster cycle

The value of this model (Figure 3.9) lies in its representation of the pre- and post-impact phases of disaster management and how they relate to each other.

Not all disasters are cyclical and some are not even recurrent. For example, virtually all of the world's flood disasters involving the catastrophic breaching of reservoir dams have been unrepeatable events. On the other hand, many extreme events in the natural world are recurrent. Meteorological disasters may be seasonal, earthquakes will recur after strain has accumulated on faults, and volcanic eruptions may follow a cycle associated with the build-up and release of magma and gases. Hence there is some utility in using a model based on the idea that disasters are recurrent events.

The 'disaster cycle' makes a basic distinction between times of quiescence, in which the emphasis should be placed upon mitigation and risk reduction, and times of emergency, in which the accent is on intervention and recovery. The five phases are:

- risk reduction (during peaceful times without disasters);
- preparation prior to impending events (assuming that forewarning is possible);
- emergency response to impacts;
- recovery of essential services;
- reconstruction.

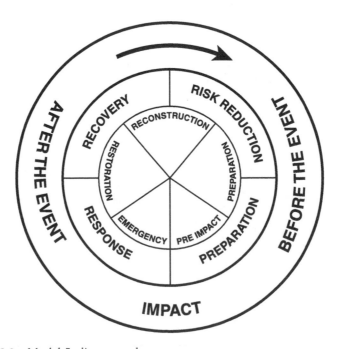

FIGURE 3.9 Model 5: disaster cycle.

The advantage of using this model is that it clearly distinguishes the phases in such a way that they can be related easily to a sequence of tasks that needs to be accomplished in order to provide, or restore, safety and security. As it elegantly simplifies the processes of managing hazard and disaster, the model has proved particularly useful in teaching disaster management to practitioners. It also helps order copious amounts of knowledge about the processes of dealing with hazards, risks and emergencies. Moreover, it helps one to draw a functional distinction between tasks on the basis of their duration; for example, emergency intervention is usually a fairly transient process; reconstruction is, in many cases, long-drawn-out; and mitigation should be a constant duty that has no ending.

The origins of the phases of disaster model are obscure and date back to the 1970s (Richardson 2005). In an influential social science book, Drabek (1986) used it as the basis of a taxonomy of sociological research about disasters. Since then, others have drawn attention to the deficiencies of the model. For example, Neal (1997) considers the asynchronous nature of the phases, and Richardson (2005) argues that they are just too simplistic to embrace and classify all the social realities of disaster. Nevertheless, the 'disaster cycle' is widely employed to characterise the process of managing major hazards and their impacts. Indeed, in its various forms, it is probably the most popular model of its kind, especially because of its ability to reduce the chaotic complexity of disaster to an elegantly simple sequence.

The shortcomings of the model can be summarised as follows.

- As noted above, not all disasters are cyclical and some are decidedly non-recurrent. Whereas the model will work for events that do not follow a regular cycle, providing they do indeed recur, it will not work for hazards that are extremely irregular. Examples of these may include certain industrial hazards, transportation crashes or acts of terrorism; these may be, in a certain sense recurrent, but they are hardly cyclical.
- The phases of the model are likely to be of widely differing duration. As a result, there may be strong discrepancies in the level of detail and the nature of the sequences of actions and events involved in each phase – i.e. the model may be inconsistent or unbalanced in its treatment of each phase.
- In many disasters, the phases are likely to overlap rather than be sequential. Hence, the idea of a sort of 'wheel of fortune', in which emergency response is followed by actions designed to promote recovery and thereafter by reconstruction planning and management, is at variance with reality. In a society that has prepared well for the next disaster, reconstruction planning may be able to take place immediately after the impact and run concurrently with emergency intervention and restoration of services (see quotation from Quarantelli, 1982, in the introduction to Chapter 9). Moreover, it is important not to separate mitigation from recovery and reconstruction processes as the latter are a golden occasion to improve the level of mitigation actions.
- The 'disaster cycle' does not make allowance for repeated impacts or those that are complex or cascading (in which one impact leads to another).

For example, in Japan after the 11 March 2011 Tōhoku disaster, recovery from the earthquake and tsunami may have taken place at a different pace to recovery from the Fukushima nuclear radiation emission crisis.

- In difficult situations, there is no guarantee that the cycle will be completed. For example, after the 1973 earthquake in Nicaragua, reconstruction stalled more or less permanently in parts of Managua.
- The model implies the existence of a status quo ante and a requirement of reconstruction to restore it. However, recovery processes should usually be progressive and should 'bounce forward' rather than 'bounce back' by providing a rebuilt environment that is safer and functions better than the one that existed before the disaster.
- While the model identifies a series of distinct disaster phases, there is no indication that they are under separate management. While emergency services may have responsibility for the immediate disaster aftermath as well as preparedness planning, they are seldom, if ever, responsible for longer-term rehabilitation and recovery or mitigation. This division is inevitable given the diversity of tasks and varied areas of departmental responsibility in government. But it means that emergency services often make decisions that have unforeseen long-term negative consequences, particularly in the shelter sector.
- There is a weakness in the graphical representation of the model in that the cycle of protective measures, such as mitigation and preparedness, appears to lead directly into a disaster when they should do precisely the opposite!
- Finally, dealing with disasters, and the risk of such events, is a pluralistic process in which, to all intents and purposes, there are many 'disaster cycles' rather than one alone. Recovery may proceed at a different pace and in different ways between different sectors: economic, medical, infrastructural, social, psychological, and so on.

We have taken the trouble to list many of the criticisms of the 'disaster cycle' because it is an important, and nonetheless valuable, model. Rather than invalidating it, the critique suggests that it has limitations, as indeed most models do.

To restore the balance, let us consider an example of the model in action. The magnitude 7.6 earthquake in Western Sumatra on 30 September 2009 seriously affected Padang Pariaman District with the loss of 1,115 lives. The tremors damaged 379,200 buildings, a third of them seriously or catastrophically. The emergency phase involved all relevant organs of the Indonesian Government plus 170 NGOs, two-thirds of which were international. The emergency phase lasted about one month.

During the early recovery phase, about 100,000 people were settled in tent camps while 13,778 temporary dwellings were constructed. Survivors were then dispersed to relatives and alternative accommodation, or they were assigned to the prefabs. The regional Yodarso Hospital of Padang suffered major damage and its functions were evacuated to tents and later to prefabricated buildings. Subsequently, as recovery got underway, deadlines were set for the repair and reconstruction of infrastructure and public buildings and for the resettlement of homeless survivors. Despite problems of policy and finance, these appear largely to have been respected.

The Padang catastrophe occurred almost five years after another Sumatran disaster, namely the earthquake and Indian Ocean tsunami that primarily affected Aceh Province in December 2004. One consequence of this for Padang is that the interval between the two disasters was used to improve preparedness. This included the beginnings of systems of catastrophe insurance and micro-insurance at the national level as well as the prevalence of disaster drills that undoubtedly saved lives in September 2009. This, then, was a good example of the importance of the mitigation phase between disasters.

Model 6: the Kates and Pijawka recovery model

The value of this model (Figure 3.10) lies in its representation of variables that express the duration of recovery; namely, level of development, availability of resources and degree of organisation.

In 1977, two geographers, Robert Kates and David Pijawka, contributed a chapter to a book on reconstruction. The title of their work was 'From rubble to monument: the pace of reconstruction'. They chose to look at recovery from earthquakes as these are the archetypical sudden-impact disasters, and in particular they studied the aftermath of three events: San Francisco 1906, Nicaragua 1972

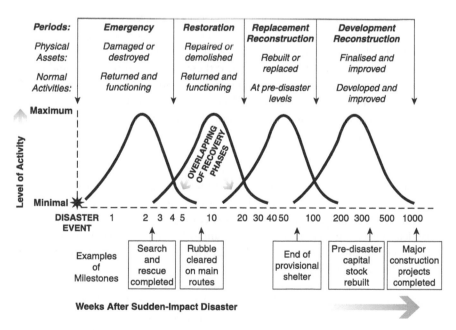

FIGURE 3.10 Model 6: the Kates and Pijawka recovery model (after Kates and Pijawka 1977).

and Alaska 1964. Their thesis was that the level of development, availability of resources and degree of organisation affected the length of time taken to achieve recovery.

The first two stages in Kates and Pijawka's model were the emergency and restoration periods. The former usually lasts from a few days to about a month and ends when the basic needs of survivors have been met. Attention is focused on pressing and fundamental needs. Major infrastructure is cleared, mass feeding programmes are set up, search and rescue are priority activities, precarious structures are buttressed and basic shelter is provided to survivors. When the majority of these tasks have been accomplished, or they no longer need to be carried out, the phase is over. The restoration period may last from two to nine months. During this time, structures that cannot be repaired are demolished, damaged buildings are rehabilitated (or entry is banned, pending reconstruction) and infrastructure and public utilities are repaired.

When dealing with reconstruction, Kates and Pijawka differentiated between the 'replacement-reconstruction' and 'developmental reconstruction' phases. They saw the former as lasting from 3 to 20 years. The Rapid City, South Dakota dam-burst floods of June 1972 led to an example of three-year reconstruction in a relatively uncomplicated situation where the scale of damage was not particularly large. In the replacement-reconstruction phase, capital stocks are gradually rebuilt, the local economy recovers to pre-disaster levels (if it can), and social equity, which prevailed in the early stages of the disaster, is replaced by social differentiation. This last observation stems from the fact that people of higher social standing tend to have greater access to capital, credit and insurance than the poor and, hence, have more opportunities to recover and more access to mechanisms that speed up the process. Kates and Pijawka noted that financial institutions are often the first to recover, as these have access to capital and credit. The model has the advantage of placing emphasis on the context of recovery. In the case of Nicaragua, corruption and civil war slowed down the process and made it a differential one in which the middle and upper classes, who had access to credit and insurance, were far more successful than poorer people.

The phase of developmental reconstruction uses monumental building to commemorate the disaster and show that the affected area has overcome the problems associated with it. This phase also marks the process of local or regional regrouping in order to launch economic growth. It usually occupies some years after the end of the replacement-reconstruction phase. According to Kates and Pijawka, the size and opulence of San Francisco's new City Hall, completed in 1929, were meant to show publicly that the problems created by the 1906 earthquake had been overcome.

The Kates and Pijawka model was qualified by work conducted by Sarah Hogg in Friuli, north-east Italy, after the 1976 earthquake (Hogg 1980). She found that the speed of recovery and reconstruction were related to the degree of geographical and political connectedness of each settlement. In other words, the pace of recovery was not uniform throughout the disaster area. Moreover, in the Friuli case, a second earthquake occurred six months after the first one, which returned the area to the emergency phase and effectively restarted the sequence of recovery. The same could be said of the 2010 and 2011 Christchurch, New Zealand earthquakes.

The Kates and Pijawka model is simple and elegant, but is it sufficiently accurate to conceptualise recovery processes as they actually are? To begin with, as in other models based on periods, the phases may overlap or they may lack an adequate conclusion. Politics, economics and social factors can influence the recovery processes, perhaps more than Kates and Pijawka suggested. Moreover, the political, economic and social context of recovery is likely to change, perhaps radically, during the process, especially if it is long-drawn-out. Effectiveness, fairness and equity in recovery processes are influenced by various forms of vulnerability, both to disasters and to subsequent processes. For example, the arrival of large amounts of relief and reconstruction money in a disaster area may lead to corruption and expropriation of resources.

A good example of long-term problems is furnished by Nicaragua. The Somoza family controlled Nicaragua from 1927 until 1979. The devastating effect of the December 1972 earthquake, which destroyed 90 per cent of the capital Managua, was one very significant element in the revolution that ended their rule. In 1973, vast amounts of relief money were syphoned off to pay for the luxury homes of the elite while the poor were constrained to live in temporary shelters of the most miserable kind. Opposition was progressively galvanised by the utterly corrupt situation until the Sandinista revolution finally occurred in the summer of 1979. This was followed by the counter-revolution of the Contras and a decade of economic ruin. Nicaragua remains the second poorest country in the Western Hemisphere, and almost half of its population lives below the poverty line. The country still has what is officially classed as a recovering economy, in which the twin depredations of civil war and natural disasters have continuously retarded development. In 1998, Hurricane Mitch killed 3,800 Nicaraguans, severely damaged 70 per cent of the country's roads, destroyed 92 bridges and caused huge losses in all sectors of agriculture. According to some estimates, this storm set back development by 20 years. The multinational fruit producers were widely criticised for exploiting the devastation to increase profits while doing nothing for laid-off workers who had been made destitute by the storm. In Nicaragua, social, economic, political and military conditions have conspired to retard or halt the process of recovery from disaster.

Model 7: cost-effectiveness (unit cost)

The value of this model (Figure 3.11) lies in its representation of the escalating unit costs of the phases of recovery in relation to the passage of time.

One question that has often been debated is the effectiveness of expenditure on mitigation. Does spending money, presumably wisely, before disaster strikes significantly reduce losses afterwards? It is sometimes written that on average for

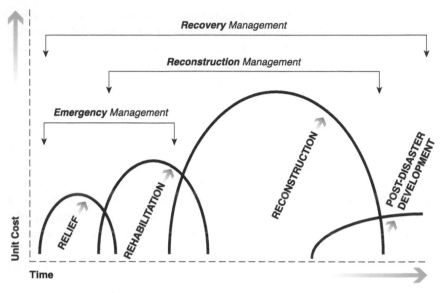

FIGURE 3.11 Model 7: cost-effectiveness (unit cost).

every one monetary unit spent on disaster mitigation, four units are saved in damage that is prevented. That may be true although it is difficult to verify in any broad, comprehensive manner. However, a very wide range of cost–benefit values have been developed, depending on the hazard, the level of vulnerability, what assets are involved, and other such details. Long ago, cost–benefit figures were compiled by Leighton (1976) for some natural hazards in California, and the figures varied from 1:1.5 for flood to 1:137 for landslides. None of the data would be acceptable as a universal generalisation.

It is sometimes also assumed that cost–benefit ratios are invariable at any stage of the expenditure, something that is not borne out by circumstances. For example, if a construction is threatened by earthquake or windstorm and has no built-in protection measures, initial expenditure on making it safe is likely to achieve substantial, even spectacular, reductions in vulnerability. For example, in wood-framed vernacular housing threatened by windstorms, the use of the 'hurricane strap' – a simple metal tie – can stop the roof blowing off, which may in turn stop the entire structure from being demolished. However, once the simple measures have been instituted, mitigation tends to become more and more sophisticated and expensive and to achieve less and less reduction in vulnerability. Fairly soon, the break even point is reached at which the cost of reducing vulnerability is equal to the value of damage avoided. At this point, in technical terms, mitigation should cease as it achieves no further net economic benefit and any further expenditure amounts to a form of economically unjustifiable risk aversion. However, it is rare that society sets its acceptable risk levels on the basis of positive cost–benefit ratios.

These observations may seem to apply only to the mitigation phase in the absence of disaster or in the periods of quiescence between impacts. However, they have a distinct resonance during the reconstruction period. At this point, there is a visible demonstration, in the form of damage, that rebuilding needs to take place to a higher standard than previously. If funds are abundant, this may take the form of risk aversion in which risk is very significantly reduced by heavy expenditure without reference to criteria that might justify the levels of protection sought.

During a long period of recovery from disaster, conditions are unlikely to remain static. Inflation will affect costs and expenditures, political priorities will change, mass media attention will fall and only sporadically be revived, and social and demographic conditions will vary. On top of this, the relative costs will be different for each phase of the recovery process. As a result of this, it is difficult to predict the outcome of recovery in advance, or what resources it will need. The vagaries of economic management mean that, in most cases, the supply of resources for recovery will vary over time. Commonly, initial costings will underestimate the expense of recovery. With inflation and the need to satisfy ever more stringent safety requirements, costs can soar. At the same time, the political will to complete the process of recovery may lapse as other problems impinge on the consciousness of decision-makers. A good example of stagnation is provided by the aftermath of the L'Aquila earthquake in Central Italy. For a brief period, L'Aquila had a pivotal role in national voting patterns and was the centre of attention. Thereafter, it returned to being a political and economic backwater. Hence, after two years, there had hardly even been clearance of rubble, and reconstruction processes continued to stagnate.

Model 8: disaster timeline

The value of this model (Figure 3.12) lies in its representation of the ebb and flow of the strands of management processes before, during and after a disaster.

The relatively high frequency of earthquakes in some parts of the world may justify the use of a cyclical model even though the cycles are not necessarily regular ones. Another way of characterising disasters using phases is to treat them individually and consider the time dimension in a linear manner. The first problem with this approach is to be able to define the conditions under which the impact of an extreme event can be classified as a disaster. Limitation of space precludes a detailed discussion of this thorny issue, but it has already been the subject of two books entitled *What is a Disaster?* (Quarantelli 1998a; Perry and Quarantelli 2005). In synthesis, quantitative definitions of the threshold for disaster (e.g. based on number of deaths or size of economic losses) tend to fail because they ignore systemic factors that betoken a qualitative change between incidents and disasters. 'Disaster' implies a need to suspend routine activities in order to cope with an entirely unusual set of circumstances involving high levels of disruption, damage and destruction. The geographical extent of the

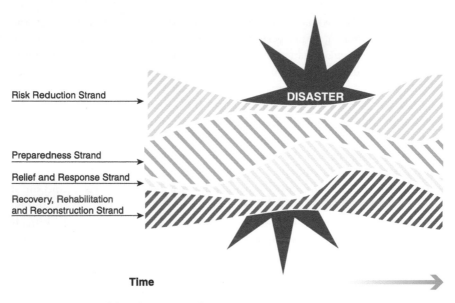

Risk Reduction Strand

DISASTER

Preparedness Strand

Relief and Response Strand

Recovery, Rehabilitation and Reconstruction Strand

Time

FIGURE 3.12 Model 8: disaster timeline.

phenomenon, the number of people affected and the size of resources involved are all variable quantities that are more or less impossible to characterise in terms of minimum thresholds.

Nonetheless, the transformation wrought by disaster to society, its assets and its activities is invariably profound, whatever the scale at which it occurs. The inception of such change may occur along a continuum that extends from 'sudden impact' to 'slow onset' disasters. Earthquakes are the archetypical sudden-impact disasters that occur without warning. Drought and accelerated erosion are examples of naturally generated events that grow, perhaps imperceptibly, to disaster proportions.

In a generalised manner, there is usually a correlation between the existence of a premonitory phase and the ability to prepare for the impact of disaster. However, it is not necessarily true that a longer lead time involves greater preparedness. Instead, there may be an optimum period in which the imminence of the impact is balanced against the availability of time to prepare and general sense of urgency.

In sudden disasters, the impact phase may begin with a period of isolation before organised assistance arrives. In fact, commonly, the first aid to be supplied is given by people who are simply at the scene when the impact occurs and have survived in such a way that they can provide help. Hence, the first rescuers are commonly people who have no training and equipment, except in the tiny minority of cases in which citizens' groups have been specially prepared for prompt disaster response. Such people commonly rescue others from under rubble, or from rising floodwaters, or from burning buildings, and so on. They may provide first aid or transport to hospital. However, unorganised assistance from spontaneous volunteers and passers-by is inefficient and can lead the rescuer into danger. Indeed, it was estimated that in the 1985 Mexico earthquake, one rescuer died for every four or five people saved (Olson and Olson 1987: 646); though bear in mind that these were spontaneous, untrained rescuers who entered precarious damaged buildings without adequate care and personal protection.

It requires a very high degree of local organisation for organised assistance to arrive at the scene of a sudden-impact disaster in a timely manner. Where building collapse has trapped and injured people, there is an imperative to rescue them within a period of less than eight hours, which represents average survival time under the rubble. Earthquakes, in particular, may cause thousands of buildings to collapse, including some that may have a high density of occupancy. Fallen beams and masonry can cause crush syndrome, which necessitates prompt rescue and dialysis in order to stop the patient dying of kidney failure. Blood loss, cranial injuries, ingestion of dust and multiple traumas all point to the need for rapid, professional urban search and rescue (USAR) and on-site medical assistance.

Commonly in major disasters, the international community mobilises its search and rescue resources over a period extending from 12 to 72 hours after the disaster. Between about 1,200 and 2,300 foreign rescuers may descend on the disaster area during this period, usually in the second and third days after the impact. For example, 1,600 arrived in Bam, Iran after the 2003 earthquake there. They may then be coordinated by UN Disaster Assistance Response Teams working under the auspices of the UN Office for the Coordination of Humanitarian Affairs. Usually, their arrival is manifestly too late to have much impact on the scale of casualties. For instance, in the Haiti earthquake of January 2010, the first country to mobilise its USAR forces was Iceland, which is more than 6,000 kilometres from Haiti. In all, foreign teams rescued only 130 people, even though the death toll may have been well over 200,000. After the fifth day, only nine people were rescued. In cases like this, it is estimated that the cost per life saved may exceed US$1 million, money that could more usefully have been spent on improving local USAR capabilities in places where entrapment and injury in disasters is likely in the future.

The duration of an emergency phase can be highly variable. It has been suggested that it is correlated with the degree of economic development (Kates and Pijawka 1977). In early 1858, travelling on mule-back, the intrepid Irish engineer Robert Mallet managed to reach Montemurro in the Basilicata Region, Southern Italy one month after the December 1857 earthquake (magnitude 7.0). In this highly isolated area, in which most of the local population had been killed by the earthquake, Mallet found the survivors to be in a state of early emergency fully one month after the tremors (Mallet 1862). On the other hand, in cases where resources are abundant, communications are robust, and there is a high degree of organisation, the emergency phase may be over in less than a week.

Emergencies can be prolonged if the impact is cascading (or compound) or repetitive. Since the magnitude 9 Tōhoku earthquake in Japan, there has been renewed worldwide interest in cascading disasters. In this case, a major earthquake caused a tsunami that led to radiation releases from some of the Fukushima nuclear reactors. This represents both a cascading disaster and a 'natech' event, composed of interacting natural and technological components (Young *et al.* 2004). The levels of uncertainty and complexity tend to increase with the number of elements in the cascade and the number of connections between them.

The emergency phase of disaster is characterised by search and rescue, medical assistance and the provision of the most basic needs, such as shelter and food. Social actions are dominated by a welfare ethos in which the normal market functions of society are temporarily suspended. The next phase is dominated by the recovery of basic services, such as utilities (electricity, water supply, etc.) and other aspects of critical infrastructure. Again, the time taken to achieve this can be highly variable according to the level of damage, the size of the area affected (i.e. the extensiveness of the damage), the complexity of the infrastructure and the availability of funds and technical resources to effect repairs.

After a major disaster that has affected a substantial area, perhaps hundreds of square kilometres, the time taken to complete the recovery process, including full reconstruction, may exceed ten years and possibly be as many as 25. Time is socially necessary in the reconstruction process. Hasty or overly rapid reconstruction runs the risk of being undemocratic, reducing the opportunities to plan the process, failing to allow adequate participation from stakeholders and postponing consideration of risks and hazards until it is effectively too late to do anything about them.

Major reconstruction should involve detailed planning based on careful consideration of the main issues. Some of these are: how to avoid replicating vulnerability, how to reduce the future impact of local hazards, how to promote community spirit, and how to preserve the cultural identity of the area in question. It goes without saying that reconstruction needs to preserve the functionality of the local area and its ability to provide employment and generate wealth. Surveys of local geological and geotechnical conditions, and of hazards, take time and so does public consultation once plans have been drawn up. Hazard avoidance and vulnerability reduction schemes also extend the time required to arrive at reconstruction.

Recovery and reconstruction are not necessarily characterised by harmony. An example is the case in which 144 people were killed, including 116 children between 7 and 9 years old, when a coal spoil heap collapsed upon schools and the urban area of Aberfan, South Wales in 1966. After the disaster, the loss of so many young, innocent lives led to a massive outpouring of solidarity; and this tight knit mining community was inundated with gifts and money while its members remained severely traumatised. There were bitter disputes about how to use the money, which proved more of a divisive than a unifying influence (Miller 1974).

In synthesis, during reconstruction, 'quick' does not necessarily mean 'efficient' as time is needed to accomplish technical and social processes such as risk assessment, measurement, planning and consultation. On the other hand, if nothing goes on for long periods of time, this could be considered inefficient. Hence, time is a linear backbone to events – an inevitable meter of progress (or its lack) – but time alone will not sort out the problems of recovery and reconstruction.

Reconstruction planning needs clearly established objectives

In terms of time, planning is a means of looking forward and marshalling resources for the future. One reason why reconstruction may be slow is the need for much preparatory work. In this context, a strong parallel is to be drawn between urban and regional planning on the one hand and emergency planning on the other. They both require considerable research and the acquisition of detailed local knowledge. Both are intimately concerned with land use and its control. Urban planning will regulate land use to ensure that incompatible functions do not interfere with one another by being too closely located. It will endeavour to improve the efficiency of the urban and regional system. Emergency planning is concerned to ensure that urgently required resources are in the right place at the right time, including manpower, equipment, fuel supplies, vehicles, specialist assistance, communications and supplies.

One would suppose that the function and utility of emergency planning ends with the emergency phase of a disaster. This is not quite true. To begin with, each disaster and its aftermath yield vital information on how to improve plans for future events since disasters reveal the extent and nature of both hazard and vulnerability. Thus, bottlenecks in the transportation system can be revealed by delays in the supply of relief. Furthermore, magnitude–frequency relationships are qualified by new information derived from impacts when they occur. Thus, probablistic seismic hazard analysis (PSHA) is based on knowledge of part of the sequence of events and their magnitude–frequency relationships, which can, however, never be perfectly known. In seismic areas, PSHA may form the basis of building codes for the reduction of earthquake damage.

Summary

The eight models described in this chapter have covered a central focus of our book – the need for recovery to become a development opportunity. So we can do no better than to repeat that objector's outburst in Ian's disaster course, noted at the outset of this chapter: 'any recovery must move forward to a better and safer future since "normality" equals vulnerability'.

Closely linked to development requirements, the phasing of recovery – as demonstrated in models 5 to 8 – is an interconnected and extended process. Model 7 describes unit cost relationships in relation to time. This shows that the reconstruction will always be the most costly phase in any disaster continuum; yet political imperatives, especially when elections are looming, may well require the allocation of relief funds in abundance with scant regard that such initial generosity may jeopardise the future where a famine of resources can cause extensive delays or easily bring the recovery process to a juddering halt.

This possibility that the funding, as well as the activities, taken in one disaster recovery phase can have negative implications for a later one was succinctly stated by Otto Koenigsberger almost 40 years ago: 'Remember that relief is the enemy of recovery, with the consequent need to *minimize* relief in order to *maximize* recovery' (cited in Davis 1978: 66; see also Chapter 2, Evolution of recovery studies, and Chapter 4, Model 18, the SWOT model).

Notes

1 Communication between Ian Davis and Dr Gustavo Wilches-Chaux, 2006.
2 Communication between Ian Davis and David Oakley, 1997.

4

MODELS OF RECOVERY

Safety and organisation

Risk Management is about people and processes and not about models and technology.

(Trevor Levine, cited in Strachnyi 2012: 6)

Our selection of models continues in this chapter as we turn to five that have a theme of risk and safety followed by seven that relate to the organisation of recovery (summarised in Appendix 1). The final model that reviews our case studies is presented here and discussed more fully in Chapter 13.

The challenge of safe recovery

Placing an emphasis on safety is an obvious and essential aspect of any effective recovery task since risks can easily be built in again with devastating consequences. While the average lifespan of a typical building is about 40 years, in highly vulnerable areas, repeat disasters can have higher levels of frequency, risking demolition, with lethal consequence, of buildings and infrastructure that were reconstructed in the relatively recent past.

For example in 1954, the town of Orléansville in Algeria – at that time a French colony – was destroyed in an earthquake that caused 1,243 deaths. After independence in 1962, this town was renamed El Asnam and 26 years later, in 1980, there was a further earthquake that killed about 5,000 and destroyed almost half of all buildings in the city. Seventy schools suffered extensive or complete damage and 95 per cent of all schools in the city were damaged to some extent. The relevance of this example is that a very large proportion of these schools were built as reconstruction following the 1954 earthquake, the suggestion being that they were shoddily built with inadequate seismic resistance – a process sometimes called the 'reconstruction of vulnerability'.

In addition to the reconstruction task, Algeria has experienced one of the highest birth rates in the world, the population growing from 10 to 30 million in just three decades. Thus the reconstruction was accompanied with an expanding population, urgently needing accommodation; and these forces gave rise to an unprecedented building boom – always a dangerous environment where shortcuts can easily occur in terms of poor quality of construction and lack of safety (Lewis 1999; Bendimerad 2004).

What's in a name?

Just after the earthquake in 1980, the town of *El Asnam* (previously *Orléansville*) was renamed *Chlef*, taking the name of the longest river in Algeria. A cynical view might regard these name changes made by the government as a political stunt, or 'cunning plan', to persuade future residents or investors that *El Asnam*, and then *Chlef*, were good places to live rather than being the site of two of the most devastating disasters in North Africa.

That form of crude official deception also appears to have been the intention of the British Government. In 1981, some years after the 1957 *Windscale* fire – the worst nuclear accident in British history – the British Government and the operating body British Nuclear Fuels Ltd (BNFL) decided to rename the site *Sellafield*.

Safe recovery models

Models 9 and 10 are designed to assist users in their diagnosis of risks. In an ideal world, this assessment will take place long before any disaster occurs, but the process is also essential in the initial stages of recovery planning. Model 10, sometimes called the 'crunch' model, highlights the progression of vulnerability from root causes, where the 'drivers of risk' are the potent forces that eventually lead to the range of unsafe conditions that give (or gave) rise to disaster destruction and deaths.

Models 11, 12 and 13 seek to explain the nature of resilience, the array of measures used to reduce risks and the way a safety culture can gradually evolve. This process may often begin with a disaster recovery operation.

Model 9: probability/consequence risk assessment

This model (Figure 4.1) is used to assess the scale of a risk relative to probability and consequence. There are many variants of the model, which is widely used. The value of this model is its simplicity as a tool to relate two of the key determinants of risk: probability or frequency of occurrence, and consequence or impact. It can be used in the form of a simple numerical risk quantification scale, in which weights are assigned to each of the severity/probability cells. Hence, to use the model, it is necessary to develop and apply indicators of probability and

TABLE 4.1 Summary of safe recovery models (9 to 13)

Model	Graphic representation	Application	Source	Location in book
9 Probability/ consequence risk assessment		Assesses the scale of a risk relative to probability of occurrence and likely consequences	Derived from multiple sources; widely used in engineering and medicine; origin unknown	Discussed in this chapter
10 Disaster 'crunch' model		Describes the causal factors that generate vulnerability and hazards	Davis (1978) and later development in Wisner et al. (2004)	The varied root cause and effects of vulnerability are discussed in Chapters 2, 5, 6 and 8
11 Resilient communities and settlements		Describes the elements of resilient communities within resilient settlements	Davis (2006)	Chapter 10: 'Resilient recovery'
12 Disaster risk reduction measures		Outlines the range of disaster risk reduction measures	New model developed for this book by the authors	Chapter 9, 'Key lessons for option 1'
13 Development of a safety culture		Indicates the stages of the progressive development of a safety culture	Davis (1987)	Chapters 2, 7, 8 and 10

consequence. One challenge, or limitation, of this model is to decide on 'what consequence and for whom?' This could involve a social, environmental, physical or economic impact assessment. Each would need its own set of indicators. The merging of all these 'elements of risk' into a single assessment is rather problematic, as it inevitably involves a highly subjective judgment.

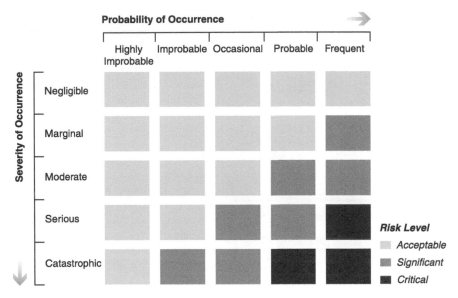

FIGURE 4.1 Model 9: probability/consequence risk assessment.

A strength of the model is its two-dimensional simplicity, but that simplicity brings risk assessment down to only two variables: frequency and impact. Other critical variables of the risk equation are missing from the vertical axis of the matrix, such as the duration of hazards like floods or droughts and the possibility of 'ramped' or repetitive impacts. However, the model is useful in introducing risk in training or public awareness programmes as it requires the users to consider these key variables and make a rough assessment of the level or scale of risks that demand action.

This model is widely used to calculate risk in business and other operations (e.g. Ayyub *et al.* 2007). Much more sophisticated versions are used in engineering. Besides the model's tendency to oversimplify risk, it is also dependent on the user's ability to estimate hazard in terms of frequency–magnitude relationships. For many threats of natural origin, these are still poorly known; and for some anthropogenic hazards, notably terrorist outrages, they are non-existent. The question of what comprises a 'critical' or even a 'significant' risk has been endlessly debated (Lane *et al.* 2012). The answers are highly dependent on context and the preferences of the user of such models. Nevertheless, risk needs to be known and thus has to be characterised. The model may be crude, but it is often effective.

For an example of flood risk assessment in the UK and the Netherlands, see the discussion of model 13, 'development of a safety culture', later in this chapter.

Model 10: disaster 'crunch' model

The value of this model (Figure 4.2) lies in its description of the causal factors that generate vulnerability to hazards. The evolution of this model is discussed

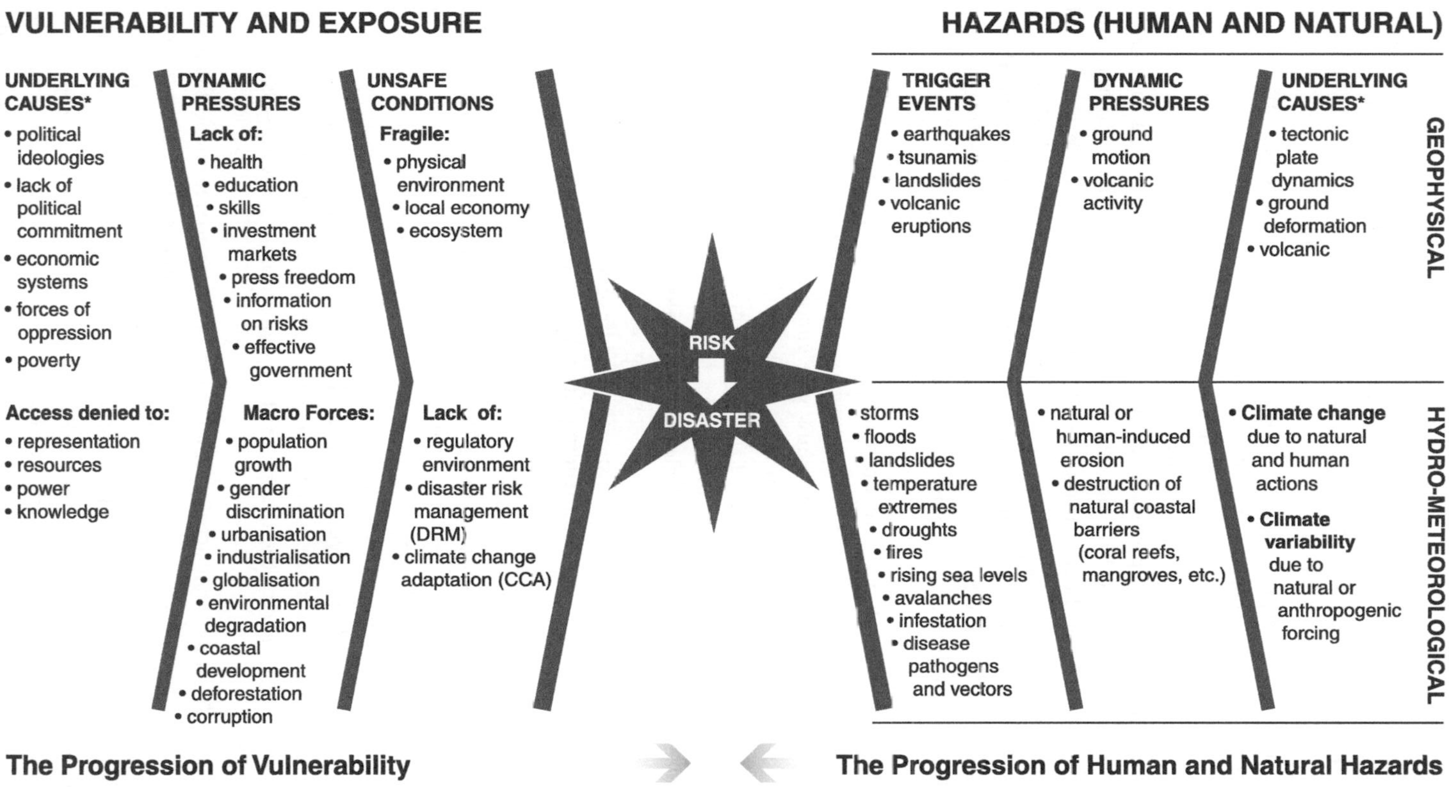

HAZARDS (HUMAN AND NATURAL)

GEOPHYSICAL

HYDRO-METEOROLOGICAL

UNDERLYING CAUSES*

Geophysical:
- tectonic plate dynamics
- ground deformation
- volcanic

Hydro-meteorological:
- **Climate change** due to natural and human actions
- **Climate variability** due to natural or anthropogenic forcing

DYNAMIC PRESSURES

Geophysical:
- ground motion
- volcanic activity

Hydro-meteorological:
- natural or human-induced erosion
- destruction of natural coastal barriers (coral reefs, mangroves, etc.)

TRIGGER EVENTS

Geophysical:
- earthquakes
- tsunamis
- landslides
- volcanic eruptions

Hydro-meteorological:
- storms
- floods
- landslides
- temperature extremes
- droughts
- fires
- rising sea levels
- avalanches
- infestation
- disease pathogens and vectors

RISK → DISASTER

The Progression of Human and Natural Hazards

VULNERABILITY AND EXPOSURE

UNDERLYING CAUSES*

- political ideologies
- lack of political commitment
- economic systems
- forces of oppression
- poverty

Access denied to:
- representation
- resources
- power
- knowledge

DYNAMIC PRESSURES

Lack of:
- health
- education
- skills
- investment markets
- press freedom
- information on risks
- effective government

Macro Forces:
- population growth
- gender discrimination
- urbanisation
- industrialisation
- globalisation
- environmental degradation
- coastal development
- deforestation
- corruption

UNSAFE CONDITIONS

Fragile:
- physical environment
- local economy
- ecosystem

Lack of:
- regulatory environment
- disaster risk management (DRM)
- climate change adaptation (CCA)

The Progression of Vulnerability

FIGURE 4.2 Model 10: disaster 'crunch' model.

Note: ★ 'Underlying causes' are often called 'risk drivers'.

in Chapter 8, 'Second comparison between 1978 and 2015: changes in vulnerability'.

The team of authors, which includes Ian, that wrote *At Risk: Natural Hazards, People's Vulnerability and Disasters*, applied this 'crunch' model to pre-disaster situations (Wisner *et al.* 2004); but its message is equally relevant to disaster recovery as a key element in any successful recovery programme is to reduce vulnerability in order to create safe conditions.

This model relates to a pressing dilemma and challenge facing recovery managers concerning the way pre-disaster deficiencies ('underlying causes' in the 'crunch' model) constrain post-disaster recovery actions. Ian recalls: in 1975, I visited an experienced urban planner, George Nez, in Denver, Colorado to interview him. He had been a planning advisor in both the reconstruction of Skopje after the 1963 earthquake and Managua after the 1972 earthquake. It proved to be a memorable encounter with a revealing discussion, well worth the long journey. Later he wrote to me to explain the persistent dilemma he had experienced in both recovery contexts. He wrote as follows:[1]

> Whenever you direct a reconstruction programme, everyone tends to blame the disaster for this or that problem. However, gradually you come to realise that 90% of the problems you encounter were all present long before the disaster event, waiting to be tackled. All that has happened is that the disaster has acted like a sharp surgeon's scalpel that has been used to expose all manner of weakness and failure, such as poor government, un-enforced building codes, lack of planning, corruption in all directions, etc. The issue poses a dilemma concerning how far it is possible to go, with the limited resources at your disposal, in addressing the residual weaknesses in a society as well as reconstructing its towns and cities.

The primary and challenging message of the 'crunch' model is this: in order to create safe conditions, actions are needed to reduce the causal factors that have created and maintained them. For example, officials responsible for disaster recovery have to look deeper and wider than merely creating a regulatory environment or instituting disaster risk reduction measures. They need to consider dynamic pressures and underlying causes and then take positive action to reduce them. If they fail to examine and tackle these generators of risk then unsafe conditions that gave rise to the disaster will return and may cause the disaster to repeat itself.

The 'crunch' model can act as a useful *aide-memoire* to remind officials to examine the scope of vulnerability by seeking answers to difficult questions through carrying out actions on the ground. One example of an unsafe condition helps explain this process: the creation of an effective regulatory system for the built environment as a vital component of reconstruction policy. The issues here are: Why did buildings collapse in the disaster and how were their occupants killed as a result? What were the underlying causal factors and, specifically, what recovery management actions are needed in order to create a safe future? This is just one of the concerns that the model highlights but it has many implications, as the next paragraph shows.

These are some typical questions for recovery officials as they consider ways to reduce underlying causes and dynamic pressures:

- Why do families occupy unsafe sites, and consequently what actions are needed in order to plan reconstruction with respect to the location of settlements?
- Is unsustainable urbanisation a major risk factor? If so, what countermeasures can be adopted in recovery policies to enable rural economies to reduce this pressure?
- Did patterns of land ownership contribute to the impact of disaster? What steps can be taken to create more equitable patterns of land tenure?
- Did the disaster casualty statistics indicate a high proportion of deaths and injuries to women and girls? If so, what steps can be taken in recovery practice to avoid gender discrimination? For further discussion on gender considerations in recovery, see Chapter 5, 'Conclusion: disaster as a starting point for recovery'. Further reference is made to gender issues in Chapter 6 under the heading 'Human rights and recovery from disaster'. Last, for a discussion concerning gender concerns in the Yolanda typhoon recovery operation, see Chapter 9, 'Option 6a: transitional shelter (temporary)'.
- Are any traditional practices in building, adapting or using buildings particularly unsafe in terms of how they relate to the impact of the kind of disaster that is expected in the future in the area? Is there a need for education designed, gradually, to effect cultural change?
- What actions are being taken to tackle corruption? The phenomenon may lie behind failure to implement regulations and will require the establishment of ethical principles at all levels of training and practice. The government will need to set examples in its own policies and practice. See Chapter 5, 'Disasters and the informal and illicit economies' (in the section 'Economic recovery and the question of livelihoods') and Chapter 12, 'Black hat – discernment' for discussion concerning the threat of corruption in disaster recovery.
- Does the government inform its citizens of the risks they face? Risk awareness is the first step in introducing community disaster preparedness into recovery plans. See non-structural risk reduction measures under model 12 later in this chapter.
- Are the national media free to report on the failings of government as well as on risks faced by citizens and emerging issues in recovery policy?
- What is the state of the local building industry? What steps are needed to improve its efficiency and place safety high on the agenda of building techniques in recovery practice? Can this be made a national priority? See Chapter 9, 'Option 7b: contractor-build permanent dwellings'.

Does this question lead one to query the state of building regulations and land use planning controls? It can be broken down into further concerns. Has the government enough ability to regulate building and planning? What improvements are needed in the design of codes? What changes are needed to the vast building programmes associated with reconstruction in the organisational and administrative aspects of regulatory review and enforcement? Finally, which national and local institutions that support regulation have the continuity and staying power to remain effective over time? See non-structural risk reduction measures under model 12 later in this chapter.

The 'crunch' model has some significant limitations. Although Ian Davis devised the original version in 1978 and it has been widely reproduced and used in education and training courses ever since, it remains unclear how the progression of vulnerability *actually* occurs, and there has been an absence of research on this subject. Therefore, it is not clear, and probably never will become clear, which underlying cause is directly or indirectly responsible for which dynamic pressure and which of these pressures leads to a given unsafe condition. The most likely explanation is that unsafe conditions derive from multiple causes, or 'risk drivers'.

Many formulations of vulnerability subdivide it by classes (e.g. physical, social, psychological, cultural, environmental and institutional – Birkmann *et al.* 2013). Unfortunately, this tends to ignore the dynamic nature of vulnerability and the tendency of one sector to influence another. An alternative approach is to classify it by phenomenon and origin (Alexander 2007):

- *total:* life is generally precarious;
- *economic:* people lack adequate occupation;
- *technological/technocratic:* due to the riskiness of technology;
- *delinquent:* caused by corruption, negligence, etc.;
- *residual:* caused by lack of modernisation;
- *newly generated:* caused by changes in circumstances.

It is clear that these categories have a degree of interplay that changes over time as conditions alter. Moreover, they are not mutually exclusive but represent different facets of vulnerability as a whole phenomenon.

One uncontroversial aspect of the 'crunch' model is the reality that there are deep-seated forces that persist in generating highly dangerous conditions. And added to this certainty is the reality that there is minimal appetite on the part of governments and international agencies to face up to the causes of disasters given the existence of acutely sensitive political issues, such as gender discrimination, rampant corruption, and denial of the human right to have information on risks. The result of this is that if they are not based on sound ethical principles and appropriate foresight, government planning decisions can place people at risk (Alexander and Davis 2012).

We will return to the important issue of corruption in two places: Chapter 5, 'Economic recovery and the question of livelihoods' in the subsection on 'Disasters and the informal and illicit economies'; and in Chapter 12 where, in the section 'Black hat – discernment', we consider the power of corruption as a risk driver.

Model 11: resilient communities and settlements

The value of this model (Figure 4.3) lies in its description of the essence of resilience to hazard and organisational pressures. The model brings together the varied concepts of resilience (which are also represented in model 3, Chapter 3).

The elements of the model can be explained as follows:

Resilience to hazard pressures: As indicated on the left-hand side of the model, communities and settlements (as well as economies, cultures, the environment, etc.) need to be resilient to extreme geophysical or climatic forces. The same is true of the technological pressures that exist in anthropogenic hazards; for example, the risks associated with explosion, crashes or toxic spills, and to a certain extent some of those connected with terrorist actions.

Resilience to organisational pressures: There is another range of forces to be resisted that come from the negative aspects of government, or the lack of positive governance. Five of these are listed in the arrows on the right-hand

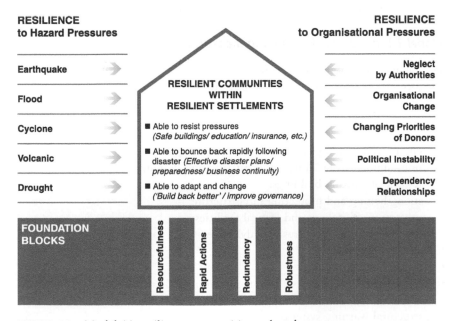

FIGURE 4.3 Model 11: resilient communities and settlements.

side of the model. However, it is rare for studies of resilience to consider these threats. Rather, they tend to focus on the need for resilience against hazard forces.

Resilient communities within resilient settlements: These elements are portrayed symbolically as the house in the centre of this model. It contains the three key elements of resilience: ability to resist pressures, to bounce back rapidly, and to adapt and change.

Foundation blocks: Resilient communities and settlements need to be firmly rooted if they are to withstand the pressures described above. Therefore this model stands on four foundation blocks that are characteristic of resilience, sometimes called the four Rs: resourcefulness, rapidity, redundancy and robustness. One might add a further block, communication – as communicating the concept and practice of resilience is an essential part of making it happen.

During the past decade, the wholesale embrace of the concept of resilience by governments, the UN and the voluntary sector indicates its popularity as an overall framework that covers the varied aspects of reducing disaster risks or adapting to climate change. However, in a similar manner to the ubiquitous terms 'development' or 'underdevelopment' – that continue to be used to describe just about everything and everyone – the catch-all term 'resilience' is already suffering the similar fate of losing its power due to overuse and overfamiliarity (Alexander 2013).

This model of resilience needs to be considered alongside model 3 (Chapter 3), which also considers the three dimensions of resilience that are located within the diagrammatic house in this model: absorbing the disruptive and damaging effects of disaster, bouncing back rapidly, and adaptation and change.

Model 12: disaster risk reduction measures

The value of this model (Figure 4.4) lies in its description of four closely linked categories of disaster risk reduction measures.

Structural measures:

- engineered buildings (construction and design);
- non-engineered buildings (construction and design);
- retrofit of existing buildings;
- strengthening and protecting infrastructure;
- protection of lifelines and critical facilities;
- flood protection measures (e.g. dykes, sea walls, barriers);
- cyclone shelters;
- tsunami refuge areas;
- fail-safe design (e.g. use of predetermined 'load pathways' to channel structural failure where it does least harm);
- structural warning systems.

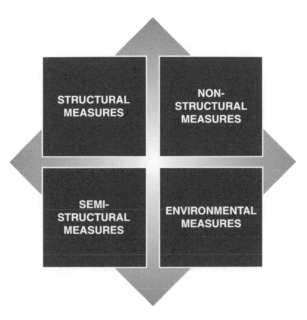

FIGURE 4.4   Model 12: disaster risk reduction measures.

Semi-structural measures:

- flood-proofing (e.g. portable barriers in entrances to buildings and sealant);
- hazard modification and control (e.g. opening floodgates to release a flood wave downstream; cloud seeding to reduce the risk of major storm damage – a highly controversial measure; snow avalanche triggering);
- redundancy (e.g. having more than one way to complete a task; duplicating staff, disaster plans, equipment, etc.).

Non-structural measures:

- disaster planning;
- emergency management;
- preparedness (alert and readiness measures);
- warning systems and evacuation plans;
- insurance and reinsurance;
- financial instruments (e.g. tax concessions);
- business continuity planning (e.g. including measures to protect the reputation of a business);
- legislation to reduce risks;
- building codes and regulations;
- land use controls in urban and regional plans;
- public awareness campaigns;
- education, training and exercises.

Environmental measures:
   In areas subject to flooding –

- flood storage areas and water meadows (sometimes classified as a semi-structural measure);
- canalisation and alternative river channels.

In areas subject to drought –

- water conservation and harvesting;
- diversification of the local economy;
- diversification of crops;
- other agricultural practices.

In areas subject to high winds –

- planting shelter belts of trees and other plants;
- using soil stabilisation materials to prevent wind erosion;
- strengthening coastal protection belts.

In areas subject to coastal surges –

- nourishing beaches with sand;
- widening the raised part at the back of a beach (the berm);
- planting mangroves;
- strengthening and rehabilitating coral reefs.

In areas subject to landslides and avalanches –

- revegetating slopes;
- protecting water courses;
- improving drainage and groundwater management.

Risk reduction in practice

A hypothetical example of some elements of these lists concerns the management of floods on a floodplain occupied by human settlements. Dredging channel beds, canalisation and armouring of channels, raising of levees and constructing dams may be relevant examples of physical measures to reduce flooding. However, it is well known that in many cases they merely shift the burden of flooding downstream, especially if increasing urbanisation creates more and more impermeable surfaces that can deliver floodwater rapidly to rivers and streams. At the same time, the presence of physical defences against flooding has been known to increase the rate of urbanisation of floodable areas under the mistaken assumption that the defences will always be effective and adequate. Setting aside floodable areas, such as water meadows (i.e. flood detention storage), can go some way to reducing the size of peak flood discharges. This semi-structural measure is seldom enough on its own. Another semi-structural measure is to provide flood prevention kits to householders or induce them to purchase such things. These include portable barriers to the incursion of floodwater and they can be effective if the levels of flooding are well under one metre above floor level. Environmental measures include revegetation, especially by planting trees, and reduction in the extent of impermeable surfaces that generate fast runoff. Finally, non-structural measures extend from land use planning (prohibiting or reducing development in floodable areas) to flood damage insurance and evacuation plans. Accumulated wisdom on dealing with flood disasters indicates that the best strategy involves multiple approaches that combine all four categories (Schanze 2006).

There is a discussion concerning the difficulty of calculating the overall costs of safety measures, as a basis for cost–benefit assessments to set against future losses, in model 13, 'Stage 3 – logic'.

The challenge of coordination

It is not difficult to list the above items in the four categories. Any consideration of these categories is a reminder of the rich and varied tools that are available to reduce risks. However, there are acute difficulties in resolving the problem of how to manage risk reduction strategies. The main challenge is the need to establish critical links between associated elements or tasks. For example, there is a clear link between safe building measures, building byelaws, the training of builders and craftspeople, the education of professional architects and engineers and the awareness of building occupants who need to know the rationale and function of the safety elements that have been used in their dwellings. Some of these measures for building safety are structural while others are non-structural. For the links to function, there has to be some level of coordination, communication between parties and shared commitment. For a variety of reasons, these attributes rarely exist in practice. The net effect is to reduce overall safety.

This elusive coordination task has been seen as unrealistic and thus likened to an aspiration to 'coordinate a culture'. But perhaps there can be an interim solution through the establishment of a multi-agency, multidisciplinary task force to ensure that vital connections are made and maintained.

Maintenance and safety

A further problem of structural risk reduction measures concerns levels of maintenance. Politicians favour high-profile capital works programmes particularly when elections are imminent, since they provide visible evidence of concern and their opening can become a high-profile media attraction. But physical measures inevitably deteriorate and require continual maintenance, and that requires an ongoing maintenance budget. However, the majority of politicians seem to perceive that there are few votes in securing such sensible stewardship of the environment and prudent protection of assets through regular maintenance.

Therefore, there is a golden rule that should be placed in large letters above the doors of treasury or financial planning ministers to the effect that: any given structural safety measure that provided security when it was installed will, without maintenance, fail to deliver the required safety in subsequent years. And rather more specifically, maintenance will certainly not occur unless there is a dedicated maintenance budget, earmarked at the time of allocating capital expenditure.

Flood protection in London

This model can describe the interaction between different categories of risk reduction measures. For example, the city of London is protected from coastal flooding by the Thames Barrier, a major example of a structural risk reduction measure. This can be regarded as a 'form that followed failure' following the 1953 East Coast flooding. This major engineering structure is linked to river protection walls that extend on either side of the Thames as far as the North Sea, for about 70 kilometres. These walls provide protection against flood levels identical to the Barrier. However, there are occasional access gates throughout the length of the sea walls. These provide essential access to the Thames estuary for emergency services or for river or estuary maintenance, boat keepers, etc. The gates are robust structures with well-engineered seals to prevent floodwaters from entering the land. However, despite the fact that they are all securely locked in severe weather conditions with potential flood risk, there is an inevitable risk of human error should a single gate or series of gates not be securely locked. This would enable floodwaters to cascade into the land and potentially to enter the London area at the back of the Thames Barrier with devastating consequences.

So this model of coastal flood protection for London requires a combination of structural, non-structural and environmental measures as follows:

- *Structural measures* – the Thames Barrier; river or sea walls; flood-proof access gates; structural warning systems, etc.
- *Non-structural measures* – training of all staff who manage the flood protection system; standard operating procedures concerning all aspects of the system, including the securing of access gates in the sea walls; public awareness for all who are involved in flood management, including those who need to gain access through the gates to secure them after use; non-structural warning systems; preparedness measures; building codes and regulations; land use controls in urban plans within the floodplain, etc.
- *Environmental measures* – nourishing beaches with sand and widening beach berms within the Thames estuary.

All these measures are parts of the flood protection system for London, but only some of the elements are managed by a single authority. For example, the management of the Thames Barrier and associated aspects of flood management will have no role in building controls or land use planning. It is also clear that the flood protection system could be compromised by human negligence, such as a failure to secure a floodgate.

In the discussion concerning model 13 (on safety cultures), there is further consideration of the flood protection problem in London.

Model 12 can have two further applications. It can be used to examine the comparative costs of risk reduction. Some are very expensive, such as retrofitting existing buildings, while others can be undertaken at much lower financial cost, such as raising public awareness. However, this is not to imply that the benefits are equal or that one can necessarily get by only with cheaper solutions.

The model can also be useful for recovery planning and has been used in consultancy assignments to show public officials the many options that are available to them. It can also be linked to model 2, the 'recovery sectors' model (Chapter 3), by recording links to relevant recovery sectors and by highlighting physical, environmental, social and economic disaster risk reduction measures.

Model 13: development of a safety culture

The value of this model (Figure 4.5) lies in its description of the stages of the progressive development of a safety culture.

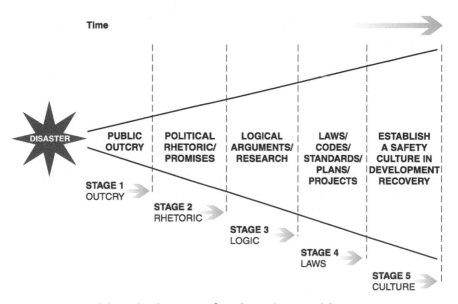

FIGURE 4.5   Model 13: development of a safety culture model.

This model suggests possible stages in the development of a safety culture over a number of years and within a given context. This may be in a community, society, town, city, country or region.

Stage 1 – inception and public outcry: Immediately after any major disaster there is an eruption of rage by the surviving and grieving public and their institutions. There may also be a pervasive, and not always rational, desire to attribute blame. Such sentiments are provoked by the waste of so many lives, sacrificed or ruined, as well as by the loss of treasured cultural heritage, buildings, infrastructure, environment and livelihoods. This outcry may be the necessary 'match' that ignites a small flame which will eventually, perhaps over many decades, help to create a stronger culture of safety.

Stage 2 – rhetoric: A familiar pattern can be deduced from mass media reaction to a major disaster. When they respond to public outcry or media pressure, political leaders are apt to make lofty promises that this disaster will never be allowed to happen again. In some cases, they go further and make unrealistic statements about relocating a damaged settlement or supplying abundant resources to the recovery process.

Stage 3 – logic: Public outcry and unsubstantiated promises may be followed by a period of hard-headed assessment and analysis. The questions asked concern the nature of future threats. To understand these, detailed risk assessments may be made on the basis of maps of hazards and analyses of patterns of vulnerability across all sectors. Economists may apply calculations to assess the benefits to be derived from the costs of protection. Although they appear logical, such enquiries are often criticised because they assess quantifiable data on the direct or indirect potential for physical or economic losses but may ignore intangible, unquantifiable or qualitative factors – for example, through the inability to place a cash value on lives, distress, disruption or cultural heritage. Albert Einstein once aptly summarised this dilemma: 'Not everything that counts can be counted, and not everything that can be counted counts.'

The logical phase often includes attempts to calculate the cost or risk reduction measures built into reconstruction. Such calculations are often made concerning the money a country will save by adopting this or that risk reduction measure to cut future disaster losses. These calculations can be applied when seeking to build safety into recovery planning. However, we believe that these calculations are normally flawed. Such calculations confidently suggest that vast sums are saved by risk reduction measures. This is probably true but our point is that it is not possible to calculate the cost of many disaster risk reduction measures; thus, it is impossible to apply these figures in making economic comparisons to encourage governments to invest in protection.

For example, Ian recalls that when he worked in architectural practice, calculations of building costs were always needed as part of the planning process; but there was *never* a breakdown cost of the various safety measures, the reason being that overall safety will relate to all manner of structural and non-structural building and physical planning factors and elements. These include aspects of the education of engineers, architects and planners; regulatory costs; possible costs associated with land use planning and siting and the choice of building materials; structural connections; constructional detailing; design of ground plans; shape of building; and foundations. Given this complexity, it is not possible to assemble the cost of such elements for the simple reason that, in many cases, these figures are not available and would be almost impossible to calculate (e.g. the costs linked to safety factors in higher education). Thus, the specific cost of building safety against earthquakes – the most significant calculation in seismic safety – is actually unknown. Published calculations that suggest safe building design adds 20 per cent to the cost are pure 'off the cuff' assumptions, not based on any reality.

Similarly, it is not possible to work out the specific costs of flood protection as part of recovery planning since such costs become totally intermingled with all the related aspects of irrigation, land drainage, water supply, etc. The international community likes to use the fashionable term 'mainstreaming' to describe the process of integration of risk reduction into all sectors of development and recovery. But when genuine integration occurs, costs as well as benefits are also merged. Thus, better and more accurate arguments are needed to promote risk reduction as a key factor in costing recovery.

Stage 4 – laws: Careful assessment and analysis can pave the way for new laws to govern safety standards, as can the creation and funding of new institutions to maintain safety, plan recovery and manage legal frameworks. By ensuring that certain actions are taken, laws lay down solid requirements and thus protect a society from the fluctuating whims of political parties. Good laws also serve to raise public awareness of the need for safety.

Stage 5 – a safety culture: Good laws are only one element in the creation of a safety culture. Other elements that are needed include vision, leadership, persistent advocacy, active public institutions, research and educational support from academia, private sector involvement, a supportive mass media and an informed and vocal public that applies pressure.

Flooding and the development of a safety culture in England, 2014

Having described the five stages, this example illustrates the model in practice. Exceptionally deep and prolonged flooding in the UK during the winter of 2014 led to an eruption of public outcry, political rhetoric and economic logic. Stage 1 involved a *public outcry* at the persistence of flooding and the perceived inefficacy of government policy both to reduce flood risk and deal with the emergency as it happened. In the words of one commentator:

> The devastating floods that have affected the UK this month are a reminder that this issue is certainly not limited to developing countries. The floods could reduce GDP by just over one per cent, knocking an estimated $20 billion off the value of the economy. These floods come off the back of

cuts in real-term spending on flood defenses, highlighting the false economy of dealing with the crisis after it has hit.

(Cabot Venton and Peters 2014)

The flooding had been caused in part by negligent public planning policies over previous decades. Nine hundred new houses had been built during the period 2002–11 within the most vulnerable floodplain areas of the Somerset Levels, in the Sedgemoor district in south-west England (Webster 2014). In addition, a further causal factor was a severe cutback in flood preventive works by a government preoccupied with being stringent in reducing the cost of public services. Despite this, amid apparent inaction by authorities, on 11 February 2014 at the height of the crisis the British Prime Minister, David Cameron, visited one of the worst affected areas and stated exactly what the affected community wanted to hear from him: 'Money is no object in this relief effort. Whatever money is needed we will spend it. We're going to lay out plans and spend what we need to.'

However, in subsequent days, Cameron's fine political *rhetoric* (stage 2) was quickly replaced by harsh economic *logic* (stage 3) as various government ministers who had responsibility for budgets qualified his assertions and backtracked on his promises. The Chair of the UK Government's Climate Change Adaptation Sub-Committee, Lord Krebs, stated on 7 March that it would be 'unfair in the long term for this flood-prone area to attract more taxpayer support (for flood protection) than similar areas. . . . It may well be the case that there are areas that are too expensive to defend' (Webster 2014).

At the time of writing, stages 4 and 5 have not yet been fully applied to the English floods of 2014. No doubt they will be in the fullness of time. However, flood defence policy was quickly re-evaluated at the national level, flood defences were comprehensively checked, and some changes were implemented in structural control measures. In the UK, flood insurance has undergone a constant process of reappraisal with each new flood disaster, and this process continues unabated – as do the floods! Ominously, the process of downsizing the Environment Agency, which has primary responsibility for many flood assessments and defences, continues apace. At the same time, there was no sign that fragmented responsibilities for responding to flood risks would be reassigned in a more rational manner. Flood policy in the UK was comprehensively reviewed in the Pitt report, which was the fruit of a public enquiry that followed the devastating and widespread floods of 2007 (Pitt 2008). The report offered 72 conclusions and 92 recommendations. After the 2014 floods, David Cameron stated that 'all but one of Pitt's recommendations have been implemented', which is difficult to substantiate as many of them referred to different tiers of government and processes of public participation that are by no means universally developed in the UK. The solitary recommendation that he regarded as unimplemented was that householders assume some of the risk for defending their own properties against floods.

This model is inevitably general and there are variables in the way a given safety culture has, or has not, been established. Examples of differing safety cultures can be seen in the following comparison between the approach to coastal flooding in the UK and the Netherlands.

Relative levels of coastal flood protection in the Netherlands and the UK

In the recovery following the North Sea flood of 1953, where 1,836 persons died in the Netherlands and 307 died in England, differing flood protection standards (or different levels of acceptable risk) were built into recovery plans.

In the Netherlands, a collective conviction exists in all levels of society that in their low-lying country, where two-thirds of the land area is exposed to flooding, there must *never* be a repeat of this disaster. Therefore, from 1953 onwards, the Dutch Government began to implement a massive flood protection system that protects against floods of a return period of up to *10,000 years*. To maintain this system, 1 per cent of the national GNP is devoted annually to flood protection. This is in line with a millennial culture, in which the Dutch have for so long defended their settlements against the incursions of sea and rivers.

In contrast, in providing protection against the same coastal flood threat, London – the best protected town or city in the UK – has been provided with defences against a *1,000-year* coastal flood through the creation of the Thames Barrier and associated sea walls along the Thames estuary. However, in 2014 following a winter of severe coastal storms, plans are being formulated for much higher levels of protection for London in recognition of its strategic importance to the UK economy (Leake and Trump 2014). It is unlikely that Britain will ever achieve the same standards of physical protection as the Netherlands, but it could be argued that the risk aversion inherent in the Dutch approach involves excessive measures and expenditures, especially on structural measures. In either case, climate change may alter the picture for both countries and necessitate a comprehensive re-evaluation of policies for flood disaster risk reduction.

On the negative side, the model discussed in this section suffers from oversimplification of the highly complex process of establishing a safety culture. The reality is that a route very different from this example may be needed if a safety culture is to be achieved. However, this model may serve the purpose of indicating a sequence for some of the social, political, economic and administrative dynamics that may or may not lead towards the acceptance by a society of the need to reduce disaster risks. The relevance to this study is that *any* disaster recovery process which seeks to build a sustainable future has necessarily to add to the tasks of treating the injured, rebuilding structures, creating jobs or planting trees. Additions represent a concentrated desire to invest in the development of a safety culture and may form the legacy of a disaster event.

Organisation of recovery models

Securing effective organisation

> In the future, countries with major housing losses in a disaster can learn
> from the experience of others and attempt to find the 'sweet spot' which
> provides the best of government management for expediency and flexibility
> and incorporates opportunities for citizens to take some control over their
> own recovery, with housing choice and participation in plans for the
> community's future.
>
> *(Comerio 2013: 37)*

The management gurus Peter Drucker and Warren Bennis have suggested that
there is a tendency for organisations to concentrate on maintaining order by
'doing things right' rather than exercising leadership and 'doing the right things'
(cited in Covey 1989: 101). The highly politicised environment of disaster
recovery after a major event in the life of a country can result in a response that
is more dictated by a highly visible show of concern to gain short-term political
benefits rather than a courageous grasping of essential and very difficult issues.
These include ensuring future safety in recovery actions (as represented in
models 9 to 13) and making certain that there is positive development, well
beyond pre-disaster norms (as described in models 1 to 4). Therefore, the models
in this section are concerned with the effective management of organisations and
the mobilisation and participation of surviving communities as well as establishing
inspired leadership.

Throughout our book, we attempt to convey the complexity and scope of
recovery due to the presence of multiple actors with their own agendas,
unforgiving time constraints, parallel multi-sectoral recovery tasks, political
pressures and continual financial demands. To respond constructively to these
challenges, the form, style and location of recovery organisations is always a
matter of critical importance. So we propose seven organisational models to
describe patterns and styles of organisation, management and leadership.

Model 14 indicates 'project planning and implementation' with a strong
ethical base, often omitted from the project cycle. A further ethical question for
recovery managers is considered in model 15 – how to maintain a balance
between trust and control.

Models 16 and 17 relate to the organisation of shelter/housing issues that are
discussed in detail in Chapters 8 and 9.

Model 18 provides a tool to assist in the design and monitoring of recovery
while model 19 is a creative tool to analyse a problem from varied standpoints.

Model 20 describes alternative approaches for governments organising
recovery. Description of the final model (no. 21) can be found in the concluding
chapter where we summarise the case studies cited within our book in a matrix
that compares the strength of governments and community participation.

TABLE 4.2 Summary of organisation of recovery models (14 to 21)

Model	Graphic representation	Application	Source	Location in book
14 Project planning and implementation		Suggests a logical sequence for planning and implementing recovery	New model developed for this book by the authors	An example of a project that follows the sequence of actions proposed in Model 14 – the case study of Malcondji village following the 1993 Latur earthquake –opens Chapter 1
15 The pendulum (after Charles Handy's trust-control dilemma)		Explores the levels of trust and control that are exercised by authorities	Handy (1995)	The Wenchuan case study in Chapter 3 and the Pakistan case study in Chapter 12, 'Yellow hat – optimism'
16 Two- or three-stage shelter and housing recovery		Compares alternative approaches to shelter to permanent dwellings	Model developed by Ian Davis and Fred Krimgold; the concept is described on page 111 of Davis (2015) and was first reproduced in its present visual form in Davis (2011a)	Chapter 9, 'Options after disaster: transitional shelter'

(continued)

TABLE 4.2 *(continued)*

Model	Graphic representation	Application	Source	Location in book
17 Modes of shelter and housing		Considers the range of shelter and housing options in pre- and post-disaster contexts	New model developed for this book by the authors	Chapter 9 is devoted to the options in this model
18 Strengths, weaknesses, opportunities and threats (SWOT)		Provides a tool to continually monitor the progress of recovery	Stanford University, USA study of Fortune 500 Companies in 1960, cited in Krogerus and Tschäppeler (2011: 12)	In Chapter 9, the strengths and weaknesses of eight options for shelter and housing are discussed
19 Edward de Bono's 'six hats' model for problem analysis		A creative analytical tool to examine a problem from six standpoints	de Bono (1985)	The 'thinking hats' model forms the basis for Chapter 12
20 Organisational frameworks of government for recovery management		Describes alternative approaches for governments organising recovery	New model developed for this book by the authors	Chapter 5 'Roles and responses of government'
21 The Mary Comerio comparison of recovery management approaches		Assesses and compares two vital ingredients of successful recovery: community participation and the role of governments	This model is adapted with permission from Comerio (2013: 37)	See Chapter 13, Principle 13 for this model as it relates to certain case studies of recovery cited throughout our book

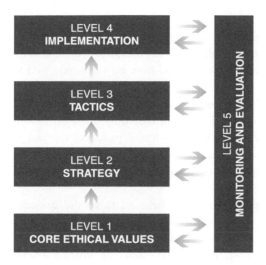

FIGURE 4.6   Model 14: project planning and implementation.

Model 14: project planning and implementation

The value of this model (Figure 4.6) lies in its description of the sequence of planning, implementation, feedback and adjustment.

By applying past experience, we seek to define a series of principles of recovery that officials will need in order to guide their actions in positive directions through all the problems and uncertainties that surround the process.

We have both sought in the past to develop and document varied sets of principles of recovery. David included recovery principles in *Principles of Emergency Planning and Management* (Alexander 2002: Chapter 7); and Ian and colleagues developed a set of principles for the first UN guidelines concerning *Shelter after Disaster* (UNDRO 1982, Davis 2015). Ian also contributed to the team that developed the principles for the World Bank guidelines *Safer Homes, Stronger Communities: A Handbook for Reconstructing after Natural Disasters* (Jha *et al.* 2010). However, such principles have previously been stated *en bloc*, often listed in a random mix of tactical concerns intermingled with those concerned with values and strategy. Instead, we suggest that there is a distinct value in disaggregating them into an orderly set of interdependent categories as described in this model in which each level flows from the one below and they all rest on a solid foundation of ethics and core values.

These principles, values or fundamental assumptions concerning how recovery can proceed in a better manner can resemble a model of project planning and implementation, as represented in the flow chart. The model is not intended to indicate a hierarchy but, rather, shows progressive stages in a sequence that builds upon what lies beneath.

Using the example of 'bounce forward' disaster recovery, level 1 involves establishing the moral imperative to preserve lives and increase safety by ensuring that recovery does not restore pre-existing vulnerability but builds safety measures into the reconstruction. Ethical values also need to permeate the entire process of recovery, such as how contracts are decided, how corruption can be avoided, how to maintain equity and fairness and how power is exercised for the good of the powerless. Level 2 involves providing the means to achieve this, including the finances, materials and principles (such as building codes). Level 3 involves distributing the resources and ensuring that they are used rationally and equitably. Level 4 is that at which practical action produces the necessary achievements, and level 5 involves a constant process of evaluating and re-evaluating to ensure the ethical principles are conformed to and the desired results are continuing to be achieved.

Level 1 – core ethical values: These relate to the underlying shared beliefs and core values of the organisation (or of society), and of the organisation's mandate, as well as those of officials charged with carrying out its work. In terms of project planning, this is the stage at which answers are sought to the questions: 'Who is likely to gain, and who might lose, in the recovery process?' and 'Who is accountable to whom?' An example of an ethical principle in disaster recovery concerns how to make the distribution of recovery resources equitable in terms of the needs of beneficiaries as opposed to their status. How to establish property rights and secure tenure for people affected by disaster are other vital concerns. See Chapter 13 for a list of principles.

Level 2 – strategic planning: This concerns the direction of a given task that will be informed by the ethical principles and the policies that arise from them. At this stage, the question should be asked: 'Why is a particular approach appropriate and what are the priorities for recovery?' An example of a strategic approach to disaster recovery involves the need to ensure that reconstruction policies and plans are financially realistic. In this context, it needs to be recognised that over time budgets will tend to decline. There is thus a pragmatic need to achieve recovery rapidly by obtaining a political consensus and the assurance that dedicated funds will remain available.

Level 3 – tactical planning: This level involves the practical application of strategies. In developing appropriate tactics, local conditions need to be taken into account. This stage in project planning must answer the question: 'How should one proceed in applying the strategy to recovery actions?' An example of a tactical approach to disaster recovery is as follows. Effective disaster recovery requires the participation of local groups. Therefore, management structures must empower local people but ensure that harmonisation is created by higher levels of government. In defining tactics, specific local groups and departments or sectors of government should be identified.

Level 4 – implementation: The approach here is guided by all the preceding levels: core values, strategy and tactics. Applications will need to be developed for each situation under the prevailing local conditions. This stage in project implementation answers the question: 'How should one move from the plans to the reality in an effective manner?' An example of a guideline for implementation of a disaster recovery practice would be: assistance imported into a disaster area should augment, complement and reinforce local initiatives, not supplant or duplicate them. Once again, in defining implementation approaches, specific local groups and departments or sectors of government should be identified.

Level 5 – monitoring and evaluation: This is the final stage and is shown sitting apart from the hierarchy of actions, though interacting with each stage. In order to indicate a two-way process of information and learning, the arrows between monitoring and evaluation and each of the stages (1 to 4) point in both directions. Monitoring will need to occur all the time in order to review progress and provide opportunities for changing the course of actions. In a complementary manner, evaluation normally takes place on the completion of a project in order to assess how effectively its aims have been achieved. Both monitoring and evaluation grow out of the first stage of ethical values as they express the need to be accountable to the beneficiaries of activities as well as to those who provided the necessary resources and those who participated in the recovery processes.

As the essence of learning is the ethical concern to be accountable and transparent, monitoring and evaluation return the process to stage 1. This stage in project implementation answers the question 'How should one improve performance in this and future projects, and what are the positive and negative lessons to be learned?' By way of example, the main purpose of the monitoring and evaluation of recovery activities is to improve recovery management policies, programmes and projects through feedback of lessons. This provides a basis for accountability, including the provision of information to the public. The test of whether lessons are learned or merely forgotten is whether measurable positive progress occurs when they are incorporated into practice.

The first case study in Chapter 1, the reconstruction of the Indian village of Malkondji, is a particularly relevant example of effective monitoring and evaluation that took place over the long term. The latter occurred immediately after the project and was followed 15 years later with a reassessment carried out by the same evaluation team. The reassessors included the original project leader, and they benefitted from a powerful learning process. The dissemination of this example of successful recovery in this book is part of a wider learning process as we extend the results to our readers.

A limitation of this model relates to an implied assumption that decision-making is a logical and direct progression from ethics to strategy to tactics and to operations. However, this is an excessively tidy view of reality, as there are many different entry points by which decisions can be made, and tactical problems may require a feedback loop into strategic thinking. Hence, this model, like all the others, needs to be used with care, discretion and flexibility.

In Chapter 6, we examine a pair of critical ethical issues that condition success or failure in recovery – namely, human rights and accountability.

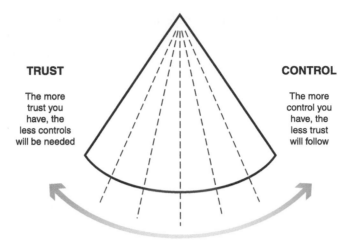

TRUST

The more
trust you
have, the
less controls
will be needed

CONTROL

The more
control you
have, the
less trust
will follow

FIGURE 4.7   Model 15: the pendulum (after Charles Handy's trust-control dilemma).

Model 15: the pendulum (after Charles Handy's trust-control dilemma)

The value of this model (Figure 4.7) lies in its exploration of the levels of trust and control that are exercised by authorities. The management thinker and theorist Charles Handy has written extensively about the relationship between trust and control in management (e.g. Handy 1995). We have taken his concept and related it to a pendulum to demonstrate that the more trust leaders place in their workforce, the fewer controls are needed. Conversely, the more that leaders rely on controls, the less the workforce will trust them.

The relevance of this model to recovery management can be seen in many of the dilemmas that the authorities face. For example, when officials provide cash grants in lieu of 'in kind' assistance, they are relying on the assumption that the beneficiaries will spend the money prudently and not on the purchase of non-essential goods such as alcohol or weapons. Conversely, if they insist on providing tangible relief items instead of cash, they are demonstrating a lack of trust. Vice versa, the beneficiaries who receive cash or goods are trusting that the groups that assist them will treat them with dignity.

In a second example, when officials initiate participatory processes involving consultation with beneficiaries over planning and reconstruction issues, those who participate are demonstrating trust that their aspirations and views will be taken into account. Where participation is denied this represents a control position. Paradoxically and despite these examples, it is unlikely that effective recovery will follow if there is *total* control or *total* trust. A balance should be struck in which authorities keep certain degrees of control and beneficiaries continue to monitor whether it is worthwhile to trust the authorities. The issue highlighted by Handy's pendulum is particularly relevant to the nature of political systems as, for instance, authoritarian regimes inevitably adopt high levels of control. Even in the most democratic societies, the relative positions of control and trust can swing from side to side over time as political, economic and social conditions change – hence the value of using the pendulum analogy in the model.

This trust-control pendulum model relates to two contrasting cases in our book. Negatively, the Wenchuan recovery case study (Chapter 3) indicates a profound *lack of trust* by the authorities in the grieving survivors of the disaster and their brave advocates. And positively, this is in sharp contrast to the *high level of trust* from the authorities in Pakistan (Chapter 12) who took the risk of providing cash grants to survivors.

The trust-control dilemma is also considered in Chapter 6 in the section on 'Accountability' and in Chapter 10 in the section concerned with 'Resilient international leadership of disaster recovery' where we discuss international agencies and the agendas and frameworks they seek to create. We suggest that these can often reflect a desire for control at the expense of trust.

Model 16: two- or three-stage shelter and housing recovery

The value of this model (Figure 4.8) lies in its description of alternative scenarios for the provision of shelter and housing. This model is a detailed consideration of the wider shelter and housing context described in model 17 and discussed in Chapters 8 and 9.

SCENARIO 1: THREE-STAGE RECOVERY

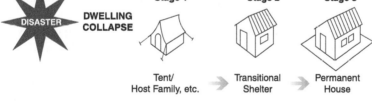

SCENARIO 2: TWO-STAGE RECOVERY

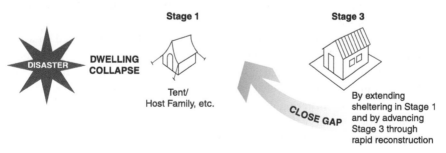

FIGURE 4.8   Model 16: two- or three-stage shelter and housing recovery.

Which scenario to choose?

Scenario 1 – three-stage recovery: In recent years, this has become the default position of many agencies, (particularly international NGOs and donors). The popularity of this approach to shelter or housing is probably due to a number of factors. To begin with, in disasters such as the 2010 Haiti earthquake or the 2004 Indian Ocean tsunami where there was an overwhelming public response to international appeals, large disaster relief funds were handled by international agencies and their national partners. Hardly any of these international agencies have any desire or competence to become involved in the complexities and demands of long-term reconstruction of housing. Instead, they need to spend their money rapidly, hence the delivery of large numbers of transitional housing units that often cost almost as much as permanent dwellings. The units are usually prefabricated or containerised dwellings, and it is common practice for each family to be allotted 40 square metres (possibly half that in poorer societies and even a quarter in the poorest cases). In both Japan and China, the floor areas tend to be smaller (19 to 33 square metres).

Governments are often unprepared for disasters. Hence, before housing reconstruction can start, building bye-laws may need revision, debris has to be cleared, titles to land and property may need to be confirmed, and building contracts need to be assigned. Long delays can ensue, and the perception is that something more substantial than tents of plastic sheeting is needed to tide the homeless over – hence the popularity of transitional housing units. This may be quite logical; for example, in areas of harsh climate, transitional housing units are needed in order to protect survivors against the elements. In industrialised societies, middle-class urban disaster survivors demand more from their governments than tents or improvised shelters, or communal sleeping arrangements, and thus transitional housing units are regarded as essential.

Scenario 2 – two-stage recovery: Where there has been pre-planning of recovery, it is possible for rapid reconstruction to occur; for example, as it did in Mexico City after the 1985 earthquake. For two-stage recovery to occur, the survivors will need to be prepared to live in tents, hotels, second homes or with host families without an interim stage of transitional units. Where this process is followed, substantial cost savings can be achieved. However, there needs to be rapid and effective reconstruction of permanent housing, which needs to be established *and maintained* as part of a dynamic, ongoing process.

The guiding principles that underpin Scenario 2 relate to the need to avoid wasting money and building materials. Available building sites for permanent housing should not be clogged up with transitional dwellings. It helps if the authorities can reconstruct rapidly by pre-planning the process. Moreover, the most effective alternative sheltering options, such as the use of host families to provide accommodation to survivors, should be strengthened. Above all, the authorities should consult the beneficiaries to determine their preferences.

Transitional shelters: should they be used?

The debate about the necessity of transitional shelters has a long history. For example, prefabricated 'temporary' housing still exists in Messina, Italy following the 1908 earthquake and in San Francisco after the 1906 tremors. After the Mexico City earthquake of 1985, the authorities rapidly requisitioned the tenement apartment dwellings that collapsed in the disaster. Then, using World Bank loans and grants, they embarked on a rapid reconstruction programme. The first apartments were completed in under a year. Meanwhile they provided rudimentary shelters for the survivors in adjacent streets with portable toilet and shower units. They were careful in their selection of which streets to use for the shelter sites: they knew that they would not become permanent slums as there would be a powerful local demand to reclaim the streets for traffic. The authorities also wished to keep the survivors as close as possible to their livelihoods, where these were still functioning. As a result, the authorities persuaded the survivors that with rapid reconstruction and extended use of emergency standard accommodation for a limited period it would be possible to avoid the additional cost of transitional shelters (Davis 2012).

The main limitation of this model is that a complex sheltering and housing process is reduced to simple headings that oversimplify the many subtle permutations that apply, as discussed in Chapter 9. Furthermore, the diversity of recovery situations in varied cultural, climatic, environmental, economic and technological contexts makes it impossible to be dogmatic about any single, preferred approach. Nonetheless, the guiding principles set out above should be applied wherever possible.

Model 17: modes of shelter and housing

The value of this model (Figure 4.9) lies in its attempt to show, in a single diagram, the wide diversity of modes of sheltering and housing in both pre- and post-disaster contexts. There is no discussion concerning the model here, other than the headings being restated below, since a detailed review can be found in Chapter 9.

The purpose of this model is to show the options and decisions that need to be made regarding shelter and the risk and occurrence of disasters. These can be listed as follows.

Options before disaster:

1   safe dwelling;
2   evacuation to safe shelter;
3   unsafe dwelling on unsafe site.

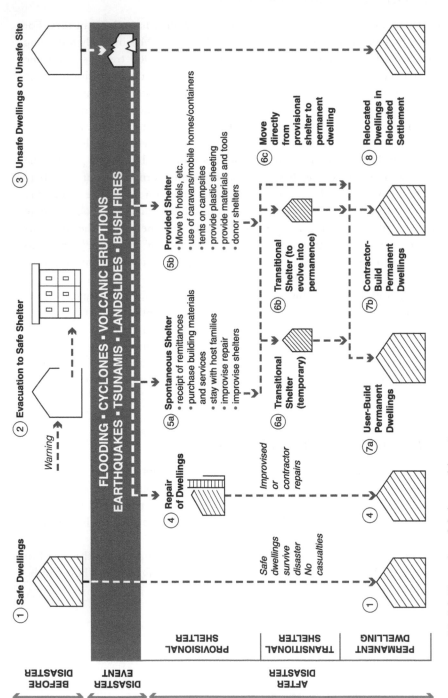

FIGURE 4.9   Model 17: modes of shelter and housing.

Options after disaster – provisional shelter:

4   repair of dwelling (repairs improvised or carried out by contractor);
5(a) spontaneous sheltering;
5(b) provided shelter.

Options after disaster – transitional shelter:

6(a) transitional shelter (temporary);
6(b) transitional shelter (to evolve into permanence).

Options after disaster – permanent dwelling:

1   safe dwelling that survived the disaster;
4   repaired dwelling;
7(a) user-build permanent dwelling;
7(b) contractor-build permanent dwelling;
8   relocated dwelling in relocated settlement.

As noted above, we return to this model in detail in Chapter 9.

Model 18: strengths, weaknesses, opportunities and threats (SWOT)

The value of this model (Figure 4.10) lies in its usefulness for the analysis of recovery options.

Introduction to SWOT analysis in disaster recovery, Managua earthquake 1973

Ian recalls that he was introduced to the application of SWOT analysis at the outset of his work in disaster risk management: this occurred in 1973 when visiting the ruined city of Managua, Nicaragua after it had been devastated by the earthquake of 23 December 1972. This *opportunity* for rapid learning came as I departed for my first period of PhD field study a couple of months after the disaster, when my supervisor at University College London, Professor Otto Koenigsberger, gave me what seemed to be the strange and quixotic advice to '[r]emember that relief is the enemy of recovery' (cited in Davis 1978: 66; see p. 78). Later, I recognised that Koenigsberger was describing the *weakness* that frequently accompanies immediate disaster assistance: the unnecessary expenditure of vast sums of money even though funds will be in short supply when they are most needed in later recovery processes (see model 7, Chapter 3). He was also referring to the manner in which misdirected aid can create passive dependency among disaster survivors.

On my way to Nicaragua, I visited Columbus, Ohio to seek the advice of Professors Henry Quarantelli and Russell Dynes, the co-founders of the Disaster

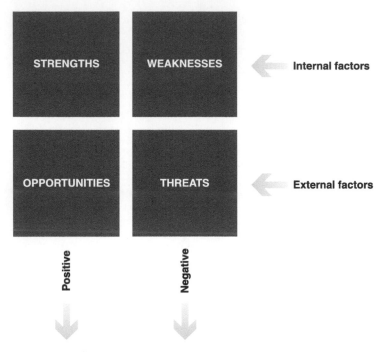

FIGURE 4.10 Model 18: strengths, weaknesses, opportunities and threats (SWOT).

Research Center that was, then, located at Ohio State University (it was later relocated to the University of Delaware). From the late 1940s, they had accumulated a vast amount of experience on the social and organisational aspects of disasters. After two days of intense study in the Center, Quarantelli offered me some further expert advice to guide my field study: 'When you get to Managua, you will find that there has been a gross *overestimation* of damage and dislocation, and an acute *underestimation* of the surviving community's capacity to manage their own recovery.' Once again, the advice was baffling as the assumption from an initial reading of the media coverage was that the damage would be overwhelming and that the survivors (or 'victims' as I used to think of them) would be dazed, apathetic and desperate for help. I had yet to learn about the *strengths* of social resilience and coping mechanisms as communities 'bounce back' from trauma.

After various visits to Managua during the recovery period, the accuracy of the advice became fully apparent. These two wise and experienced men, Quarantelli and Koenigsberger, were able to anticipate the *threats* to recovery that four decades later still lurk in the collective mindset of certain officials. These dangers derive from biased information, stereotypical assumptions and an acute inability or unwillingness to understand the needs and capacities of the affected population.

The SWOT model, a familiar and deceptively simple method of analysis, is used to gain a clearer understanding of the effective design and implementation of projects. Typical SWOT characteristics can usefully be related to the sectors of disaster recovery described in model 2, as shown in Table 4.3.

The relevance of the SWOT model to disaster recovery situations lies in the need to keep asking the following four critical questions at regular intervals

TABLE 4.3 The SWOT model applied to recovery sectors

SWOT categories	Recovery sectors				
	Psychosocial	Environmental	Institutional/ governmental	Physical	Economic/ livelihoods
Strengths	• The coping abilities and resilience of survivors and their institutions • Religious faith can provide support	• The ability of the natural environment to recover	• Local governments and institutions rise to the challenge of recovery • Recovery pre-planning is a major asset	• The creation of safe buildings and infrastructure can strengthen skills and the local private sector	• Regional and international solidarity can provide valuable support • Financial instruments may help share costs of recovery
Weaknesses	• Where there have been high casualties, the will and capacity to recover will be inhibited • Mental health problems are often a serious concern	• Damage to the environment causes severe indirect economic impacts • The supply of large quantities of timber for new building can cause deforestation	• Pre-disaster political and administrative constraints will seriously inhibit recovery actions • Recovery demands will stretch already weakened government capacities	• Corruption will damage recovery progress and actively contribute to unsafe buildings	• The influx of external agencies and contractors can weaken local resources when they need to be built up
Opportunities	• Support the therapeutic value of work for survivors	• Protect and restore natural environment • Protect floodplains, coastal areas and steep slopes	• Strengthen capacities of government • Make full use of local institutions and their leaders	• The disaster will have cleared out many substandard physical elements, offering an opportunity to secure improvements	• Create jobs through use of local resources • Use cash grants for assistance and develop individual banking by survivors

(continued)

TABLE 4.3 *(continued)*

SWOT categories	Recovery sectors				
	Psychosocial	Environmental	Institutional/ governmental	Physical	Economic/ livelihoods
Threats	• Long delays in recovery can prevent or retard the mental and physical recovery of communities	• Trees, plants, crops and coral reefs can take many years to recover, and the economic consequence can be serious	• Political rivalries and interests will negatively affect recovery plans	• The scale and speed of building work can pose a threat to safety, quality and participation of affected communities	• External financial loans can result in serious future national debt obligations

throughout the recovery process. They can usefully be asked for micro-scale projects as well as for the overall macro level of operations.

- How can the various participants in recovery make full use of their strengths? Given the scale of problems and the daunting challenges to officials, the need constantly to accentuate the positive is particularly important.
- How can one compensate for, or rectify, weaknesses?
- How can one maximise any opportunities that can appear in recovery management?
- How can the recovery process be protected from possible threats? For example, planners in Haiti who were engaged in the management of earthquake recovery recognised that new disasters might occur that would threaten the progress of reconstruction. It did indeed happen in the form of a tropical cyclone and a cholera epidemic.

The main value of this model lies in the fact that it induces officials to probe deeply into the problems and opportunities they face throughout the recovery process. The model's virtue lies in its simplicity, versatility and applicability at all scales.

Model 19: Edward de Bono's 'six hats' model for problem analysis

In Chapter 12, this model is used as we attempt to set out the conclusions of the book. Hence, it is dealt with in detail there and described only in outline here. Model 19 is a complement to model 18, the SWOT analysis model. In his famous 'six hats' model (Figure 4.11), Edward de Bono discusses creatively the need to see a problem or an issue from varied standpoints (de Bono 1985). He identified six distinct directions for decision-making and assigned each one an appropriate

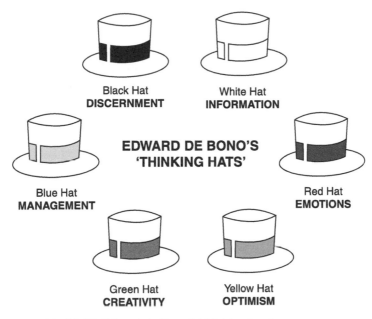

FIGURE 4.11   Model 19: Edward de Bono's 'thinking hats'.

colour. This model has often been used in group brainstorming sessions in which participants are invited to choose a hat or are assigned one of a particular colour and then asked to see the problem that is set before them from that standpoint. The hats are as follows:

Black hat – discernment:

- considers a judgment, being devil's advocate – an invitation to challenge, to be cautious and conservative and to state why a given approach will not work;
- monitoring and evaluating each stage in shelter and housing to make policy changes where approaches are not working.

White hat – information:

- considers what information is known or needed – what are the facts?;
- assessments of shelter and housing damage and needs – what resources are available?

Red hat – emotions:

- considers instinctive feelings – what can we gain from our hunches, emotions and intuition?'
- personal identity in shelter and housing – 'home' not merely 'house';
- understanding *genius loci* and the value of tradition.

Yellow hat – optimism:

• considers whether there are any positive opportunities, benefits or harmony;
• looking beyond shelter and housing to its wider values and implications for health, well being, safety, the environment and community building.

Green hat – creativity:

• considers, with opportunities to rethink radically, alternatives and new ideas;
• an openness to alternative and innovative ways to achieve and maintain shelter and housing from traditional approaches.

Blue hat – management:

• considers the subject and asks what we are we thinking about and what the goal is;
• defines short- and long-term goals, strategies and tactics.

Model 20: organisational frameworks of government for recovery management

> Best practice is to have defined a reconstruction policy and designed an institutional response in advance of a disaster. In some cases, this will entail a new agency. Even so, line ministries should be involved in the reconstruction effort and existing sector policies should apply, whenever possible. The lead agency should coordinate housing policy decisions and ensure that those decisions are communicated to the public. It should also establish mechanisms for coordinating the actions and funding of local, national and international organizations and for ensuring that information is shared and that projects conform to standards.
>
> *(Jha et al. 2010: 1)*

The value of this model (Figure 4.12) lies in the way it compares alternative approaches to the organisation of recovery at the national level. Options 1 and 2 in the model have both been adopted in various attempts to manage disaster recovery by national governments.

Option 1: form a dedicated disaster recovery organisation (India example in Chapter 12: Yellow hat – Optimism)

This approach was adopted by the Government of India following the Gujarat earthquake of 2001, when it set up the Gujarat State Disaster Management Authority (GSDMA). The GSDMA was established straight after the earthquake

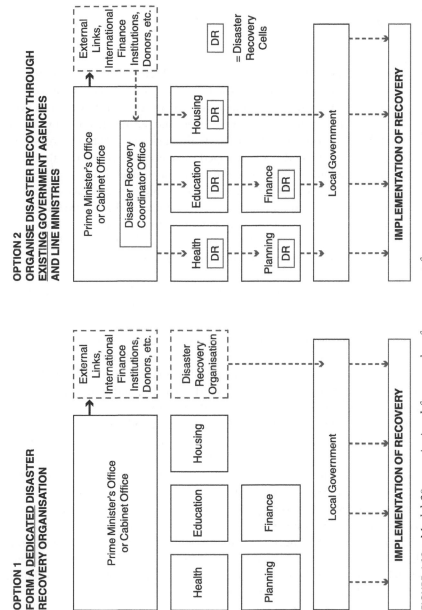

FIGURE 4.12 Model 20: organisational frameworks of government for recovery management.

as it was intended to act as a single focal point for reconstruction management that could negotiate the terms of loans and manage the flow of recovery grants from international financial institutions (IFIs), such as the World Bank and the Asian Development Bank. Praveen Pardeshi, the UN Development Programme's Director of Earthquake Recovery, described the creation and functions of the GSDMA as follows:[2]

> Gujarat struggled initially to find the right balance between utilising the existing district administration structures through the Collectors and devising a specific body to manage the disaster recovery. Ultimately the GSDMA . . . was set up to manage the entire recovery and reconstruction efforts being funded by Government of India, the World Bank and the State Government of Gujarat. The GSDMA was managed by senior state government officials, was linked to line departments, but had an independent financial and executive authority to disburse funds and review progress and take corrective policy measures based on field assessments. However, the GSDMA used the existing field agencies of the state governments, such as the Collectorates, the district councils and the line Departments of Public Works, Education, Health and Water Supply to implement the programme.

When the comprehensive earthquake recovery was completed, the GSDMA continued to exist with a wider remit, as defined by the Government of Gujarat, to 'act as a nodal agency to plan and implement pre-disaster preparedness and mitigation activities including training and capacity building of all the stakeholders involved in disaster management' (GSDMA 2014a); and '[t]o go beyond reconstruction and make Gujarat economically vibrant, agriculturally and industrially competitive with improved standards of living and with a capacity to mitigate and manage future disasters' (GSDMA 2014b).

This administrative option has the following strengths. First, the workload of recovery management is massive and this option provides a dedicated management that focuses exclusively on recovery. Second, a specialised body can provide a 'one-stop entry point' for external donors to provide support to the recovery efforts. They prefer this to having to deal with multiple contact points located in individual ministries. However, option 1 has some weaknesses. The first of these is that establishing a dedicated body is very expensive with a high cost in financial and human resources. Second, there is an inevitable and serious risk of duplication with confusion over who is responsible for what; for example, will the rebuilding of schools come under the recovery agency or the Ministry of Education? Third, the staff of a dedicated agency will normally be seconded from line ministries, which will weaken their capacities beyond the effects of damage and human losses sustained in the disaster. Finally, disaster recovery agencies may exist for many years, taking on a life of their own, and thus be difficult to close down.

Where a separate agency is created, it should follow the World Bank guidelines for the national management of disaster recovery, which include the clause:

'The reconstruction agency, even if it is new or temporary, must work closely with existing line ministries and other public agencies to provide efficient and effective post-disaster reconstruction' (Jha *et al.* 2010: 197).

Option 2: organise disaster recovery through existing government agencies and line ministries

Two examples are given here.

Mozambique example

This approach has repeatedly been adopted by the Government of Mozambique when it manages cyclones, droughts and flood disasters. There have been several reasons for the adoption of this approach. The government considered and rejected the idea of establishing a dedicated agency as it believed that this concept had a number of inherent weaknesses. These were summed up as follows:

- A dedicated agency would have been an expensive approach for a country with limited resources, given the need for a large team of officials as well as new accommodation, equipment and vehicles.
- Such an approach would take away responsibility for disaster recovery and the creation of safety from relevant line ministries. For example, the Mozambique Government believed that responsibility for school reconstruction should lie with the Department of Education and not be delegated to a specialised recovery agency.
- A dedicated agency would be likely to become permanent and be difficult to dismantle after recovery was completed.
- It would inevitably need to be staffed by key personnel drawn from existing departments, which would be weakened in the process.

The Government of Mozambique therefore uses existing line departments, and instead of creating a special recovery agency, it has given them a mandate to manage long-term recovery. The National Institute for Disaster Management (INGC) was created in 1999. It is only allowed to coordinate short-term interventions; for example, in the case of floods, only during the emergency period. Once the floodwater recedes, it is left to line ministries to repair the damage and plan for long-term recovery, such as rebuilding dykes, bridges and schools.

One should ask how effective the system is in practice. In 2005, the World Bank conducted a study of disaster recovery in Mozambique. Its authors conclude that:

> Recovery responses were generally managed and coordinated by the line ministries with the oversight of the CCGC [Coordinating Council for Disaster Management].

Most post-flood recovery was undertaken within the framework of existing development programs. On a practical level, as the community survey found, this meant that reconstruction could be facilitated using standard ministry drawings and specifications for schools, health posts/centers, and hospitals.

(Wiles et al. 2005: x)

Recurrent annual floods reveal that some unaccountable and ineffective ministries have yet to understand the concept of 'building back better'. In terms of development programmes, the lack of coordination among line ministries increases the vulnerability of communities to floods. Decentralisation could have reduced the drift of population to urban centres, but it only exists as an administrative concept on paper with minimal action in practice. Due to lack of essential facilities, such as water and electricity supplies, long-term disaster recovery resettlement programmes designed to move people away from river basins failed. This occurred because of lack of coordination between the relevant ministries responsible for providing basic amenities. This has further increased the dependency of communities on immediate relief after each flood[3] (Spaliviero 2007).

Chile example

The Government of Chile also made a decision to use existing ministries, programmes and budget lines in the management of the earthquake recovery (2010–14) rather than to create a 'super-minister' in a dedicated recovery agency. Therefore the Ministry of Housing and Urban Development (MINVU) was allocated responsibility for urban and housing reconstruction, the Public Works authority was responsible for roads and infrastructure, and Departments of Health and Education managed programmes within their own sectors.

Initially a Committee of Emergency worked directly under the President, but this was replaced by a Committee of Ministers that coordinated reconstruction polices at the national level with Governors coordinating local level recovery (Comerio 2013).

A hybrid approach combining options 1 and 2, adopted in Indonesia following the 2004 earthquake and tsunami

In the guidelines on reconstruction developed by GFDRR in the 2010 publication *Safer Homes, Stronger Communities*, the advice offered concerning the national management of disasters favoured the creation of a dedicated body in government to take overall control of recovery (option 1):

In certain situations, especially after a large-scale emergency, government may establish a dedicated organization or taskforce to coordinate, reinforce, or in some cases temporarily replace the responsibilities of line ministries. The taskforce can sometimes better coordinate tasks among ministries and departments. The taskforce is usually created for a specific period of time and will return responsibilities to the relevant line ministries, either gradually or when specific objectives are met.

(Jha et al. 2010: 10)

However, this policy was modified by September 2014 when a disaster recovery framework was launched at the Second World Reconstruction Conference, organised by GFDRR. Three arrangements or approaches to national recovery management were outlined:

1   Strengthen and coordinate existing line ministries to be the reconstruction leaders, sector by sector.
2   Create a new institution for recovery management.
3   Adopt a hybrid arrangement combining aspects of 1 and 2.

The hybrid approach was evident in the Indonesian Government's recovery programme following the Indian Ocean tsunami in December 2004 and the Nias earthquake in March 2005. The GFDRR report that: 'In adopting this model, a sunset clause existed from the outset. The four-year mandate of BRR [Agency for the Rehabilitation and Reconstruction of Aceh and Nias] maintained urgency for reconstruction and encouraged a handover strategy to existing administration in Indonesia' (GFDRR 2014b: 35).

The World Bank, the UN and the EU are clear that there is not one preferred option, and that there are significant advantages using existing structures, addressing sustainability concerns and allowing pre-disaster work to be done within existing institutions. (GFDRR 2014b: 34–5).

The alternative approaches to recovery management are discussed in Chapter 5 'Roles and responses of government'.

Summarising the pros and cons of model 20

These are the arguments for creating a *new* organisation for recovery management concerns (option 1):

1   The need to recognize the unprecedented demands of the situation on a weak central government or a damaged local government require a radically different form of administration with strong links to central as well as local governments.
2   The workload of recovery management is massive and this option adds all these tasks to the existing everyday responsibilities of line ministries. Additionally, resources are limited to those required to realise the 'building back better' concept.

3   Having a dedicated agency can assist governments of developing countries who are often dependent on external agencies, such as the World Bank, the UN, NGOs and bilateral agencies, which are needed in order to assist them in repairing the damage caused by disasters.

These are the arguments for using *existing* bodies for recovery management concerns (option 2):

1   Working through existing line ministries to manage recovery is much cheaper than creating a new body that requires salaries, premises and vehicles.
2   This option has the advantage of leaving responsibility where it naturally belongs. Thus, the reconstruction of health facilities is the responsibility of the Ministry of Health, reconstructing safe schools is that of the Ministry of Education, and so on. The rationale is that existing line ministries and departments have the ultimate long-term responsibility, hence it is desirable to strengthen existing bodies rather than to wastefully duplicate or replace them.
3   It is important that disaster recovery is 'normalised' in order to relate it to established protocols, policies and future responsibilities. This is more likely to occur when using existing line ministries than by working through a newly created agency dedicated to recovery.

The argument for using a hybrid approach that combines options 1 and 2 is that in combining the two options, the efficiency and power of a dedicated body can be melded with a time fuse that will place a finite limit on its duration and, thus, return authority to line ministries. In order to select an option, three approaches are needed. First, define the precise organisational need. Second, assess what can be done though governments and gauge local government capacity. Third, assess what can be done via existing institutions and structures.

International, national and local patterns of leadership of recovery situations are discussed in Chapter 10, 'Resilient international leadership of disaster recovery'.

Afterword: reality and realism

The previous chapter began with a model (no. 1) that was designed to capture the essence of recovery. This model concluded with a final target: 'Scenario 4: developmental recovery'. Such words sound persuasive and logical when set down in the model, but we are aware that the terminology can also appear facile and idealistic. Under this heading, there are the following descriptions: 'build back both better and safer', 'complete all sectors', have an 'inspired vision' and have a 'strong, persistent and wise leadership'. These expressions are all aspirational, depicting the best possible conditions, since it is in the nature of models to be optimistic. But they do not indicate how immensely challenging it will be for *any* government – whether rich or poor, authoritarian or democratic – to achieve such targets.

Given the persistent lack of resources, skills, knowledge and political capital, any study of a recovery situation will reveal the scale of this immense challenge; hence, it is unsurprising that these intentions are rarely achieved in full. While officials certainly need to set their sights high in order to meet the exacting demands of recovery, they should not be dismayed or discouraged when full success eludes their best efforts.

For all the reasons stated at the beginning of the previous chapter, we regard the models described in Chapters 3 and 4 as being both useful and necessary as they relate to various topics to be explored in later chapters. However, as we have noted, all models are limited and require continual review and development, new models are always needed, and models always need to be used with caution. In this respect, it is important to specify clearly the conditions under which a model is to be used and to what it applies. This will help avoid the misuse of models by applying them to inappropriate circumstances, possibly without the necessary modification and qualification.

The next chapter considers the impact of disasters on people, economies and livelihoods, urban environments, buildings, local governmental capacity and the natural environment. This is to provide an essential basis for the subsequent detailed examination of the elements of recovery.

Notes

1 Letter from George Nez to Ian Davis, 1975.
2 Private communication to Ian Davis from P. Pardeshi, Director of Earthquake Recovery, UN Development Program, India in 2006.
3 Private communication between Ian Davis and Dr Titus Kuuyuor, Chief Technical Advisor, Disaster Risk Reduction, UN Development Programme, Maputo, Mozambique in 2014.

5

RECOVERING FROM WHAT?

The impact of disaster

In model 2 (Chapter 3), we develop a pentagonal representation of five aspects of disaster impact and recovery: environmental recovery, institutional and governmental recovery, recovery of the economy and livelihoods, psychosocial recovery and physical recovery. These five sectors are by no means a comprehensive list; for example, we discuss 'cultural recovery' in this chapter as we look in some detail at these varied yet closely related recovery sectors. However, we do not dwell on physical recovery here as the reconstruction of the built environment with specific reference to shelter and housing is covered in considerable detail in Chapters 8 and 9.

The impact of disaster as a starting point for recovery

In considering the phenomenon of disaster, we are dealing with extreme events whose impacts vary from being instantaneous to very slow. This can involve single or multiple events, ramped, variable or cascading impacts, and secondary or collateral vulnerabilities. Hence, there is no single diagnostic of what disaster is and what it is not. There have been two compendia of essays on the subject (Quarantelli 1998a; Perry and Quarantelli 2005). There have also been attempts to quantify the magnitude of disaster, not in terms of physical forces (i.e. hazard magnitude) but of human impacts (Foster 1976; Keller et al. 1992). None of these works has settled the definitional debate or prescribed an adequate way of measuring the overall impact of disaster. However, there is a pervasive sense that particular magnitudes of event are separated from one another by thresholds that relate to how profound or widespread the impact is as well as the ability of society to cope with it. A possible typology is given in Table 5.1, but it should be borne in mind that terms such as 'incident', 'disaster' and 'catastrophe' are used very loosely by academics, administrators and emergency managers.

TABLE 5.1 A tentative classification of the size of impacts

	Incidents	*Major incidents*	*Disasters*	*Catastrophes*
Impact	Very localised	Generally localised	Widespread and severe	Extremely large
Response	Local efforts	Some mutual assistance	Intergovernmental response	Major international response
Plans and procedures	Standard operating procedures	Emergency plans activated	Emergency plans fully activated	Plans probably overwhelmed
Resources	Local resources	Some outside assistance	Interregional transfer of resources	Local resources overwhelmed
Public involvement	Very little involvement	Mainly not involved	Very involved	Extensively involved
Recovery	Very few challenges	Few challenges	Major challenges	Massive challenges

Source: partly after Tierney (2008).

Moreover, there is no precise, scientific way of distinguishing the meanings of words such as 'major' and 'massive'.

Despite these misgivings, to some extent disaster is defined by society's ability to cope with the impact and consequences of disaster. The thresholds of coping may vary considerably by sector (administrative and institutional, environmental, social, etc.) and with the resources of particular classes and groups in society.

For ease of discussion, we will restrict ourselves to sudden-impact disasters that have limited or no forewarning. Earthquakes, tornadoes, hurricanes and explosions are examples of these phenomena. One expects that events with slower onsets and longer warning times would engender more elaborate preparations and mitigation measures.

The first task is to understand the geographical extent and seriousness of the impact. In the case of a localised major incident or small disaster, this may be a relatively simple task as the area is readily circumscribed. However, major storms, floods or earthquakes may require more than a week of systematic reconnaissance and collection of reports from the field before the entirety of the disaster is clear. During this period, there is a serious risk that remote and outlying settlements may be ignored even though their need for assistance may be as great as that of more centrally located urban areas. A persistent problem with the assessment of needs concerns the difficulty assessors have in distinguishing between chronic, pre-disaster gaps and needs and those created by the disaster event. After the 1983 Popayán earthquake in Colombia, Ian recalls being taken to see a reconstruction programme: it became clear to me that none of the beneficiaries of the new housing reconstruction project had suffered losses in the disaster. They were poor and enterprising families who had seized a unique opportunity by migrating to the recovery area in the confident expectation of securing support. When I asked about their qualifications to be provided with new houses, I was told that while they may not be 'victims of the earthquake', they were certainly 'victims of life', and since there was surplus aid available it was appropriate to make the houses available to them.[1]

Assuming that local resources have been overwhelmed, and in some measure destroyed, it will be necessary to organise the importation into the disaster area of first necessities, such as medical supplies, bedding, emergency response vehicles and trained personnel. One hopes that this is achieved in relation to an assessment of needs on the ground. Sadly – indeed, tragically – this is often not the case, and comprehensive needs assessments tend to be cursory or non-existent. This is usually because their importance is underrated, officials are too busy with directing the arrival and distribution of relief supplies, or no one is available or trained to conduct them. The results of this include piles of redundant relief goods, duplicated activities, urgent tasks that are not carried out, and areas and problems that are neglected (Gerdin *et al.* 2014). In cases of massive disasters where there has been extensive media coverage and a consequent flood of public donations, such as the Southeast Asian tsunami of 2004 or the Haiti earthquake of 2010, assessments tend to *follow* rather than *determine* patterns of assistance. After the tsunami: 'the slow-moving humanitarian needs assessment did not drive the initial humanitarian response. The availability of an enormous amount of funds in search of activities was the driving force for the earliest decisions' (de Ville de Goyet and Morinière 2006: 37).

It is easy to give the impression that needs assessments should only be carried out at the start of the emergency phase. Nothing could be further from the truth. The needs of survivors and their communities change over the duration of the disaster and hence they should be reassessed constantly. This requirement is emphasised in 'Model 14: project planning and implementation' (Chapter 4). In level 5 of the model, 'monitoring and evaluation' are continual processes, applicable to all stages of recovery and implementation. This is the first step in the process of humanitarian supply chain logistics (Tatham 2012) which can be extended to embrace the management of intercontinental transfers of goods, equipment and personnel.

The most basic human needs are safety, lodging, food, hygiene and medical care. Hence, in disaster, the survivors need to be protected against further impacts (e.g. the damaging effects of aftershocks in earthquakes). Those who have lost their homes, either temporarily or for the long term, need to be given shelter from the elements and provided for in terms of the other needs. Injured and sick people should be cared for by trained, equipped medical personnel. Supplies of essential medical drugs and consumables should be reinforced and guaranteed. Efficiency in providing relief ensures that wastage is minimised and resources go where they are genuinely required. Needs assessment should take account of those survivors who are able to provide for themselves in any of the ways mentioned above.

In the evaluation of needs assessments following the 2004 Southeast Asia tsunami, two highly experienced authors — Claude de Ville de Goyet and Lezlie Morinière — were perplexed as to why the surviving communities were not involved in assessing their *own* needs, followed by cash grants from assisting groups. Their recommendation No. 6.3 made the radical proposal to

> empower the affected individuals or families to assess and prioritise their own welfare needs by using cash subsidies whenever possible.... The need for thematic assessments would be considerably reduced if, when possible, the affected people were given the financial means to decide whether they want a better shelter, a boat, food or any other welfare item brought at high cost by expatriates. This approach would go a long way toward compliance with the Sphere principle of 'respecting the dignity of victims' in countries with active market economies, such as those affected by the tsunami.
>
> *(de Ville de Goyet and Morinière 2006: 64)*

Excessive aid can destroy local markets and have a generally debilitating effect on the recovery. Aid dependency is a common and highly negative syndrome. Allowing or inducing survivors to rely on aid reduces their ability to provide for themselves, to organise their own lives and to return to a situation of comparative autonomy. To encourage trade, markets may need to be regulated in order to avoid profiteering and inflation in the cost of essential items.

There is commonly a very strong discrepancy between immediate and short-term aid and long-term recovery. This is the result of a presumed disjuncture between disaster relief and development aid. How to achieve a transition from surviving the early aftermath to the rebuilding of settlements, productive capacity and society is thus a difficult issue. Generosity in the short term can turn to neglect in the long run. During a long and complex process of recovery from disaster, much depends on both the commitment and the ability of governments to look after their citizens. Self-organisation of communities has a role in this. Emergent groups and consultative processes can have a substantial impact in making the recovery process more democratic and vigorous, but national support is bound to be needed. To have a functioning government that is willing to tax its citizens in order to provide aid to a recovering disaster area is to ensure that reconstruction will, in some form, go ahead. The two main considerations in the minds of leading politicians are the strategic importance of the recovery process to the national economy and the value of votes obtained from the populations of the disaster area. None of this necessarily adds up to a rational, sensible recovery process, but it helps describe motivations, regardless of how successful the process is eventually.

We now consider the starting point of recovery and its potential in terms of the main categories of human life in which it occurs, starting with economics and employment.

Economic recovery and the question of livelihoods

Simple descriptions of disaster tend to focus on the casualties and direct losses. Indeed, in the more distant past, these were the only effects that were measured. Clearly, direct losses are easier to measure than are many indirect ones.

The model of unit costs (no. 7) shows that in most cases, the largest expenses are likely to be incurred in the reconstruction phase, not in the emergency intervention or repair of basic infrastructure nor even in the developmental reconstruction phase (see also 'Model 6: the Kates and Pijawka recovery model' in Chapter 3). Particular problems can arise if the planning process for funding the recovery does not take into account the fact that more money will be required to reconstruct buildings that have been lost than to carry out activities such as emergency relief, hazard protection and monumental or developmental reconstruction.

Death and injury: The value of a human life is generally calculated on the basis of lost earnings and foregone economic worth caused by premature death. As such, the figure has no moral implications, and it is inevitably as generalised as it is approximate. For people who are injured, especially those who have long-term impairments to their health or suffer permanent disablement, the costs of medical treatment, health insurance and reduced productivity as workers need to be taken into account.

Direct damage: Assets that can be damaged consist of housing, commercial and industrial premises and physical infrastructure (roads, bridges, airports, railway lines, power lines, water distribution networks, and so on). Less tangible assets include the amenity value of sites and services. The total destruction of an asset involves costs based on its economic value and the expense of substituting or replacing it (including demolition and waste disposal costs). The value lost by being unable to utilise the asset is usually counted as an indirect cost. Damage that is less than total may involve pro rata losses but, in many cases, the repair costs are not directly commensurate with the value of the asset. For example, if one takes a historic building that is part of an area's cultural heritage, repair may be very challenging and expensive as complex restoration techniques need to be employed. A further complication is added by the fact that most repair does not involve simply restoring an asset to its previous condition but also ensuring that it is upgraded as it is rebuilt. For instance, housing that is damaged by an earthquake because it is not resistant to seismic forces should be rebuilt with anti-seismic retrofitting incorporated into it. This is the 'bounce forward' approach to creating resilience, rather than the static 'bounce back' approach that merely recreates vulnerability (see Chapter 9, 'Option 4: repair of dwellings').

Infrastructure: Infrastructure is defined as the basic facilities, networks and services required for society to function. It consists of interconnected assets that support the operation of services, economic activities and welfare. Critical infrastructure delivers the essential services on which life depends: food and water, wastewater treatment, transportation, food production, distribution and sales, banking, electronic networking, health care and welfare. Disaster can disrupt, damage or destroy critical infrastructure of local or national significance, although systems are generally resilient to most kinds of impact. In part, this is because of the existence of redundancy; for example, backup equipment or alternative routes through a network. The cost of damage to infrastructure, and disruption of the services it provides, may be difficult to estimate in that it depends on damage to assets, lost revenue from services, knock-on costs of disruption or lost services, and the time and expense of repairing the infrastructure. In today's globalised economy, loss of infrastructure may have international consequences. However, generally, infrastructure has been repaired quite quickly after major disasters.

In 1998, a large area of eastern North America suffered about 80 hours of continuous freezing rain – a major ice storm with consequences for both the USA and Canada. More than 1,000 steel towers for electricity transmission lines collapsed under the weight of accumulated ice, and so did 35,000 wooden utility poles. Although the overall costs may have reached US$4–6 billion, electricity supplies were restored to all consumers within a matter of weeks, although this required the deployment of more than 20,000 workers to make the repairs. Interruption to the food supply can cause perishable and refrigerated goods to become unsaleable while interruption to banking services can cause loss of income or failure to complete vital transactions, many of which are time-sensitive. Hence, a variable proportion of the economic losses caused by the failure of infrastructure may be knock-on (i.e. indirect) effects. Although it is usually unlikely that a full 'domino effect' occurs, some of the more expensive losses caused by knock-on may, in a certain sense, be disasters in their own right; for example, if they lead to proliferating bankruptcy through unexpected loss of revenue (Harris 1998).

Indirect losses: In the second half of the twentieth century, worldwide losses in disasters apparently increased fifteenfold (Munich Re 2000). Part of this increase was the result of more comprehensive and consistent reporting of disasters and the losses that they incur; and part resulted from gradually increasing inclusion of indirect losses, which in 1950 were hardly considered at all. In point of fact, indirect losses can outweigh direct ones. They include loss of business and revenue, the cost of legal proceedings associated with the disaster, and loss of image. Regarding the latter, the tourism industry is notoriously volatile and mass cancellations are common in areas affected by disaster, regardless of whether tourist facilities are significantly affected or not. Much, then, depends on the degree to which the area in question is economically dependent on the revenue from tourism. In areas of cultural heritage, areas where people come to participate in winter sports, areas where there are coastal amenities, and so on, the dependency may be very strong, and disaster may severely depress the local economy whether or not tourism remains feasible. The only saving grace here is that the industry tends to recover quickly. Disaster can also graphically reveal the risks incurred by living in a particular area, which may depress property prices and cause the local economy to suffer deflation, at least for a certain length of time.

The globalisation of production has two possible effects when disaster strikes. One is that multinational companies can transfer production – or purchasing – elsewhere, and the other is that they may instead be critically dependent on production that is highly integrated across regional and national boundaries. For example, if components suddenly cease to be available, production of finished products will be affected. Thus, the earthquake and tsunami in north-east Japan in March 2011 affected vehicle production in Europe, and the floods in Thailand in October and November 2010 led to a worldwide shortage of computer components.

Livelihoods: Personal and family resilience are usually highly dependent on wealth generation. Unemployment after disaster can easily induce dependency on subventions, which may be unreliable as well as being a poor way to encourage people to take the initiative in reconstruction. Many scholars and public administrators see recovery from disaster as being critically dependent on employment since livelihoods need to be protected and stimulated. The destruction of industrial and commercial premises, infrastructure and agriculture can leave people without income or prospects of work. On the other hand, unless reconstruction stagnates, there will probably be a post-disaster boom in the building trade and all activities that support it. For a period of months or years, this may sustain the local economy with its multiplier effect, but there is obviously a need to avoid stagnation when the rebuilding process either ends successfully or loses its impetus.

In developing countries, NGOs have experimented with cash-for-work schemes in order to support disaster survivors' income generation and, at the same time, support the reconstruction process. The example in Chapter 12 ('Yellow hat – optimism') of the rural reconstruction project in Pakistan after the 2005 earthquake highlights the value of the cash donation approach to housing reconstruction. For long, it was widely believed that cash handouts would induce aid dependency. That may be true, but it depends on what the cash is used for. In the vicinity of Tacloban in the Philippines, many survivors of the November 2013 cyclone received cash handouts which they used to buy food and materials to construct shelters but, where possible, they used the money to buy the instruments of gainful employment, such as fishing boats. Meanwhile, the problems with cash-for-work schemes were their lack of longevity and their inability to lead to stable employment once the NGO money had been spent. In disaster relief circles, there is a general feeling that it is better to sell people goods and services than to provide them for free. At the same time it is recognised that, if such a strategy is to work, people must have the means to purchase what they need and, to some degree, this must involve a distribution of wealth that is not severely inequitable. Flooding an area that has been devastated by disaster with cash can lead to increases in corruption, black markets and expropriation of wealth. For example, in Afghanistan and parts of Africa, the financial responses to complex emergencies have funded militias and thus increased the level of conflict. Elsewhere, the situation regarding livelihoods has much to do with the macroeconomic impact of disaster.

Macroeconomic impacts: Large disasters tend to cost between 0.2 per cent and 10 per cent of annual gross domestic product (GDP). The proportion varies according to both the size of the national economy and the destructive power of the disaster agent in terms of direct and indirect losses, knock-on effects and the cost of recovery. Large, diverse economies, such as the USA, can sustain high losses with macroeconomic effects that are relatively small and of short duration. In contrast, when Hurricane Mitch struck Nicaragua in 1998, it was estimated that losses had set back development by two decades. In 1995, wildfires in Mongolia caused losses that amounted to 120 per cent of GDP, which probably only shows that GDP may be a very poor estimate of national wealth.

At the national level, the economic effect of a disaster is usually transient, however large the event is (here we must exclude events such as a major asteroid impact, whose economic impacts are unknown). The multiplier effect of reconstruction efforts may cause a temporary boom in the local economy, although it may be accompanied by considerable indebtedness as people and firms take out loans to fund their recovery. Alternatively, at the local level, some losses may be permanent. For example, the January 1995 earthquake in Japan severely damaged the port of Kobe, one of the largest ports in the world. As a result of liquefaction caused by the earthquake, Kobe lost trade permanently to Tokyo and Yokohama. By 1997, the port had been fully reconstructed with the most modern handling equipment in Japan, but the container port only recovered to 85 per cent of 1994 trading levels.

Generally, economic stagnation after a disaster is a sign of the marginalisation of a community, settlement or area. It indicates that the area's population and leaders lack political and economic influence in the national arena. Seen from the other side, if governments fail to stimulate the economy of an area then local attempts to recover may be thwarted by lack of resources to revive economic activity. In the Province of L'Aquila, Central Italy after the 2009 earthquake, there was a net loss of 16,000 jobs within a year as the local economy contracted. Very little was done to revive it and many workers, particularly professionals deprived of accommodation for their activities, left for greener pastures elsewhere. Recovery was, therefore, interminably slow.

Disasters and the informal and illicit economies: It is possible that about 20 per cent of the world economy exists 'below the surface' and is thus not registered in official statistics. In some places, up to 70 per cent of economic activities may be 'informal' or illicit. The scope of such activities varies from simple failure to declare taxes when products or services are sold to highly organised forms of criminal activity which, at their most insidious, comprise multinational crime syndicates that are effectively investment companies operating outside the law. Illicit trade includes trafficking in narcotics, armaments, endangered species, toxic waste and human migrants. It is facilitated by the world's 78 tax havens, through which, it is estimated, passes half of world trade, whether legitimate or not.

On the positive side, illicit activity provides employment in places where there may be few legitimate sources of work. Hence, it keeps people alive. On the other hand, it provides no tax revenue with which to fund recovery and reduce the threat of future disasters. It also tends to avoid health and safety controls, or to constitute a threat to safety in its own right as one sees with the drug cartels in Mexico that have massacred tens of thousands of people. There are many instances in which disaster is a source of, and opportunity for, corruption to thrive. Organised crime may see the exceptional circumstances and relative chaos of a disaster aftermath as an opportunity to infiltrate political administrations and dominate the recovery process. As the construction trade is the basis, or starting point, of much organised crime (Saviano 2008), reconstruction after disaster may represent a golden opportunity for criminals to extend their reach. At the very least, this will distort the macroeconomic processes of recovery. It may lead to the perpetuation of vulnerability on several levels, from the proliferation of unsafe construction techniques to the effect of violent crime on local businesses and employment patterns to the fact that organised crime is under no obligation to treat the recovery process as a welfare function (Alexander and Davis 2012).

We further discuss the issue of corruption in recovery in Chapter 12, 'Black hat – discernment'.

The economics of development and humanitarian aid: In recent decades, humanitarian assistance to disaster areas has become a major international industry. It is estimated that US\$4.5 billion were donated by the world's general public after the Indian Ocean tsunami of December 2004. Most aid is supplied to provide immediate or short-term assistance, and international agencies have been reluctant to develop long-term commitments to areas. Despite a need that has long been demonstrated, there have been relatively few attempts to link disaster assistance to development aid. This reflects the reluctance of disaster aid agencies to become involved in development issues.[2]

Flooding disaster areas with free food and goods can lead to distortions in, or the collapse of, local markets and in turn the destitution of local producers. Hence, there has been a change towards more selective use of aid. Damaged infrastructure has been repaired through cash-for-work and food-for-work initiatives, although the drawback of these is that they tend not to provide stable, long-term employment despite the need to regenerate livelihoods in areas in which disaster has destroyed many of the sources of paid work. There has also been a re-evaluation of cash handouts. Though conventional wisdom suggests that these can end up in the wrong hands due to corruption, or may be used inappropriately, they do have the merit of stimulating trade and the circulation of money in the local area. Small amounts of cash tend to be used wisely by families and individuals, but the combined effect of cash handouts may be to drive up prices and inflation in the disaster area.

The providers of humanitarian aid in disaster areas have been accused of subsidising narcotic drug production, funding militias and mafias, and creating aid dependency among the survivors. All of these things have happened, but they are by no means the rule and neither are they inevitable outcomes of aid and assistance. However, there is no guarantee that international aid will solve problems rather than create them in such places. Generally, economic help is beneficial if it helps forge a stable society that is capable of generating its own wealth and in which strong efforts are made to avoid the marginalisation of any sector of society. Aid that contributes to local autonomy tends to be the most successful.

The pattern of international aid after disasters is heavily dependent on strategic alliances. Some of these are postcolonial, such as British and French involvement in West Africa; other forms of alliance are modern, such as the use of European Union funds. Whether or not a country receives external aid, there is a need to assure the long-term recovery of the disaster area and avoid stagnation or other economic ills (deflation or excessive inflation, for example). Businesses need to be stimulated and sources of employment created and maintained. Regeneration programmes need to be created carefully and realistically as there is no guarantee that a marginalised area can be brought into the economic mainstream merely by providing post-disaster aid. The result may instead be empty factories and offices and jobs that very quickly cease to exist.

Studies by economists are equivocal about the effect of natural disaster on economic growth (Albala-Bertrand 2007). Moreover, the effects differ considerably from one country to another, depending on the available assets and wealth levels. Very large or recurrent disasters can reduce economic growth. On the other hand, Japan recovered well from the loss of 4 per cent of GDP in the 2011 earthquake and tsunami, and New Zealand weathered the costs amounting to 10 per cent of GDP in the Christchurch earthquake. Countries with smaller, poorer economies would not have sustained such losses as easily. Shabnam (2014) found that death tolls in natural disasters had little impact on economic growth, but if a disaster affects a large number of people then this can reduce growth significantly, presumably because it removes workers, and sources of taxation and purchasing power, from the local or national economic arena.

Disaster is also an opportunity for growth in terms of the renewal of assets and stimulus provided by public expenditure on recovery (Greenberg *et al.* 2007). In such cases, higher taxation and rising public sector debt may counterbalance the positive effects. Moreover, there is abundant evidence that the poor have to bear a disproportionate economic burden in the aftermath of disaster because they lack access to credit, have low or no financial resources and lack political influence over economic decision-making. It is striking that there is very little literature on welfare economics in the context of disasters. When Shughart II (2006) coined the term 'Katrinanomics' after Hurricane Katrina struck the US Gulf of Mexico states in 2005, he was referring to an *absence* of welfare economics, not something like Roosevelt's New Deal public works programme being adapted for the economic doldrums of an area laid waste by disaster.

We believe that the term 'welfare' should be rigorously defined in relation to disaster. In essence, it represents *the provision of care to a minimum acceptable standard to people who are unable adequately to look after themselves.* In Chapter 8, Table 8.2

compares a 'welfare' with a 'developmental' approach to shelter and housing provision. There is an indistinct dividing line between welfare and unjustified largesse with public funds. In the hope that political corruption can be eschewed, we recommend that welfare be rigorously defined in relation to people's needs after disaster and what, from savings and assets, they are or are not able to provide for themselves. This is not an argument for neoconservatism and the shrinkage of aid to the needy: it is a plea for more effective use of funds to help those who really need to be aided in their recovery. Moreover, welfare should not be used as a means of inducing aid dependency.

In this context, it is evident how economics interacts with politics – the focus of the next section of this chapter.

Roles and responses of government

The politics of disaster can be exceedingly complex, subtle and subject to rapid or abrupt change. Disaster aid is attractive to politicians because it is a vote winner, but unfortunately this does not encourage governments to behave responsibly or to commit themselves to supporting viable long-term recovery. In a multiparty democracy, it is all too easy to blame disasters on previous administrations and defer investment in disaster risk reduction until future ones. However, good government is a key element of recovery from disaster.

Two basic elements distinguish the role of government, at all levels, in disaster risk reduction. One is the rule of law and the other is citizens' trust in the institutions that represent them and manage basic services. Contrary to the impression given in some Hollywood films, disaster does not usually cause the rule of law to break down. On the contrary, it has been used as an opportunity to introduce repressive or restrictive legislation; for example, that which restricts the right of free expression. When disaster contributes to the erosion of democratic values and the rule of law, alternatives to government may appear, such as militias and mafias.

Government response to disaster therefore needs to be intelligent, robust and sustained. Intergovernmental relations during the aftermath of disaster may resemble a barter market in which lower levels of government argue for more assistance and higher levels, faced with the need to apportion scarce resources among competing needs, try to ration them. It is a sad fact of life that the politics of disaster are largely the politics of aid, not those of prevention. The main functions of government, vis-à-vis disaster, are to protect citizens against harm, which may require housing and feeding them; to relaunch the local area economically, which may involve providing employment or the means of generating it; and to manage or provide basic services and infrastructure. Government intervention in recovery from disaster is essential and thus accepted without cavil. However, it is not necessarily the only source of organisation and response.

Most countries have at least three tiers of government: national, regional and local. However, the degree of centralisation or devolution varies considerably from one country to another and so does the way in which functions are distributed between the levels. Thus there is a considerable difference between a country composed of federated states (such as India, Germany, Mexico or the USA) and a unitary country with a centralised administration. Hence, there is no single model of recovery that is valid for all nations.

If one assumes that after a major disaster the lead will be taken by the national government, then a basic choice has to be made between creating institutions to deal with disaster recovery or else using existing ministries. This dilemma is discussed in detail in model 20 (Chapter 4). In New Zealand, the Canterbury Earthquake Recovery Authority was set up to manage the recovery from the Christchurch earthquakes of 2010 and 2011. In contrast, Italy used the existing national Department of Civil Protection to coordinate the short- to medium-term recovery from the L'Aquila earthquake of 2009 and has left much of the long-term management to local authorities, backed by the Ministry of the Interior. There is no particular criterion which indicates that one method is better than another, but all recovery requires coordination between ministries since it requires sustained input to deal with health, social security, economy and employment, public works, public security, and so on.

In Haiti, the earthquake of January 2010 damaged many state-owned assets and effectively destroyed the fragile government in Port-au-Prince. Institutions had to be rebuilt and restaffed, and civil servants had to be trained. Lack of international commitment to this process weakened the role of government, the key decision-maker in the recovery process.

Part of economics is about people's preferences for expenditure and decision-making on monetary matters. This ushers in a wider consideration of how people react to disaster socially, psychologically and culturally.

Psychosocial recovery

Two landmark studies chronicle the mental suffering caused by sudden-impact disaster. One, by eminent sociologist Kai Erikson, describes the aftermath of a flood in a coal mining district of West Virginia (Erikson 1976); and the other, by eminent anthropologist Tony Oliver-Smith, describes the aftermath of a massive landslide that overwhelmed the town of Yungay in Peru (Oliver-Smith 1986). The mental and emotional suffering caused by disaster can be highly debilitating and can last for a very long time indeed. In fact, in 2006, David met one of the survivors of the Aberfan (South Wales) landslide of 1966, which overran two schools at assembly time and killed 128 small children. After 40 years, she was still manifestly enduring the effects of the disaster, which was one of great tragedy and suffering.

The main psychological effects of disaster are depression, anxiety and post-traumatic stress disorder (PTSD). The last of these tends to be an umbrella term for physical, intellectual, behavioural and emotional symptoms of psychological distress. Hence, to an extent it includes the other two conditions, but its most common effects are arousal, tension, anxiety and a tendency constantly to relive the worst moments of the impact.

As a rule of thumb, in clinical tests one might expect up to 10 per cent of the affected population to show the symptoms of PTSD or anxiety. In severe cases, the figure may be higher, and generally women are more at risk than men. On the other hand, men are more likely to resort to antisocial behaviour such as substance abuse or domestic violence as an outlet for their anguish and frustrations. Some persistent cases of PTSD are prolongations of critical incident stress (CIS), the effect of close experience of casualties and destruction at the height of the disaster. CIS is most likely to be experienced by front line rescuers.

Natality tends to increase after large disasters, possibly from a reduced population base. It is not clear whether this is the fruit of an instinctive desire to 'repopulate' or of changes in social relations caused by the disaster. Alternatively, outmigration and the fragmentation of communities may reduce intimate contact between men and women.

The destruction caused by disaster can be interpreted by a person's subconscious as a sort of symbolic 'end of the world'. Many of the normal referents and coordinates of daily life may have been swept away or changed out of recognition, including people's homes, which are usually their principal source of physical sanctuary. This is one of the main causes of psychological distress. Another may be the fragmentation of community that commonly occurs when groups of people are dispersed during the allocation of temporary housing. As noted elsewhere in this book, communities are not uniformly positive and therapeutic entities, but at the most local scale, they do tend to contain familiar faces, helpful neighbours and supportive family members. To have such a social network dispersed can take away a person's sense of coping. Hence, to be cast into temporary housing far from familiar people and surroundings can be a devastating psychological blow to a person who is already struggling to come to terms with losses, destruction and possibly bereavement.

In the short-term aftermath of disaster, rates of occurrence of suicide or murder do not necessarily rise. The enhanced social cohesion of the post-disaster period, sometimes known by sociologists as the 'therapeutic community', tends to act as a control on violence within it. In the short term, disaster brings greater consensus on what is just and right: there is a moral, as well as a practical, imperative to deal with the problems caused by the disaster, and this tends to impel people to work more closely together and to enhance the welfare function of society. As a result, it is common for suicide rates to drop. However, there may nevertheless be a surge in mortality, particularly among the aged and infirm. Destruction and disruption can cause premature death by influencing the psychological reaction of those who are precariously hanging on to life, regardless of whether or not physical problems are encountered. The surge can last for weeks or months.

Disasters are complex phenomena that create intricate relationships between physical forces and human societies. Some people find that working in the aftermath of disaster allows them to 'realise themselves' in that they acquire clear goals, benefit from a new and enhanced social matrix of support and interaction with their peers, find that helping others is rewarding, or otherwise reap the benefits of the new reality. Many others have to live instead with suffering, deprivation, physical or psychological impairment, bereavement or inconvenience and thus do not view the fact of having survived so positively. Whereas the short-term reaction to a one-off, sudden-impact disaster may enhance social cohesion, in many cases the longer-term reaction involves worsened social tensions or increased divisions in society. This may drive up rates of crime and violence. Above all, worsening inequality, lack of social justice, deprivation of human rights, a climate of uncertainty and exclusion from decision-making processes can all contribute to psychological problems. However, there are few indications that disaster worsens or causes permanent mental illness.

Cultural recovery

The earthquakes that affected the regions of Marche and Umbria in central Italy from September until December 1997 damaged some 1,200 religious buildings, most of which had the status of historic monuments. Even in the relatively small earthquakes that occurred further north in the Emilia-Romagna and Lombardy Regions in May 2012, the most spectacular damage occurred to churches, towers, castles, city walls and other relics of the distant past. In any area that has been settled for centuries, historic monuments tend to be a highly visible part of daily life. They embody the continuity of time between the generations and, as noted elsewhere in this book, help define the *genius loci*, or spirit of place, of a settlement. Hence, when they are damaged or destroyed by disasters, both tangible and intangible resources are depleted.

Culture is a very difficult phenomenon to define. Obviously, it is far more than the historic monuments of a city; indeed, some forms of culture are more or less instantaneous, or at least thoroughly modern. Culture is multifaceted and, in effect, we all participate in or belong to multiple cultures according to what we do and with whom we associate. In this sense, whether it is popular or high-brow, of long gestation or invented yesterday, culture is the sense of shared beliefs, values, behaviour patterns, spirituality and traditions (in whatever combination) that arise from the social part of life. It provides guidelines for living and for interacting with other people. It enables a shorthand form of communication, in which another person with the same culture will be assumed to have some degree of common experience, beliefs and attitudes. Hence, many of the coordinates of life are cultural in nature, or at least in origin. At the same time, culture is a dynamic phenomenon in which those elements that are inherited from our forebears or developed over time are cross-fertilised with external influences. Many of these are part of modern, globalised consumer culture.

There are various ways in which culture can be damaged, degraded, destroyed or mutated by disaster. Besides monuments, works of art and architecture, museums, artefacts and other visible signs of cultural activity in the present or past, many intangible elements are at risk. Traditions and practices may be suspended, either for lack of a place in which to practise them or because they are judged to be inappropriate in the aftermath of a disaster. Population fluxes may cause some people to leave an area affected by disaster or others to settle in it, thus disrupting the social activities that enable people to adhere to practices and beliefs. On the other hand, religious observances may be reinforced by disaster; for example, mourning for the dead or praying for deliverance.

Culture may enter into in a state of flux as a result of a disaster, or if it is already influenced by outside forces, those effects may intensify during the aftermath. The worst situation here is one in which people feel that their culture is threatened by outside influences and this demoralises them or makes them intolerant (IFRC 2014).

In preparing to recover from disaster, cultural aspects should not be ignored, however intangible and, therefore, esoteric they may seem. Plans that are not culturally compatible will be rejected, perhaps not for very rational reasons, while plans that are culturally acceptable will, one supposes, be welcomed. Hence, planned actions should be scrutinised for their cultural compatibility. Culture in its more recognisable forms should be promoted, not ignored, as it can boost morale among the survivors of a disaster. For instance, street theatre and music were vigorously promoted in Christchurch, New Zealand after the earthquakes of 2010 and 2011. Far from being a frivolous attitude to recovery, it helped boost local morale and remind people that there is more to recovery than simply physical construction.

Failure to rebuild damaged symbols of a local culture may have a deleterious effect on morale. No matter how well infrastructure and housing are being reconstructed, if people have to look upon the failed symbols of their identity every day, this may convince them that recovery is failing. Indeed, according to model 6, the Kates and Pijawka model (1977), there is a strong element of triumphant monumentality in the final stage of recovery. Not only will it symbolically record the end of the reconstruction period but it will probably also commemorate the disaster in monuments such as that shown in Figure 5.1.

This issue is further discussed in relation to housing in Chapter 8, 'Shelter preferences and functions'; and the theme recurs as one of the most important threads that runs through each of the sheltering and housing options in Chapter 9.

Although the days are gone in which people's social and psychological make-up were considered to be determined by their physical environment, there is nonetheless a relationship between culture, perception and environment that is as strong as it is subtle. Limitations of space demand that the following account of disaster and environment is very condensed. In reality, there are many studies of flora, fauna, soils and landscapes in relation to the impact of disaster. There is room here to bring out a few essential elements of them.

FIGURE 5.1   Monumental bass relief panel in the doors of the church of San Rocco in Lioni, depicting earthquake impact in a town in which 228 people died in the magnitude 6.8 earthquake of 1980 (photograph by David Alexander).

Environment

The devastation of 432 square kilometres of largely urbanised land by the tsunami of March 2011 in north-eastern Japan is an indication of how disasters can affect the built and natural environments – in this case an impact worsened by radioactive contamination from the Fukushima Dai'ichi power plant. There are, of course, many different kinds of environmental impact that result from disasters. The loss of crops and natural vegetation in drought; the loss of soil as well in desertification and accelerated erosion, sedimentation, salinisation and seismic landslides; and direct destruction, such as wind-blow damage to forests, are all examples of environmental impact. So is the destruction of buildings and physical infrastructure.

Left entirely to its own devices, the natural environment will usually recover from disaster of its own accord. In many forms of biosphere, wildfire is even beneficial as a stimulus to seed germination, enrichment of soils and recolonisation of burnt areas by vegetation. Beaches will rebuild their own equilibrium after storm surges and tsunamis, and landslides will stabilise after intense rain or earthquakes. Hence, some of the greatest environmental impacts come not from natural disasters but from the recovery process after them.

As noted by Kates and Pijawka (1977), reconstruction tends to use more space than the built environment that it replaces. This often requires converting more land from rural to urban uses. Without an adequate survey of hazards and adequate protection measures, reconstruction can extend urbanisation into areas of high risk. Moreover, measures such as embankments against landsliding or levees to contain floodable rivers can be fallible and eventually will be dismantled or overtopped by natural forces. Recovery may destabilise fragile ecological balances more than disaster itself is capable of doing. Development may fragment habitats, reduce slope stability, increase river discharges, pollute ground and lead to more intensive use of land that lacks adequate carrying capacity. Hence, it is opportune to include in recovery some measures designed to restore natural balances. Managed natural areas can have high amenity value and be useful buffers against future impacts; for example, in the case of coastal parks that act as buffers between storm surges or tsunamis and urban development further inland. However, it is as well to recognise that land of high amenity value becomes very attractive to developers and possible residents, which may create a demand to reduce the level of environmental protection.

Having considered the impact of disaster thematically, it is now time to draw the threads together.

Conclusion: disaster as a starting point for recovery

This chapter has looked at disaster as a starting point for recovery processes, and it has done so in a sectoral manner, considering economics and livelihood, government, psychological, cultural, and environmental aspects. It is important to engage with experts in each of these fields and to give these issues the attention they need. However, it is also important to treat recovery from disaster holistically and to look hard at the linkages between the sectors. Physical and psychosocial or cultural recovery are intimately linked. Failure to study the connections and take them into account can create problems for the future because complex strategies will have been thought through only in part and will lack the dimensions that they need in order to ensure complete success. One way to ensure that this problem is tackled is to involve the beneficiaries in the recovery process and listen to their points of view (see Chapter 8, 'Voices of survivors').

For example, in all societies there is a gender perspective on disaster. In those cultures in which the roles of men and women, or boys and girls, tend to differ more substantially, it is particularly important to look at the recovery process from both male and female perspectives. In this, no assumption of superiority or dominance should be made. Priorities may differ, but failure to take into account the roles, needs and aspirations of women, for example, will have negative repercussions for the whole of society (Enarson 2012). Further reference is made to gender issues in Chapter 6, 'Human rights and recovery from disaster'.

Managers of recovery operations need to consider ways to avoid the varied sectors of recovery becoming silos in which officials in a given sector build walls around their area of concern and fight for resources – in the all too familiar arrangements experienced in so many of our governments. Thus, cross-cutting teams drawn from the varied sectors need to be formed to require a fully integrated response, and the process is helped when they are all under the authority of the prime minister or president, who has the best chance of 'banging heads together' to secure effective and sustained integration. A further important implication relates to the subject of the higher education curriculum and the Disaster Risk and Recovery Management Masters courses where social workers, economists, architects and engineers, environmentalists and health professionals may meet. Teamwork in course design and specific teaching are needed to emphasise the interdependence of linked disciplines in disaster recovery, just as in all aspects of society.

Although recovery from disaster needs adaptive management, it also needs a recognisable model to be followed, albeit not with blind rigidity. The magnitude and impact of the disaster will be one set of influences on the form that the model should take. Another set will consist of the potentialities in each sector of recovery. These are made up of resources, assets, skills and social consensuses. Constraints may be environmental, political or cultural. Finally, although all disasters embody some unique characteristics, all have common ground. Hence, it is possible to search for analogies in good practice elsewhere providing this is adapted to local circumstances. In the rest of this book, we give some suggestions and ground rules for establishing viable models of reconstruction to follow.

In our next chapter, we explore some of the ethical dimensions of recovery management, asking how the multiple decision-makers can expand their normal obligations and accountability to their boards of directors to become 'downwardly accountable' to the supposed object of their concern: their 'clients', the survivors of disasters. We also return to the cultural recovery theme discussed above with a focus on vital yet intangible dimensions of recovery that provide sense of place, belonging and identification between people and their new environment.

Notes

1 Conversation between Dr Gustavo Wilches-Chaux and Ian Davis in 1994.
2 However, there are notable exceptions in which organisations regard themselves primarily as 'development agencies' such as CARE International, Oxfam, World Vision, Tearfund, Practical Action, Habitat for Humanity, etc.

6

SOME KEY ELEMENTS OF RECOVERY

In the previous chapter, we considered the starting point of recovery in terms of different sectors in the physical, social and psychological realms. These aspects are summarised in the second model, presented in Chapter 3, in which the main sectors of recovery are portrayed in a hexagonal diagram as facets of a holistic process. In this chapter, we examine some of the issues arising from disaster that are directly pertinent to recovery because together they form the context in which it occurs. They include human rights and responsibilities, the accountability of institutions and their leaders and the preservation of a sense of place, as well as issues of risk and safety in the recovery process. It is evident that these elements differ from each other very substantially. However, they illustrate the breadth of issues that must be addressed in recovering from disasters. Although it is very difficult to tie such disparate elements together, the last section of this chapter gives some thoughts on the possible role of a holistic approach to recovery – one that does endeavour to unite the disparate elements as much as this is possible. We begin with what may be the most fundamental of all elements – human rights.

Human rights and recovery from disaster

In countries where corruption and bureaucratic incompetence are rife, certain individuals and groups may manipulate their political connections to receive or distribute aid at the expense of others. Still other groups may receive little or no aid because of their ethnicity, religion, gender, age, or social standing. These abuses can leave individuals and families at risk and prolong the time they have to stay in poorly built and even dangerous camps and shelters for internally displaced people.

(Human Rights Center 2005: 1)

Each citizen of the world is, or should be, endowed with both rights and responsibilities. This is a crowded planet and we have a responsibility to live in peace and cooperation with each other, to promote good stewardship of resources for the future and to respect the rule of law. In return, we have a right to enjoy basic freedoms – of speech and expression, for example. Our rights are limited by economic scarcity and what is commonly acceptable in moral and ethical terms. Restrictions on rights are imposed by the political, social and cultural conditions under which we live. On that basis, there is ample scope for different definitions of what is seemly, permissible, acceptable and appropriate. There are also plenty of opportunities for conflict, oppression and deprivation of human rights. Definitions of what is acceptable in terms of 'the common good' are easily influenced by power relations and political expediency. Given the importance of creating and maintaining a good human rights situation, the less benign influences may need to be contested and resisted. One hopes that this can be done peacefully, ethically and within the law; that is, if the law is favourable to human rights.

Fundamentally, disaster risk reduction should take place within a framework of human rights. We should all enjoy the right to live peacefully, peaceably and in relative safety. Many of us do not. Time and time again, disasters have been used as an opportunity to repress the population, or elements of it, and to introduce measures that restrict human rights, usually under the guise of 'maintaining public order'. These are sins of commission. Sins of omission involve neglecting the rights of particular groups, usually minorities in society, and thus leaving them at a disadvantage (e.g. see the complex situation in Myanmar: Barber 2009; Thawnghmung 2013).

For years, it has been recognised that there is a gender dimension to disasters (Enarson and Dhar Chakrabarti 2009). A report by the Japanese Women's Network for Disaster Reduction was issued in the wake of the 2011 earthquake and tsunami in north-west Japan, and this described in some detail how decision-making during the aftermath and recovery was managed by men to the absolute detriment of women's issues (Domoto et al. 2011). When cyclone Haiyan (Yolanda) struck the eastern Philippines in November 2013, many men did not evacuate and thus drowned in the storm surge, leaving a substantial number of women as widows with families to bring up under harsh conditions (Faure Walker and Alexander 2014).

There are many instances in which women are more versatile and resourceful than men after disaster. They may, by aptitude or necessity, have greater ability to rear children and carry out multiple tasks. Their business acumen and resourcefulness are not in doubt. However, there are indications that in many disasters, women bear a heavier psychological burden than men, having higher rates of post-traumatic stress disorder, anxiety and depression (Richter and Flowers 2008). In various traditional cultures, their range of options and opportunities for action may be reduced by restrictions placed on them by the male leaders of society (van der Gaag et al. 2013). Hence, there are many instances in which the key to recovery from disaster is the empowerment of women.

If there are too many obstacles in the way of giving them parity with men in decision-making, the key decision-makers need to be induced to listen to their concerns and take proper account of them. In our view, parity is the only morally acceptable option, and while we recognise that it may take time to achieve, we trust that reforms will be introduced where they are needed.

In conducting fieldwork in the Tacloban area of the Philippines after cyclone Haiyan (Yolanda), David found women to be more articulate than men and, in many cases, to be more resourceful as they faced the daily struggle to survive and rebuild. Providing microcredit and other opportunities to women can enable them to start or revive small businesses – in clothing manufacture and the retail trade, for example. In the longer term, much depends on the educational and employment opportunities that are available to women and girls. In Middle Eastern countries, there is a predominance of women among university students. However, whether they are discriminated against in the employment market will determine their ultimate success. Employment is needed that not only favours women on equal terms with men but also takes account of their specific needs – for example, in raising children without having to abandon their careers.

In disaster recovery, as in other aspects of life, gender parity is an issue of human rights; and one hopes that disaster can become a catalyst for improving rights, not restricting them – in other words, an occasion for introducing reforms. Another aspect of this is the treatment of people with disabilities in disaster. There are many forms of disability, which may be physical or cognitive; may be short term, long term or permanent; and may involve varying degrees of autonomy or dependence on life support equipment or carers. People with disabilities may be disadvantaged in disaster by limitations of their ability to hear or react to warnings, to evacuate and to protect themselves against harm. At least one in seven people in society is likely to be disabled. This very heterogeneous group of people overlaps with the elderly; though not all people of great age are disabled, and not all people with disabilities are aged (Stough and Mayhorn 2013).

In countries that have made some effort to improve the lot of people with disabilities, there is a trend towards enabling them to live at home, wherever this is feasible, and in any case with as much autonomy as possible. There have also been some improvements in accessibility and mobility. Indeed, the concept of 'design for all' is gaining traction (Alexander and Sagramola 2014). Nevertheless, there are very many instances in which no provision is made for people with disabilities in disaster or during the recovery process. The result can be discriminatory and leave people with disabilities in a highly disadvantaged state. Worldwide, neither emergency planning nor recovery planning is routinely broadened to include provision for the disabled; and yet morally and ethically, under the provision of 'design for all' principles and practice, it should be. In seeking to improve this situation, there is a fundamental antagonism between knowing who is disabled and where people with disabilities live and work and respecting their right to privacy. It can be resolved by encouraging people with disabilities, and their carers, to participate in voluntary programmes in which their needs are assessed and accommodated in recovery plans (Abbott and Porter 2013). Sadly, this approach is far from universal, and people with disabilities have to contend with shelters that are not wheelchair accessible, evacuation plans that take no account of reduced mobility, and alarm systems that make no provision for those who cannot see or hear the warning signs. Civil protection and recovery plans are usually made for and on behalf of groups of citizens whereas provisions for people with disabilities require individual attention, which needs more resources but is by no means infeasible.

It is a principle of planning for minorities that they should not be lumped together and treated as a single category. The needs of particular minority groups should be considered separately, even in cases in which economies of scale can be achieved by combining initiatives. In and after disaster, ethnic and cultural minorities and people of diverse sexual orientation should not be discriminated against. Moreover, it will probably be necessary to monitor social relations in order to ensure that the aftermath of disaster does not become an occasion for the introduction of repressive and discriminatory measures. This requires a participatory approach in which decision-makers know who are the representatives of such groups, listen to their concerns and introduce measures to avoid discrimination if it materialises or looks likely to do so. It is vital that these issues are tackled during the recovery rather than being postponed until after it as, by then, the problems will be entrenched and difficult to remedy.

In summarising the issue of human rights in relation to recovery from disaster – if one can! – we can view it as something that requires constant vigilance and the application of fairness and moral principles. It is difficult to define the boundary between rights and freedoms on the one hand and, on the other, the responsibility to make concessions to social living and the rights of our fellow human beings. For instance, legislation to counter the threat of terrorism may be intended to protect citizens but can end up sacrificing the very freedoms that it is designed to safeguard. Infringement of human rights can be linked to issues of corruption, exploitation, nepotism, organised crime and other forms of exploitation (Boyce 2000). Governments that are serious about protecting citizens' rights should consider appointing a guarantor or at least a senior official who monitors the situation and reports on possible infringements so that corrective action can be taken. The rights of neglected minorities will require liaison with their representatives and, potentially, measures to reduce conflict or at least to guarantee action to redress any wrongs.

We consider human rights violations against protesting survivors and their advocates following the Wenchuan earthquake in our discussion of model 1 in Chapter 3.

Accountability

Accountability is part and parcel of human rights. It means taking responsibility for decisions in a demonstrably honest manner. In practical terms, accountability is perhaps more easily defined by its nemeses: lack of transparency, unclear decision-making processes, and refusal by key officials to take responsibility for their decisions and actions. If circumstances exist in which these issues are likely to be a significant problem then oversight by an independent authority may be needed – one that has the power to investigate and is judged to be relatively immune to undue influence. Secrecy is the enemy of accountability. Granted, a modicum of confidentiality is necessary in politics, statesmanship and emergency planning. However, overemphasis on this aspect can be used as an excuse to shield bad decision-making from necessary scrutiny. The consequences may be unnecessary suffering, a persistent sense of unfairness, subversion of the recovery agenda towards personal, political or factional gain, and inefficiency in relief and rebuilding.

Who intervenes in disaster, and to what extent are the organisations accountable? Ian recalls his experience of leading the group that was developing *Shelter after Disaster* guidelines in 1977, the first UN study of the subject (UNDRO 1982): at that time, I was working with Fred Cuny (an urban planner), Fred Krimgold (an architect) and Everett Ressler (a social scientist). During the course of this demanding project, the four of us began to realise that one of the fundamental problems with the humanitarian system was the almost total absence of accountability that prevailed in the 1970s. This applied to outside agencies and individuals, including ourselves, as well as to the beneficiaries of assistance. As researchers, consultants, UN representatives or NGO staff, we all felt financially and morally accountable to whoever employed us, but that seemed to be the sum total of our accountability. Thus we were able to discover that inappropriate shelter goods could be provided, emergency campsites could remain half empty, post-disaster dwellings could fail miserably, yet the occupants or recipients of the aid had no ability whatsoever to obtain redress from those who had failed them just at the precise point when support was most needed.

We therefore decided to write about this unsatisfactory situation in the radical magazine *New Internationalist*, for which we developed the chart shown in Table 6.1. Rather uncharitably, we called the piece 'The Interveners' as we thought that this was an accurate description of the roles of all assisting groups who were outside the immediate community of survivors: they literally 'intervened' in the survivors' own recovery situations (Davis 1977a). We noted then – as we can confirm now, 38 years later – that while various mechanisms have developed in public life in our countries to make academics, doctors, politicians and others accountable to their 'clients' through inspections, evaluations, appraisals, terms of reference, social audits, and so on, there has been exceedingly slow progress in 'downward accountability' in the relief, recovery and risk management sectors. Thus, poor-quality assistance can persist without suitable checks and remedies. This helps explain the many examples of waste and inappropriate assistance that we chronicle in this book.

TABLE 6.1 The interveners in disaster

The interveners in a disaster					To whom are they accountable?		
The interveners	Official reasons for their presence	Self-interest in the disaster	Underlying beliefs which will govern their actions	Activity within the disaster area	Officially	In practice	To the survivor
United Nations organisations, such as the World Health Organisation, UN Office for the Coordination of Humanitarian Affairs, and UNICEF	• Responsibility to member nations embodied in their terms of reference	• To show how valuable they are • To ensure their future growth and funding	• Inadequacy of locals • Their help is wanted • International cooperation is essential from the earliest days of recovery	• Coordinate other groups • Allocate UN money • Work with UN development programme agencies, the national governments, and relief organisations	• To their UN superiors • To the national governments that contribute to the affected country	• Hardly anyone, but possibly a journalistic exposé	• No accountability
Overseas governments, particularly those in the more wealthy nations	• To assist less fortunate nations – often formalized in official treaties	• To fulfill their foreign policy aims • To boost home business • To win votes at home • To reward faithful allies	• Inadequacy of locals • Their help is wanted • There is a shortage of materials and expertise in the disaster	• Allocate their financial help and other supplies • Often to take control of the disaster relief operation, in proportion to their influence • Work with the national government	• To their home government • To the local government	• Hardly anyone, but possibly a journalistic exposé • To their home voters if near election time	• No accountability
Overseas charities concerned with relief and development	• A major reason for their existence is to aid disaster survivors	• To show their humanitarian concern • To show to their supporters, and to rival agencies, how	• Inadequacy of locals • Their help is wanted • The official response is slow and inefficient	• Direct foreign helpers and experts • Distribute aid • Work with both government relief agencies and those at the grass roots	• To the director of their charity • To their charity's financial supporters, including their	• To the director of their charity • To their charity's financial supporters	• Through the local grass roots organisations when they work with them,

(continued)

TABLE 6.1 The interveners in a disaster (continued)

The interveners	Official reasons for their presence	Self-interest in the disaster	Underlying beliefs which will govern their actions	Activity within the disaster area	To whom are they accountable?		
					Officially	In practice	To the survivor
		valuable they are • To ensure their future growth and funding	• They are uniquely suited to work with the poor • They are politically neutral		home government	• Possibly a journalistic exposé	otherwise no accountability
Foreign experts in relevant subjects	• To use their expertise in conjunction with one of the above organisations	• To use their superior knowledge • Rivalry with others who have similar skills • To impress these rivals	• Inadequacy of locals • Their help is wanted • Their skills can not be found locally • Their skills are relevant	• Provide appropriate advice • Work with the foreign organisations who are providing help	• Possibly to their superiors in home university or agency • To those that have sponsored their work	• Hardly anyone	• No accountability
Local elites – who often function like the foreign experts	• To help the poor	• To preserve their local power and authority	• Inadequacy of locals • Their help is wanted • They have the knowledge of the local situation and what is wanted	• Work with local business and the national government	• To local business	• Hardly anyone	• Occasionally through the local grass roots organisations, otherwise no accountability

Source: developed by Ian Davis, Fred Cuny, Fred Krimgold and Everett Ressler. This chart was originally used in Davis (1977a) and is reprinted by kind permission of *New Internationalist*. Copyright New Internationalist.newint.org

Accountability in the sheltering and housing sector

We must recognise that in recent years, serious efforts have been made by concerned bodies to develop effective patterns of accountability to survivors. An early example came in 1995 with the publication of the significant 'Accountability in disaster response' (Humanitarian Practice Network 1995). Two years later, in 1997, a key development occurred following the Rwandan genocide when the Active Learning Network for Accountability and Performance (ALNAP) was created by the NGO community. This was given the ambitious mandate to create a 'system-wide network dedicated to improving humanitarian performance through increased learning and accountability' (ALNAP n.d.). Since then, ALNAP has expanded to include a wide spectrum of members, including UN organisations and the Red Cross. It has produced a stream of important publications which aim to improve project evaluations, distil key lessons and generally strengthen accountability in the humanitarian system. In the mid 1990s during a period of questioning, Peter Walker, currently Director of the Feinstein International Center at Tufts University, expressed his concerns as follows: 'Agencies rarely ask themselves the question: "what quality of assistance have disaster survivors the right to expect from us?", going beyond rations delivered, where most present standards concentrate' (Walker 1996). Work such as this by ALNAP and Walker has provided particular solutions to the unaccountability described in Davis (1977a).

Despite such admirable efforts, is the message getting through? In our recent travels to review disaster situations in China, Haiti, India, Japan, Italy, New Zealand, Pakistan, the Philippines, Sumatra and the USA, we have both repeatedly observed governments, UN agencies and NGOs continue to act in a cavalier, unaccountable manner without adequate participation by survivor groups. When will *interveners* become genuine *partners* with the surviving community, thus responding to the need for 'downward accountability'?

The issue of accountability is closely related to trust – as demonstrated in model 15's trust-control pendulum, described in Chapter 4 – since it is not possible to become genuinely accountable to someone, or some community, without there being a high level of mutual trust. In Chapter 9, 'Option 3: unsafe dwellings on unsafe site', we record the example of a lack of accountability of an Indian NGO to the residents of a failed housing reconstruction project following a cyclone.

In 2006, Ian recalls meeting the director of a housing reconstruction project carried out by a major Australian NGO in Aceh in the wake of the 2004 tsunami disaster: I enquired about his background in the building reconstruction sector. He responded that his background was that of a military officer and that this was his second career and he had no previous training or background in building or housing. But that did not seem to dent his confidence in the slightest, as he expressed the view that 'housing reconstruction is a pretty straightforward business'. I told him that I did not share his view, having spent over 30 years grappling with the topic, which seems to become more demanding and less straightforward as every day passes.

When considering 'provided' shelter or housing, we remain baffled over the question of who is competent to make informed decisions. In an example in the medical field of humanitarian operations, it would be totally unthinkable for persons without medical skills to conduct surgical operations on injured patients; or within the water and sanitation field, authorities would certainly not permit any 'man or woman in the street' to design and manage a safe water supply or sewerage system. So why is there a persistent pattern in which NGO officials – or in some instances, government staff – with no training whatsoever in architecture, engineering, housing design, building contracting or settlement planning are entrusted by their directors to design shelter projects or permanent housing? We discuss this official neglect further in Chapter 8, 'Shelter and housing experience'.

A further symptom of missing leadership in the shelter and housing sector within the donor community can be found in the absence of expertise within key agencies. For example at the time of writing in 2015, the European Commission Humanitarian Office (ECHO), the largest funding body of humanitarian shelter programmes, does not possess a dedicated shelter expert while USAID has a single expert in this field (Gray and Bayley 2015).

This lack of professionalism and failure to be accountable to survivors lies at the heart of many of the failed projects that we have described in this book. The consequence is long-term misery for survivors who have to suffer inadequate provision. However, some further fine-tuning is needed for we are also aware that there have been failures where architects and other professionals have designed and directed projects. So in the final analysis, when appointing the managers and leaders of projects, *competency*, *capacity* and *experience* are more important than *qualifications*.

Managing and leading shelter and housing programmes are demanding tasks that require substantial knowledge and a range of skills. It is most unlikely that any individual will possess all the qualities needed (see Table 6.2), but it is essential that the full range of skills is present among those who make up the recovery project teams.

We have discussed the accountability of governments and other 'interveners' in local affairs. In a sense, we all need to learn and conserve the lessons of history so that we can be accountable to the future, the world we will pass on to our descendants. The next section considers this in terms of *genius loci*, the sense of place – a phenomenon that is often severely threatened by the destruction that disaster causes and that requires specific measures if it is to be conserved.

Genius loci and preservation of the identity of places and human settlements

In Chapter 5, we discuss cultural recovery. We now sharpen the focus to consider *genius loci*, literally the guardian spirit of a place. This term is a Latin phrase that

TABLE 6.2 Team and leadership requirements for the management of shelter and housing projects

Knowledge	Skills	Attitudes
. . . of the relevant field and its application to recovery planning: architecture, engineering, planning, construction, etc.	*Interdisciplinary teamwork*	Empathy, listening skills and accountability to survivors
. . . *of disaster and development principles and practice*	Creativity, improvisation, ability to make much out of limited resources	*Leadership, vision and integrity*
. . . *of project management and financial management*	Understanding and respect for local culture, social patterns and building traditions	Ability to see both short- and long-term needs and macroscopic and micro-level concerns, and to merge them in project design
. . . of working with low income groups and of advocacy	Training, mentoring and educational skills	Political awareness and sensitivity
. . . of disaster risk reduction	*Social skills and communication skills needed for participatory management*	*Patience, tenacity and perseverance*
. . . *of the given disaster situation, of the multi-sector recovery plan and of the key players*	*Coordination skills*	A willingness to learn and adapt as recovery proceeds

Note: credentials required by leaders and directors are given in *italics*.

expresses the concept of the cultural identity of a geographical locality. There are many different individual phenomena that can contribute to, or even singly constitute, the essence of *genius loci* and make it distinctive (Norberg-Schultz 1980). Architectural character might be one ingredient, such as the palazzi in Rome or Florence or the Eiffel Tower in Paris; features of the natural environment are another, such as Sugarloaf Mountain in Rio de Janeiro; and monuments or artefacts, such as particular sculptures or memorials, also contribute. Often, such details are a shorthand way of remembering a place. What would Sydney be without the Harbour Bridge and the Opera House, or Rome without the Colosseum? Some examples show that the defining ingredients of *genius loci* do not need to be old and sanctified by time, but can be of recent construction. Hence, one could argue that the Opera House, designed by the Danish architect Jørn Utzon and located on Bennelong Point in a prominent location in Sydney Harbour, is an attempt to endow Sydney with yet more *genius loci* than it already had – an attempt to define its character in a building.

There are several reasons why *genius loci* is important. It helps make places distinctive and memorable, and this is usually in a positive way as we have little inclination to think of the concept in negative terms, as a concentration of ugliness. It embodies the spirit of a place and thus engenders loyalty to it by its citizens. Thus it is a shorthand expression for the sense of belonging. Finally, *genius loci* is most commonly manifest in terms of heritage, defined as the artefacts and constructions of the past that are precious to us because they embody art, architecture and the best of human endeavour, as well as the talent of former citizens who were inspired by their home territory.

Rikuzentakata in Iwate Prefecture, Japan was a town of 23,300 inhabitants but was largely razed to the ground on 11 March 2011 by a tsunami that killed more than 1,000 inhabitants (Ogasawara *et al*. 2012). A symbol of the town is the *Cryptomeria japonica* cypress tree, one example of which remains standing on the seafront of the former urban area. This tree, much the worse for wear, is carefully

FIGURE 6.1 The coast at Rikuzentakata, Iwate Prefecture, Japan and the lone *Cryptomeria japonica* tree, survivor of the March 2011 tsunami and a symbol of hope and reconstruction (photograph by David Alexander).

supported by scaffolding but remains symbolic of the town and its desire to be reborn. It died and was later replaced, but the roots of the original tree were conserved in a museum.

In many instances, it is the cityscape – a total environment built by people for their own use and enjoyment – that defines *genius loci*. Thus the painted houses of Copenhagen, the airy vastness of Tiananmen Square in Beijing, or the Art Deco buildings of Napier, New Zealand (built during reconstruction after the 1931 earthquake) all embody a spirit of belonging and a respect for human endeavour. However, a significant number of these essential areas of built environment are seriously at risk of destruction by disaster or conflict, or simply by neglect, abandonment and slow decay. Thus, the Arg-é Bam (citadel or castle) of Bam in Iran was more than 80 per cent destroyed by the earthquake of 2003. As this building is the defining feature of the city of Bam, as well as the world's largest adobe structure, much attention was given to its rebuilding (Khatam 2006). The Arg-é Bam had been inserted in UNESCO's register of World Heritage Sites but, in view of the destruction, was placed on its list of sites at risk. In mid 2013, after much reconstruction work by international teams, it was removed from the latter.

Disaster clearly represents a threat to *genius loci*. What happens when the latter is not upheld? Cultural heritage protection and restoration are expensive, complex and time-consuming processes that are seldom among the immediate priorities of recovery from disaster. Ethically, they should not take precedence over the provision of food, shelter, medical care and other basic assistance to survivors. However, in many cases, loyalty to a place, and therefore to a recovery process, is intimately bound up with *genius loci*.

Christchurch, New Zealand was badly damaged by the earthquakes of 4 September 2010 and 22 February 2011, the second of which killed 185 people. One casualty was the city's Anglican cathedral – a work built over the period 1864 to 1904 and designed, in the first instance, by the celebrated British neo-Gothic architect Sir George Gilbert Scott. Incidentally, Scott was concerned enough about earthquakes to include some anti-seismic provisions although they were largely ignored in the execution of the design. Damage in the February 2011 earthquake was very substantial, including the loss of the tower and steeple and the west front (Sibley and Bulbulia 2012.). A design for a temporary cathedral was commissioned from the Japanese architect Shigeru Ban, and this was constructed in late 2012 and early 2013. Meanwhile the controversy over the fate of the damaged cathedral continued.

Questions about the insurance premiums for historic stone buildings, the structural stability, the relative costs of partial or total replacement versus reconstruction of the original building, and the future seismic stability were hotly debated. It should be noted that this Christchurch cathedral was located at the centre of the (now largely demolished) city centre and that it formed a harmonious, if spatially discontinuous, grouping with other neo-Gothic buildings, such as the former Canterbury University complex and the city Museum.

FIGURE 6.2 Christchurch Cathedral, New Zealand after the 2010 and 2011 earthquakes (photograph by David Alexander).

In terms of *genius loci*, much depends on the extent to which the Church of Christ is considered representative of the spirit of the city and whether the original building is regarded as indispensable to its rebirth during the reconstruction. Opinions differ among residents and other New Zealanders. However, it is to be remembered that the very name of the city reflects its Anglican mother church and tradition. To some extent, it is evident that multicultural immigration and rehabilitation of the indigenous population has diluted Christchurch's Anglican tradition; and while this has endowed the city with new riches, it has also introduced new opinions about what is best for the city.

My own view is that world heritage and local tradition would both be poorly served by not rebuilding Christchurch Cathedral as a restored version of the original, which was designed by Scott and executed by the New Zealand architect Benjamin Mountfort. Christchurch faces the problem of outmigration, the cost of relaunching its economy and urban functions, and the severe challenge of recreating eminently liveable urban spaces. Some observers of this process point to the recovery of Napier on North Island after the 1931 earthquake, in which the destroyed city centre was rebuilt in a delicious Art Deco style that led it eventually to be declared a UNESCO World Heritage Site. That is as may be, but there are few indications that Christchurch would be able to follow a similar pattern. Napier, pre-earthquake, lacked a significant architectural heritage whereas

Christchurch had one and now has no guarantee that it will necessarily acquire another. Interestingly, Christchurch's Catholic cathedral – an equally impressive building – was also devastated by the earthquakes, but there was remarkably little controversy about rebuilding it.

Next in this brief meditation is the question of how to absorb the disaster into the *genius loci* of an affected area. During recovery and reconstruction, there is usually a debate about how to commemorate the event. Commonly, the stakeholders divide into two factions. There are those who wish to ensure that the disaster is properly commemorated with monuments, indicators (e.g. of the vertical level that flooding reached), museums and documentation centres. On the other hand, there are those who have no wish to be reminded daily of the horror represented by the disaster or who feel that memorials of any kind damage the image of the affected place. It can be quite difficult to reach a compromise between the two sides. One is reminded of the Herculean struggles of Giovanni Tosatti to provide a documentation centre and memorial to the 268 people who died in the 1985 Stava Valley mudflow in the Italian Alps (Tosatti 2003), work that was strongly resisted by many people in power. However, one thing is clear: one way or the other, a large disaster will be absorbed into culture because it represents a milestone and defining point in the culture of a locality or region.

Monuments can have a practical purpose as well as a symbolic one. For example, in Rikuzentakata, eastern Japan, there is a project to plant a long line of trees at exactly the elevation reached by the March 2011 tsunami. This is one of many examples where the magnitude of a disaster is commemorated in practical terms that will serve as a benchmark and warning for future generations. Another way is to leave some remnants of damage *in situ* without either demolishing or repairing them. Thus when the historic walled town of Venzone in the Friuli Region, north-eastern Italy was rapidly rebuilt following devastation by two earthquakes in 1976, its Augustinian convent church was left as a ruin to be a lasting monument to the disaster. Elsewhere, monumental constructions can embody the successful conclusion of reconstruction processes; for example, the Rua Augusta triumphal arch to the Praça do Comércio in Lisbon – a symbolic entrance from the Tejo estuary to the city and an embodiment of the reconstruction after the 1755 earthquake; or the rebuilding of San Francisco City Hall in 1929, almost a quarter of a century after its predecessor was destroyed by the 1906 earthquake and fire (Kates and Pijawka 1977).

Finally, it should be borne in mind that *genius loci* can be deliberately destroyed by malevolent interests. The Bamiyan Buddhas were two rock and stucco statues, 35 metres and 53 metres tall, which were sculpted in a cliff face located 230 kilometres from Kabul in Afghanistan between AD 507 and AD 554. In March 2001, they were dynamited by the Taliban who regarded them as idolatrous. Subsequently, there has been much debate on the future of this UNESCO World Heritage Site, including plans to reconstruct one of the Buddhas using anastylosis – the recomposition of original material with new additions (Margottini 2014).

Among elements of the context of recovery from disasters, we have considered rights, accountability and the need to conserve the spirit of place. In rather broad terms, we now consider human safety in the conclusion to this chapter.

Conclusion: holistic and integrated perspectives

If human rights were very strictly observed, there would be relatively little difference between people's levels of safety. However, the increasing gap between the rich and the poor ensures that the latter are far more at risk of disaster and far less able to devise remedies than the former. Safety and risk reduction are issues that cut across the segments, facets and categories of recovery from disaster. With careful planning, they can be unifying themes. Reconstruction should not take place in unsafe areas when there are safer alternative sites. Communities should not remain highly vulnerable to hazards when there are well-known affordable measures that could protect them. Recovery and reconstruction plans should place an accent on fairness in the use of resources, especially regarding safety considerations. Recovery should be carried out with adequate attention to the safety of workers. Staff and citizens should be encouraged to draw attention to safety risks and these should be openly discussed in order to find remedies. If such provisions are implemented, and there are no political or cultural barriers to doing so, then there will be good prospects for creating a 'safety culture' in which hazards and risks are taken seriously and responsibilities are shared among members of the community and officials. Commonly, such a culture takes many years to achieve, but this should not discourage leaders from advocating it.

Because of the multifaceted nature of recovery, it is not easy to design strategies that are truly holistic. The diagram in Figure 6.3 depicts the complexity of the interplay between vulnerability and resilience in the light of the many facets, scales and relationships involved.

In recovery, as in disaster risk reduction, there is a need to achieve a consensus on what is necessary to sustain human life, health and well being (Becker and Tehler 2013). McEntire (2011) sees this as a matter of liabilities and capabilities, ensuring that the former are shouldered and the latter are fully exploited. McEntire *et al.* (2010) note that vulnerability reduction needs to concentrate on both external and internal stressors; i.e. shocks such as earthquakes or storms that affect the community from outside and those, such as dissent and anxiety, which affect it from within. Some authors (e.g. Aldrich 2012) put their faith in social capital. However, the concept has been criticised as being too narrow, too awkward and too restricted (Inaba 2013). Nevertheless, if it is broadened to embrace *human capital*, it does convey the idea that skills and abilities can effectively be utilised in concerted action to rebuild well and reduce future risks of disaster.

McEntire (2005) offers his own *vade mecum* of holistic disaster recovery. It has four requirements: to avoid the restoration of past vulnerability while concomitantly seeking to reduce present vulnerability; to avoid promoting or

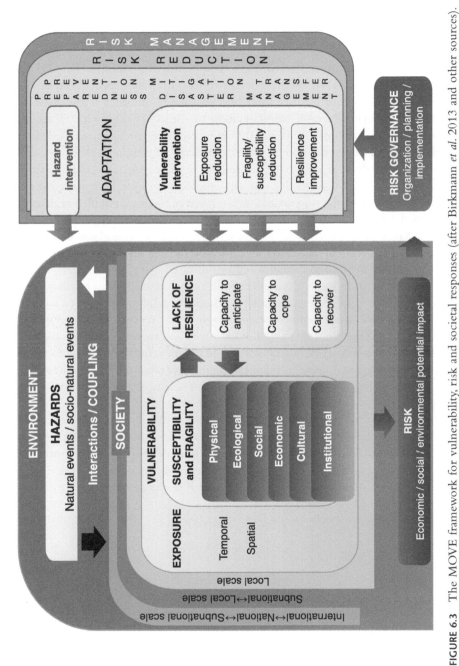

FIGURE 6.3 The MOVE framework for vulnerability, risk and societal responses (after Birkmann *et al.* 2013 and other sources).

contributing to the probability that disaster will strike; to promote social, political and economic progress and try to ensure that this is not nullified by future disasters; and to acknowledge the centrality of the community's response to disaster. Like many other scholars (e.g. Cardona 2004), McEntire sees vulnerability reduction as an essential root of the problem of how to achieve recovery while ensuring safety.

Recovery and associated disaster risk reduction should not ignore interdependencies. One of these is the relationship between disasters and development, both of the economy and of society. Another is the need to integrate disaster reduction with adaptation to climate change, with an agenda for general sustainability (Reniers 2012). Hence, one requires sustainable disaster risk reduction and sustainable lifestyles and usage of resources. Recovery and disaster risk reduction need to be sustainable in their own right and as part of the wider agenda. To ensure that this happens, recovery must be practised at the local level and have the support and involvement of the local population. It must be based on coherent plans that are accepted by their users, stakeholders and beneficiaries. The plans must be sensitive to local needs and be flexible in execution. Disaster risk reduction must be considered a fundamental service on a par with the provision of utilities and health care.

Last, integration does not mean applying the same model and techniques to every problem. It means ensuring that no fundamental dimension of recovery is ignored, including the connections between facets, sectors and geographical scales. This will require leaders with both vision and integrity. An essential skill is the ability to project developments and interconnections into the future and envisage progress in different elements of the recovery; in other words, the ability to build and test scenarios. This requires experience and perspicacity. It also requires political savvy and adaptive management techniques.

In Chapter 7, we consider a series of dilemmas that face decision-makers in recovery management. We continue some of the themes discussed in this chapter; for example, the first dilemma considers the issue of reform versus continuity, and we look at issues of identity and continuity in the reconstruction of Skopje as seen by a resident who grew up in the city as it was being rebuilt.

7

DILEMMAS IN RECOVERY MANAGEMENT

This book repeatedly demonstrates that disaster recovery is a highly complex process that is characterised by widespread confusion, contradictory needs, continual pressures and multiple planning dilemmas. In 1943 at the height of World War II, in his Mansion House speech on 9 November, Winston Churchill spoke of one of the fundamental dilemmas of recovery. His concern still confounds every recovery manager facing the challenge of political expediency versus practical needs: 'no airy visions, no party doctrines, no political appetites, no vested interests must stand in the way of the simple duty of providing beforehand for food, work, and homes.'

In this chapter, we examine some of the more common dilemmas that are likely to perplex officials. Although they may appear to be in conflict, opposing or alternative options are often vital elements of successful recovery. Therefore, it is necessary to find ingenious ways to resolve the dilemma by satisfying *both* sides. Hence, this chapter sets out arguments for and against particular approaches in four selected dilemmas that are judged to be typical of the field. This is followed by a listing of key steps that will resolve the conflicting demands (Davis 2007; Jha *et al.* 2010: 10).

First dilemma: reform vs continuity

This dilemma has a long history:

> There is nothing more difficult to plan, more doubtful of success, nor more dangerous to manage than the creation of a new system. For the initiator has the enmity of all those who would profit by the preservation of the old system and merely lukewarm defenders in those who would gain from the new one.
>
> *(Niccolò Machiavelli c. AD 1500)*

This issue is discussed in Chapter 3 in 'Model 1: progress with recovery'. Here, we note the serious risk of returning to a status quo ante form of vulnerability. However, returning to the status quo is certainly not the same as seeking continuity with the past. When writing about the reconstruction of towns and cities following World War II, Professor Nicholas Bullock perceptively noted that: 'reconstruction was not just about building a "nobler" new world. Planning for the future was inseparably mingled with a desire for continuity with the past' (2002: 6). He was referring to a persistent dilemma that faced planners in the post-war reconstruction of European cities – a dilemma that still persists today in reconstruction after natural disasters. He noted the 'window of opportunity' created by the war to 'sweep away what was bad, unjust, worn out and inadequate', but he also noted that 'the same process of change also created a longing to return to what was familiar, to the happier images of a pre-war world fondly remembered' (Bullock 2002: 6).

In one of the most significant buildings to be reconstructed in Britain after WWII, this dilemma was resolved in a manner that pleased the architectural reform lobby as much as it pleased as the conservation lobby. In November 1940, the English city of Coventry was devastated by bombing. The cathedral was shattered and rebuilt during the 1950s to a competition-winning design by the architect Sir Basil Spence. The design succeeded in combining continuity and reform by linking the remains of the nave of the destroyed cathedral to a completely new structure.

Perhaps the contrasting attitudes of the reformers and the conservers can be likened to 'herbivores and carnivores'. When reflecting on post-WWII recovery in Britain, the social historian David Kynaston described the herbivores (reformers) as:

> progressive liberal-minded, somewhat left of centre, eager to build a New Jerusalem after the chaos and destruction of War: the carnivores (conservers) were very different, seeing themselves as practical men, impatient of woolly do-goodery, usually right of centre, and giving the highest priority to restoring the nation's vitality.
>
> *(Kynaston 2015: 5)*

Reform vs continuity in the reconstruction of Skopje, Yugoslavia (1963–90) following the earthquake of 1963

This dilemma of whether to adopt a new vision with widespread reform or to conserve old urban patterns is a familiar debate in many reconstruction programmes, and it was particularly evident while the city of Skopje was being reconstructed after the 1963 earthquake.

In 1974, Ian first visited Skopje, in what was then Yugoslavia (now Macedonia), to review progress with reconstruction 11 years after the earthquake that had devastated the city. It caused more than 1,000 deaths among a population of

200,000 and destroyed 40 per cent of the housing stock. The project manager of the reconstruction was Polish architect and planner Adolf Ciborowski – one of the pioneers of disaster reconstruction (see Chapter 2, 'Evolution of recovery studies'). George Nez, an American urban planner, was also an advisor to the project. We include a quotation from him in our discussion of model 10 in Chapter 4 (see Model 10: disaster 'crunch' model) in which he describes a dilemma he encountered in his work on the recovery plan for the city. The reconstruction plan included a grandiose design called a 'city gate' with a centrepiece composed of a transportation hub for buses and trains in the form of a vast structure designed by the renowned Japanese architect Kenzo Tange, who in addition to the transportation structure, won the international competition for the urban plan (Ladinski 1997).

Professor Olga Popovic Larsen is an architect who grew up in the reconstructed Skopje. She now lectures at the School of Architecture of the Royal Danish Academy of Fine Arts in Copenhagen. The dialogue that follows is from an interview with her that was carried out by Ian, focusing on her experience of the reconstruction of her home city. First, she describes her family's experience of the earthquake:[1]

PROF. POPOVIC LARSEN: I was born in 1963, before the earthquake. We were on a summer holiday in Struga [a small town on Lake Ohrid] and were due to come back home to Skopje the day before the earthquake. I was a baby and very ill. The doctor in Struga asked my parents to stay few more days. As a result, we were not in Skopje when the disaster occurred. It was felt in Struga, and when the news came through, my father went back to Skopje to help out. Our building was severely damaged but did not collapse. Had we been in Skopje we would have been fine. It took one year to restore the building. During that time we lived in Veles with my mother's parents. My parents were commuting to work. However the building adjacent to ours collapsed and everyone was killed. We were fortunate that none of the extended family members were killed, but many of my parents' friends, colleagues were less fortunate.

IAN: Reform or evolution? Do you recognise the following conclusion concerning a dilemma in reconstruction that I wrote almost 40 years ago?

Seen from any standpoint, the reconstruction achievement is a remarkable example of efficiency and co-operation following a disaster. But with internationalism and all its benefits, there come inevitable problems of cultural 'appropriateness': at the small scale, the suitability of the Dexion-framed temporary house for a climate as severe as Skopje, and at a large scale, the Japanese 'Tokyo scaled' transportation centre for a modest town in need of gentle structures. But, more important, the rebuilding of Skopje raises again the fundamental question of the appropriateness of a 'revolutionary modern architecture' as opposed to an architectural evolving within a local tradition to satisfy local needs.

(Davis 1975b: 663; see also Figure 7.1)

FIGURE 7.1 Skopje in 1974, 11 years after the earthquake. This indicates the contrast between traditional Macedonian architecture and the rebuilt slab blocks that reflect a desire for reform and modernism (photograph by Ian Davis).

IAN: Was I correct in what I wrote [Davis 1975b]? And is the grandiose transportation centre in the middle of Skopje [Figure 7.2] out of scale and does the ambitious 'City Wall' urban plan by reconstruction director Adolf Ciborowski and his colleagues work?

PROF. POPOVIC LARSEN: I recognise your conclusion very well and it is correct: the rail/coach transport interchange was totally out of scale. A decision to elevate the train station and position it on the first floor was done so that the coach station could be placed on the ground level. The separation in levels meant that train and coach departures/arrivals could run simultaneously and without any interference – a great idea allowing for punctual functioning of a transport interchange. However, lifting the train platforms and trains off the ground meant constructing a forest of massive concrete pillars to support the heavy weight of the trains, to deal with the movement

FIGURE 7.2 Transportation Centre, Skopje, under construction in 1974. It was designed by Kenzo Tange (photograph by Ian Davis).

and vibration-induced effects; and all this in an earthquake zone meant that it was a very expensive solution. It took years to complete the interchange. I remember as a child walking around the unfinished station year after year following minimal progress.

Yet, perhaps for a Japanese or a Western European person it is essential that trains and coaches run on time and delays do not happen. It is not that in Macedonia time does not matter, but this is the Balkans after all and any local architect would have known that perhaps keeping the trains and coaches running on the same level (on the ground) would not have been such a bad solution, even if it caused small delays. Time is measured in days and hours. Minutes and seconds are not a reason for concern.

The 'City Wall' urban plan was inspired by the ruined town wall in the centre of Skopje [see Figure 7.3]. It was a valid inspiration, but one that was difficult to read. The story was told that one could only read the resemblance to the old city wall from a bird or aircraft perspective. But for the time the scale was also wrong. Skopje was a relatively small town with several public buildings in the centre of a larger scale. Most people lived in small houses or apartment blocks up to max. 4–5 floors. The 'City Wall' consisted of towers with 15–16 floors. However the design and layout of the City Wall housing blocks did have some very positive features. The buildings created pedestrian areas, also the internal building planning and the planning of apartments was quite good. Even today the apartments are known for their good layout.

FIGURE 7.3 The concrete towers and slabs represent the 'City Wall' concept of the reconstruction master plan of Skopje in a photograph taken in 1974 (photograph by Ian Davis).

IAN: *Loss of tradition* [Figure 7.1]. Do you have additional overall conclusions about the reconstruction as seen by a resident, and architect, rather than as an external visitor?

PROF. POPOVIC LARSEN: I only know Skopje 'from after the earthquake', but looking at old plans, drawings and photos as well as talking to my family who remember Skopje from before the earthquake brings sadness to me, a sadness felt as an architect and citizen. In my view it is sad that the history and tradition of a beautiful city was completely erased. It was not that the earthquake did that on its own. In my view this could have been achieved if more had been built using local materials, local architects working within local traditions to create something new but something growing out of the locality. I knew no different – it was the only city I knew as a child, but people who lived in the 'old' and 'new' Skopje felt a loss. The modernist architecture of the reconstruction had been changed too fast and was not rooted in anything the general public knew or understood. The new architecture, especially some of the public buildings, had no local references.

IAN: *Long-term impact of temporary housing* [Figure 7.4]. I noted in my paper [Davis, 1975b] that the location of the 'temporary' houses in the suburbs of Skopje, planned for about 9–12 months but remaining to this day, had a negative effect on the future plan of the city. Do you agree – and if so, what was that negative effect?

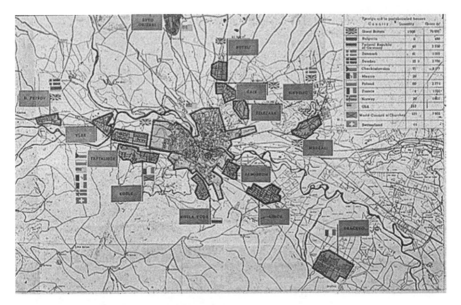

FIGURE 7.4 Map of planned temporary settlements provided by donor governments in 1963. Fourteen thousand houses were donated and many have become permanent, thus influencing the overall layout of Skopje (plan given to Ian Davis in 1974 by the Skopje City Planning Department).

Note: most of the temporary settlements shown on this map were positioned along the River Vardar. The old Skopje extended along the River Vardar and connected to these new 'suburbs'. As a result, the new city forms an extended linear plan about 40 kilometres long.

PROF. POPOVIC LARSEN: In some locations temporary housing of very low quality was erected, but as the government allowed for huge numbers of people to move into Skopje, the need for housing was desperate. As a result, some of these temporary houses were never taken down. Some became slum-like parts; for example, in Suto Orizari, where a large number of Roma (gypsies) still live in them today. [See Figure 7.5; an estimated 25,000–50,000 Roma live in Skopje.] In other parts of the town, as in Vlae, Scandinavian houses were constructed as temporary emergency housing. These houses were of timber frame construction, with good flexible planning. Also the individual house plots were larger. This allowed for extending, upgrading and altering them into permanent housing. This remains to be one of the most sought after, very pleasant and green parts of the city.

IAN: *A child's view of recovery.* What was it like to be a child growing up in a reconstructed city?

PROF. POPOVIC LARSEN: I had a happy childhood. I lived in a part of Skopje (Debar Maalo) that at the time was a protected environment. It was one of the few parts of the town where the old urban fabric with small narrow streets and low housing had survived. Many buildings had collapsed in the

FIGURE 7.5 Roma children playing in the suburb of Suto Orizari, Skopje in 1974. They lived in US Army prefabricated huts, which are present in the background (photograph by Ian Davis).

earthquake, but the new ones had to respect the old scale. Everything was in walking distance: my school, gymnasium, later even the School of Architecture where I studied for my degree. The place was very green and it was like living in a small town.

However, the huge new public buildings constructed almost overnight, without any reference to anything local, also the temporary housing that was not really temporary . . . brought lots of questions about architectural values, and about the meaning of tradition and scale. I felt grateful for all the international aid that was sent to Skopje both in saving lives and also, later, in rebuilding our community. Yet, I wished so much I could go and visit my grandparents' house, or houses in that scale . . . but parts of town were wiped out and replaced with new architecture that at times felt very alien to all of us.

IAN: *Communist vs democratic governance of recovery*. Finally, what was the effect on reconstruction of Skopje being set within a communist country, as compared with its present status in a democratic capitalist state? Did this bring significant advantages?

PROF. POPOVIC LARSEN: This is difficult for me to answer because as you can see I have some criticism [of] how Skopje was rebuilt after the earthquake in 1963. Yet, although at the time it was a communist country with decisions made in a centralised government, there was also some efficiency in it.

After the initial and urgent humanitarian help, with help from foreign experts, a procedure was defined how to assess damaged buildings and decide which ones to 'save' and which ones are beyond repair. I think that the scale of the disaster was huge and this process was efficient. However, when things are done quickly, the likelihood of making a mistake is greater. Maybe some of the public buildings could have been saved. Also the urban planning was changed with no or little respect for the scale of the old city planning.

I strongly believe in democracy and in respect for a multitude of opinions. Since the dissolving of Yugoslavia, the new Macedonian state has been moving towards a democratic capitalist state. Yet when I go back to Macedonia these days, I am astonished by the new architectural developments. The city planning is a development of the existing one, but there is a great trend of building [in] styles that have no relevance to our building culture, traditions and architecture. There are many sculptures in the city that often sit uncomfortably next to each other because they represent very different historical moments, are built in different materials, some are made to look old . . . buildings are re-clad with facades in 'neo styles' that have never been present in these parts of the world. We even have a triumphal arch that is positioned leading on one side to a large public square and on the other side onto a dead-end road. It makes me sad that democracy allows for such architectural developments.

The 'City Wall' and the transport interchange perhaps did not respect or understand locality enough, but the architectural intentions and values in them are clear and honest. There is complete lack of architectural values and clarity in the new Skopje democratic architecture.

The final statement by Olga Popovic Larsen relating to the transition from reconstruction to 'normality' is important to note. She regarded the reconstruction of Skopje, which under a communist government took about 30 years to complete, to be 'clear and honest' in contrast to the ongoing development and expansion of the city under a democratic government that lacks architectural values of 'relevance to our building culture'.

The example of Skopje is a clear attempt by the authorities to manage the recovery with total commitment to reform in the shape of modern architecture for a modern socialist city. There was certainly no dilemma present here as the opportunity to reconstruct with continuity and respect for the rich architectural traditions of the region was probably not even considered and was probably regarded as 'backward'.

There are two main arguments for reform. First, disasters reveal widespread failure that has to be rectified in effective reconstruction to avoid a future repetition of catastrophe. As model 1 (Chapter 3) demonstrates, a return to the status quo is not a viable option. The failures that require reform include inadequate governance, weak planning systems, unsafe buildings, unsuitable infrastructure and failures to implement risk reduction measures. Second, disasters inevitably accelerate the normal process of urban renewal. Worn-out buildings and infrastructure, as well as inadequate planning structures or substandard institutions, can be replaced and upgraded in what amounts to a unique 'window of opportunity'. In Managua, the capital city of Nicaragua, a policy of widening streets to improve safety, to enable safe evacuation of buildings, and to provide access for emergency vehicles virtually paralysed reconstruction planning policies. This standstill lasted for a decade after the 1972 earthquake, and its negative consequences can still be observed more than 40 years later.

Conversely, there are three main arguments in favour of continuity. First, all plans for reform have to compete with another vivid plan. This is memory, embedded in the minds of many citizens, of the pre-disaster settlement layout and of the functions and appearance of buildings. This concern is further discussed in Chapter 6 under the heading '*Genius loci*'. Second, socially acceptable reconstruction must take account of pre-disaster norms or else risk the alienation of residents and communities who need cultural continuity in order to recover. Third, a universal demand exists to reconstruct rapidly so as to minimise social hardship, to facilitate economic recovery and to capitalise on political will to gain access to available funds. But rapid reconstruction implies the minimum adoption of time-consuming changes, including safer street patterns and structural changes in building, etc.

Three strategies enable one to resolve the dilemma. First, accept the need for *both* reform and continuity in recovery planning and management. Second, plan recovery and reconstruction *before* the disaster as an element in pre-disaster planning, thus anticipating the dilemma. Third, define criteria for policies that include both reform and continuity.

Second dilemma: use existing government line ministries vs create new organisations to plan and manage the recovery

This dilemma is discussed in detail in relation to model 20, 'organisational frameworks of government for recovery management', in Chapter 4 where we discuss the competing arguments and suggested resolution.

Third dilemma: reconstructing existing unsafe settlements vs relocation to safer sites

As noted by Jha *et al.* (2010: 2):

> Relocation of affected communities should be avoided unless it is the only feasible approach to disaster risk management. If relocation is unavoidable, it should be kept to a minimum, affected communities should be involved in site selection, and sufficient budgetary support should be provided over a sufficient period of time to mitigate all social and economic impacts.

Many attempts have been made on safety grounds to relocate complete settlements after disasters, but there are places in which new settlements were created that failed to replace the original unsafe towns, as the following two examples show. In the first example, an attempt was made in Belize, South America after 1967 to relocate the exposed coastal settlement of Belize City to higher ground at Belmopan, 82 kilometres away. This was on account of repeated past hurricanes and, in particular, severe storm and surge flooding damage sustained when, in 1961, Hurricane Hattie destroyed 75 per cent of the dwellings. Eventually, the parliament buildings, government ministries and embassies moved, but commerce, officials and most residents preferred to remain in Belize City. Both places now exist and prosper in parallel. The population of Belize City is about 71,000, while Belmopan has about 14,000 residents.

The second example concerns the town of Gediz which, from 1970, relocated to New Gediz in western Turkey. Gediz has suffered a long history of earthquakes, among which damaging events occurred in 1886, 1896, 1944 and 1970. The last of these was a magnitude 7.2 earthquake that resulted in 1,086 deaths, 1,260 injuries and the destruction of 9,452 dwellings. Shortly after the disaster, the government passed a resolution to rebuild Gediz on a new, safe site, and reconstruction took place 7 kilometres from the original site at a location called Yeni Gediz (New Gediz). The new site may not have been any safer than the original site, and Professor Nicholas Ambraseys – a seismologist with extensive experience of Turkish earthquakes – declared in a briefing to students that the original site of Gediz was not set on any earthquake fault but the new site certainly was. However, as the earthquake affected an area of 13,000 square kilometres, the distance to the fault line is not the only relevant issue. The decision to relocate could have been based on many factors, such as soil conditions, landownership patterns, expropriation of land, politics or ignorance, and not merely on the grounds of safety. A further factor could have been the fire that broke out after the earthquake in an area of wooden houses, contributing to damage in the Kayalar area of Old Gediz – the worst damaged part. If the intention of the relocation was to replace the old settlement, this did not succeed as, in a similar manner to Belmopan, both old and new Gediz coexist side by side. It appears that those who left after the earthquake for Yeni Gediz and those who remained are both satisfied with their towns.[2]

Wherever global pressures of urbanisation and population growth are present, it is clear that new towns will always be needed to accommodate these expansions. Therefore, relocated towns, such as Belmopan and New Gediz, will probably eventually take root and evolve into viable towns. However, even if such relocated towns are built on safe sites, the decision to relocate will not have solved the residual problem of existing unsafe settlements as they continue to exist and expand.

Two further examples relate to situations in which the pressure for relocation came directly from the disaster survivors. The first of these concerns the Latur earthquake that struck Maharashtra State in Western India on 30 September 1993 (see the Malkondji case study in Chapter 1), after which 52 villages in the Kilari region of Latur, India relocated approximately 10 kilometres away from their original sites. In total, 35 villages were demolished in the earthquake, almost 8,000 people died and 16,000 were injured. In the early weeks after the disaster, influenced by villagers concerns, a decision was made by the Government of Maharashtra to relocate all the villages in the Kilari area that suffered more than 70 per cent damage. These villages had been built on what was locally called 'black cotton soil', which was about two metres deep. Survivors believed that this soil explained the heavy damage in this area. Their discontent was additionally fuelled by the fact that these areas had been cremation sites and burial grounds, and they had no wish to remain with such cultural and religious associations in addition to the lack of safety.

The authorities were attracted to the relocation option because of the sheer volume of earthquake debris that filled the destroyed villages; they believed that it would not be possible economically to rebuild rapidly on the destroyed sites given the problem of clearing away the debris. They were advised by their engineers that the soft cotton soil would require expensive deep strip foundations or piles to support future dwellings. A further difficulty involved getting all participants to agree on the demolition of commonly owned walls and re-establishing property boundaries.

A visiting team of experts from the Earthquake Engineering Research Institute (EERI) in California concluded as follows:

> popular sentiment considered relocation an opportunity to provide earthquake victims with well-planned and neatly laid out new villages at new sites without any segregated compartments for different castes and communities. Villagers, reinforced by prominent social science institutions, pleaded for relocation. The government responded politically to such strong sentiment by agreeing to the relocation. More than 27,000 houses were ultimately relocated in over 52 villages. The new villages were located in close proximity to the old villages.
>
> *(Greene* et al. *2000: 3)*

In the next example, the town of Beichuan in Sichuan Province, China relocated 24 kilometres away to New Beichuan over the period 2008–13. This major relocation of a large town received widespread support from the surviving community. It is discussed in detail in Chapter 3, 'Government relocation of communities and settlements'.

Unlike the examples of major relocation of towns discussed above, there are examples in which *localised sections* of settlements have been successfully relocated due to landslide or flood risks. For example, land where settlements once stood may have been eroded as a result of the impacts of floods or coastal erosion and, therefore, there may be no alternative but to relocate to a new site (Jha et al. 2010: 77–86).

There are two main arguments for this sort of relocation. First, the small-scale relocation of highly vulnerable local sections of a town or settlement may be necessary to avoid replicating the risks that caused the destruction. Such relocations must be as close as possible to the pre-disaster location. Second, where there is massive loss of life and where bodies remain in the ruins, which become an impromptu cemetery, the survivors have a strong reason for wanting to relocate that authorities will be wise to respect. This is what occurred in Armero, Colombia after the lahar (volcanic mudflow) of 1985; in Kilari, India (see previous page); and in Bechuan, China (see Chapter 3, 'Government relocation of communities and settlements').

In contrast, two arguments militate against relocation. First, as a guiding principle, whenever possible, resettlement of disaster-affected communities should be avoided for social, cultural, economic and environmental reasons. Second, without social and environmental impact assessments of new sites and the involvement of the affected communities, people may abandon the new sites because they do not support their socio-economic and cultural needs.

Four solutions may help resolve this dilemma. First, social and environmental assessments should be conducted as a prelude to recovery and reconstruction planning. Second, all assessments should be participatory. Third, one should recognise the significance of various hazards to local relocation, and these should be investigated thoroughly and, if necessary, mitigated before the relocation takes place. Finally, where relocation has taken place, it is essential that authorities maintain services, including disaster preparedness plans and provision for residents who decide not to relocate.

Fourth dilemma: speed of reconstruction vs vital requirements in reconstruction planning

W. S. Morrison was first Minister of Planning in the UK Government, in which capacity he was charged with responsibility for post-war reconstruction. In 1943, he argued as follows:

> There are some problems of Government in which speed of decision is the great thing, in which it is essential that some decision, even though it be not the ideal decision, should be taken quickly ... you do well to ask yourself two questions. First, is the damage that would be done by some delay in reaching a decision more serious than the damage that a wrong decision would entail? Secondly, is the material that is the subject of your deliberation such that a decision found to be defective in practice can be readily amended?
>
> *(cited in Madge 1945: 43)*

The revision of building codes does not always have a positive impact upon the recovery process. Professor Ken Hewitt, a leading geographer who specialises in disaster studies, visited the Greek Island of Kefalonia in the late 1970s. This had suffered a devastating earthquake on 12 August 1953. He met the mayor and asked about the overall recovery following the disaster. The mayor recounted how, following the disaster, the Greek Government decided to require that every building was reconstructed to satisfy stringent seismic safety standards. The consequence was a series of protracted delays while the codes were revised and earthquake-resistant plans were formulated. The survivors, who were living in miserable emergency accommodation, then decided virtually *en masse* that the extended delays were intolerable, so they emigrated to Montreal. Thus the safe houses, when completed, were occupied by families who were not survivors of the disaster.[3]

In such contexts, one argument for rapid recovery is that many stakeholders will exert pressure for speedy action. These include politicians, contractors, beneficiaries and international agencies. Many fear that political will, and consequent funding allocations, will evaporate rapidly if there are delays in reconstruction (see 'Model 7: cost-effectiveness'). Conversely, an argument against rapid recovery is that adequate time is needed in order to revise building codes, set up the means of ensuring safety, legislate in favour of improved land use planning controls, and train professionals and builders. Adequate time is also needed to ensure that participation is responsible, the quality of reconstruction is controlled and planning is carried out with care.

To resolve this dilemma, there are four possible approaches. First, one should recognise the reasons for satisfying the demands of both sides of the debate. Second, it may be necessary to explain to all the stakeholders who are pushing for rapid reconstruction (especially the beneficiaries) that severe costs are associated with overhasty reconstruction. Sophisticated decision-making tools will be needed in order to cope with these conflicting pressures. Third, where possible, and where reconstruction has been pre-planned before the disaster, one should adopt techniques that help meet the demands of safety, quality control, participation and planning. Fourth, in order to accelerate recovery, 'action planning' techniques will enable decision-making to occur while the planning design process goes on in parallel.

Summary recommendation for resolving the dilemmas

In recovery management, these four dilemmas occur again and again. But, as we note, their resolution is rarely a matter of 'black or white' alternatives. Rather, solutions that are viewed according to shades of grey, involving balanced compromises, are more appropriate and perhaps more likely. There are many other dilemmas and some that specifically relate to the shelter and housing concerns discussed in Chapter 8. When such complex and finely balanced dilemmas surface, it may help to undertake the following actions. First, conduct a SWOT analysis of each alternative option (see model 18 in Chapter 4). This may help broaden the picture of the situation and suggest how to resolve the dilemmas. Second, if the two parties to a dilemma are maintaining deeply held positions, plan a course of action to address the issue and bring both parties together to seek a measured consensus. Explain that it is imperative in reconstruction to minimise conflict and seek a collective commitment to the noble cause of rebuilding lives, livelihoods and property.

In the next two chapters, we focus our attention on sheltering and housing, which are perhaps the most visible and problematic aspects of recovery planning.

Notes

1   Olga Popovic Larsen: personal communication to Ian Davis, 24 October 2014.
2   Communication between Ian Davis and Yasemin Aysan, 15 February 2014.
3   Kenneth Hewitt, personal communication, 2006.

8

LESSONS RELATING TO SHELTERING AND HOUSING

HERR report on shelter

> Providing adequate shelter is one of the most intractable problems in
> international humanitarian response. Tents are too costly and do not last
> long enough. Plastic sheeting can be good but most often is low quality
> and falls apart immediately. Rebuilding houses takes years, even when land
> issues are not major obstacles.
>
> *(Ashdown 2011: 25)*

This quotation is a particularly dismal conclusion from the *Humanitarian
Emergency Response Review* (HERR), an authoritative report by a distinguished
international body convened by the UK Government in 2011 to examine the
state of international humanitarian assistance.

Why is sheltering regarded as a problematic element in humanitarian
assistance? Ann-Margaret Esnard, Alka Sapat and their colleagues have conducted
a detailed international research project on what they term 'population
displacement'. In one chapter of their outstanding study, they agree with the
conclusion of the HERR report, believing that post-disaster housing:

> remains an area riddled with problems, both in the United States and in
> other countries. While Hurricane Katrina put the issue of disaster housing
> policies into the national limelight, post-disaster housing has been a
> lingering problem for decades.
>
> *(Esnard and Sapat 2014: 120)*

They discuss a range of problems that include official neglect, 'messy policies',
the gap between survivor expectations and bureaucratic procedures, lack of

advocacy, and neglect of renters. They conclude that a way forward is to take a holistic approach to shelter that will 'yield more dividends than piecemeal attempts to find solutions' (Esnard and Sapat 2014: 139). We attempt to make the same point concerning the need for integrated recovery in model 2 (the 'recovery sectors' model in Chapter 3).

The commonly used acronym SAD (shelter after disaster) has proved to be rather apt. One further explanation for the sadness in the overall state of the shelter sector, with its entrenched problems, may result from a fundamental mismatch. This is the gaping hole that exists between official apathy and neglect of shelter and housing and their vital importance. The priority need for strong shelter policies and practice was described in the HERR report as being: 'critical to health, employment, family and safety. Without adequate shelter, in all but the most benign climates people are terribly vulnerable' (Ashdown 2011: 25).

We fully agree about the importance of shelter and its implications for related sectors (as noted in model 2) and we recognise this official and unofficial neglect; but we question whether shelter is such an 'intractable problem'. We are fully aware of weaknesses in provision and problem areas, but we are also aware of examples of highly successful shelter and housing provision and agency programmes that may have eluded the attention of the HERR commission. We opened Chapter 1 with an example of effective housing reconstruction in India, which is not unique, and in Chapter 12 ('Yellow hat – optimism'), we describe an outstanding example from Pakistan. Perhaps the failures noted in the HERR report were more directed at the immediate shelter or transitional phase. Other examples of success and failure are described throughout the options presented in the next chapter.

In this chapter, we aim to set the scene for this critical part of disaster recovery by undertaking some stocktaking. We decided to start by reflecting on the changes that have occurred since the early 1970s when Ian began to study the sheltering theme. We review five comparisons of changes that have affected the shelter sector. This is followed by a discussion concerning eight common dilemmas that face officials who work in this field.

Shelter and housing lessons: 1972–2015

In 1972, Ian embarked on a study of shelter following disaster. This was initially PhD research (see Davis 1985) but it later developed into the first book on the subject – *Shelter after Disaster* (Davis 1978) – and later into his role as editor of the first UN guidelines in shelter, published in 1982 and currently refreshed for publication by IFRC as the second edition (UNDRO 1982; Davis 2015).

This close involvement in shelter and post-disaster housing reconstruction has been continuous, with Ian's view of the subject being mainly from an architectural and physical planning standpoint (Davis 2011b). In addition, David has also been associated with shelter and reconstruction for more than three decades – though seen from his complementary geographical standpoint (Alexander 1984).

Therefore, this chapter provides a unique opportunity for us to jointly reflect on our direct personal experiences of these subjects.

The issues set out in this chapter have been constantly recurring throughout our careers as we have thought about these matters, conducted research, lectured, advised on policies, debated them with colleagues and continually written about them in a steady stream of publications (Aysan *et al.* 1995; Aysan and Davis 1992; Aysan and Oliver 1987; Davis 1975a, 1975b, 1977a, 1977b, 1978, 1981, 1985, 2011a, 2011b, 2014a, 2014b; UNOCHA 2006; UNDRO 1982; Charlesworth 2014; Alexander 1984, 1989; Alexander *et al.* 2007).

As well as drawing on each other's knowledge, our starting point has been to apply our own experiences to each option. Although many of the events described in this and the next chapter occurred decades ago, we recall them vividly and accurately as we documented our experiences at the time. These encounters have often been the formative elements in shaping our views on recovery, and we are able to return to our extensive writing about these issues. These direct experiences far outclass any lessons we have gained from reading reports, from the experiences of colleagues, or from second- or third-hand conference case studies.

Varied hats, varied perspectives

To avoid our reflections having a detached academic bias, sometimes remote from the complexities that face practitioners on the ground, our views also derive from wearing hats other than those of lecturers and researchers. These have been as directors of NGOs working to respond to shelter and housing needs as well as consultants to governments and international agencies; and briefly we have both been disaster survivors from war bombing or from being in caught up in a major earthquake, as we describe in Chapter 9. The issue of different 'hats' leading to different viewpoints is discussed using Edward de Bono's 'six hats' model (model 19 in Chapter 4) as the theme for Chapter 12.

A further influence on the perceptions we express in this book has come from the variety of countries in which we have lived or worked: Britain, the USA, Canada, Italy, Switzerland, Sweden, Denmark and Japan. Each location, with its distinct culture and political-administrative framework, has required us to wear new hats, bringing an enriched set of perspectives and understandings concerning disasters and recovery processes.

Revisiting *Shelter after Disaster* (1978)

In preparing this chapter, we have been rereading *Shelter after Disaster* to see what has changed in the intervening 37 years since it was published. This was the first book that Ian wrote and, as noted above, the first book on the subject (Davis 1978).

Perhaps this initial lack of competition may explain why this modest book has latterly been described in inflated terms (Charlesworth 2014: see p. 5 and p. 242).

A leading American architectural critic, Witold Rybczynski (2005), also generously referred to the book as a 'classic' when reviewing emergency housing following Hurricane Katrina in the light of the policy suggestions contained in *Shelter after Disaster*.

First comparison between 1978 and 2015: the expanding shelter and housing problem

Population growth: The most obvious fact to notice in revisiting this book is that there have been dramatic global changes during the period 1978–2015. One aspect has been escalating population growth. In 1978, the world population was 4.3 billion, but this had reached 7.2 billion by 2014. This has clearly been a major factor in the growth of vulnerability and in the consequent shelter challenge. For example, the Guatemala earthquake, which occurred two years prior to publication, tends to loom large in *Shelter after Disaster*. That disaster resulted in 22,000 deaths. In sharp contrast, this book is being written four years after the Haiti earthquake of 2010, in which an estimated 220,000 deaths occurred (i.e. ten fatalities for every death that occurred in Guatemala).

Disaster statistics – rising damage, declining casualties: The EM-DAT database of the Centre for Research on the Epidemiology of Disasters records that 137 natural disasters were reported globally in 1978, with 38,096 deaths and damage that amounted to US$6.2 million. In 2013, 184 natural disasters were reported globally, causing 10,192 deaths and creating damage of US$61.8 million – a tenfold increase in damage, but not in casualties. Lloyd's of London stated in May 2014 that damage and weather-related losses around the world have increased from an annual average of US$59 billion in the 1980s to an annual average of almost US$200 billion in the past ten years. They report that the year 2011 was the most expensive on record for natural disasters, with insured losses costing the industry US$126 billion. In 2012, a single event, Superstorm Sandy, caused US$35 billion of insured losses, making it the second costliest hurricane in US history, after Hurricane Katrina (Kollewe 2014). Figure 8.1 illustrates the escalating patterns of damage from 1975 to 2013.

Aside from the massive loss of lives in mega-disasters such as the South Asia tsunami of 2004 or the Haiti earthquake of 2010, there has been a gradual decline in disaster deaths. This is probably due to the impact of improving economies in Asian and Latin American countries as well as improved disaster preparedness, such as effective warning systems.

Oxfam's expanding involvement in disasters:[1] We have chosen to highlight contrasts within Oxfam over this time period in its response to the growth of disasters and given the expansion of this agency as the largest development agency in the UK. In 1978, Oxfam did not possess an emergency department. Two or three people worked in the head office to handle disaster response. In addition, there was a manager in charge of stores of emergency supplies. There was also a small health unit, an industries officer and an engineer working in the

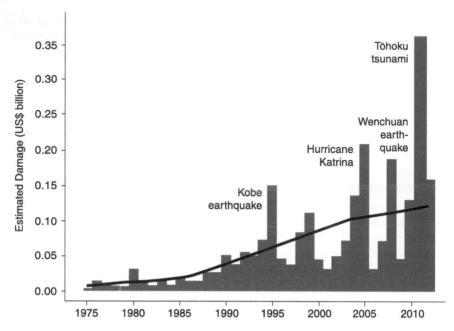

FIGURE 8.1 Estimated damage in US dollars (billions) caused by natural disasters, 1975–2012 (data provided by the Centre for Research on the Epidemiology of Disasters, courtesy of Debarati Guha-Sapir and Philippe Hoyois).

education department. That comprised the sum total of the agency's disaster response capacity. Oxfam's Technical Unit, to become so important in developing their work on water and sanitation capacity, was not formed until 1984.

In 2015, 37 years later, Oxfam's humanitarian capacity is rather different. Ninety-three staff work in their Humanitarian Department within Oxfam's head office, which embraces health and technical units. They also have a logistics team responsible for the stores that are reserved for emergencies. Additional staff work overseas as 'emergency support professionals'. The Humanitarian Department is now part of Oxfam International, integrating the work of five key large Oxfams: UK, America, Australia, Holland and Spain. Whenever a large-scale disaster occurs, all five Oxfams are involved, and Oxfam UK takes the lead in major emergencies.

Shelter and housing experience: A survey carried out in 1976 just after the Guatemala earthquake indicated that out of the 40 international agencies involved in reconstruction, only 5 had prior housing experience in Guatemala. Of the remaining 35 agencies, only 7 had staff with low-cost housing experience (Thompson and Thompson 1976). Despite the exponential growth in disaster incidence and organisational response in the past 37 years, this neglect of shelter and housing by the international community has hardly changed. Graham Saunders, Head of Shelter and Settlements at the International Federation of Red Cross and Red Crescent Societies (IFRC), stated that the sector has failed to develop because the institutions have not advanced their own understanding of the subject despite progress at field level.[2] Thus, out of 522 national and international signatories (as of March 2014) of the Code of Conduct in Disaster Relief prepared by the International Red Cross and Red Crescent movement, a mere dozen employ full-time expert advisors with experience in shelter or housing reconstruction. This official neglect is also discussed in Chapter 6, 'Accountability in the sheltering and housing sector'.

The gap in professionalism: There is no escaping the urgency of the need to tackle the problems of administrative incompetence in the shelter, housing and recovery sectors, as is discussed in the section on 'Accountability' in Chapter 6. We are tempted to cite examples of major housing reconstruction projects undertaken by agencies without a track record in reconstruction yet with sublime confidence by their directing staff, who have no qualifications whatsoever in construction management, architecture, planning, urban design and engineering, and who have no field experience upon which to draw. This concern poses a complex dilemma that is discussed later in this chapter in 'Eighth dilemma: professional vs lay knowledge and expertise'.

Expanding levels of interest: By far the most significant contrast between *Shelter after Disaster* and the opportunity that this book provides to review the issues three decades later is one of expanding disasters and a consequent shelter and housing problem that has expanded vastly and produced a growing need to face the challenge. Esther Charlesworth's groundbreaking study *Humanitarian Architecture: Fifteen Stories of Architects Working after Natural Disasters* provides encouraging evidence of the depth, scale and variety of concern from the architectural profession in 2014 (Charlesworth 2014).

In 1978, when shelter after disaster was a mere backwater concern of a few dozen individuals, it would have been totally inconceivable to have imagined the level of international attention, let alone recognition, of the subject. In 2014, one of the 15 architects highlighted in Esther's book – Shigeru Ban, the Japanese architect of post-disaster structures, including shelters in Japan, Rwanda, Taiwan, Haiti, Sri Lanka and New Zealand – was awarded the 2014 Pritzker Prize, the world's most prestigious architectural prize in global architecture. Ban made imaginative use of redundant shipping containers as temporary housing for survivors of the Tōhoku earthquake and tsunami in Japan.

Second comparison between 1978 and 2015: changes in vulnerability

In the third and fourth chapters, we present 21 models of disaster recovery. One of these – model 10, the 'disaster crunch model' (see Chapter 4) – began its life on page 3 of *Shelter after Disaster* as a simplified diagram, as shown in Figure 8.2. This was originally included in the text to challenge an assumption that prevailed widely in the 1970s that disasters were caused by earthquakes or floods or storms, without recognition of the significance of the vulnerability of dangerous conditions.

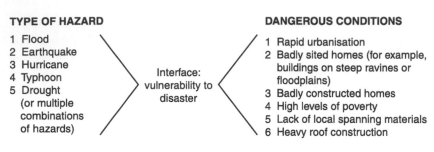

TYPE OF HAZARD

1 Flood
2 Earthquake
3 Hurricane
4 Typhoon
5 Drought
(or multiple
combinations
of hazards)

Interface:
vulnerability to
disaster

DANGEROUS CONDITIONS

1 Rapid urbanisation
2 Badly sited homes (for example,
   buildings on steep ravines or
   floodplains)
3 Badly constructed homes
4 High levels of poverty
5 Lack of local spanning materials
6 Heavy roof construction

FIGURE 8.2 The 'pressure and release' model (Davis 1978: 3).

It took 36 years of experience of vulnerability as well as the contributions of colleagues for the above conceptual diagram to evolve into what is often called the 'crunch' model in the form of model 10. This model has been applied to varied disaster risk contexts in a study of disaster risk management in Asia and the Pacific (Davis 2014b).

Poverty reduction: Since there is a strong correlation between vulnerability to disaster impact and poverty levels, it is worth noting the significant decrease in global poverty levels from 1978 to 2015. The Millennium Development Goals, established by the world's governments in 1990, set the target of halving the proportion of the global population whose income is less than US$1.25 per day by the year 2015. By 2010, five years ahead of schedule, 700 million fewer people were living in conditions of extreme poverty than in 1990 (UNDP 2010). Elliot (2015) reports a decrease in the proportion of the global population living in extreme poverty from over a third in 1990 to 14.5 per cent in 2011, equivalent to around 910 million people having risen out of extreme poverty.

Rising inequalities: During this period, there was a dramatic increase in inequality. In 2015, OXFAM published research findings which indicated that the monetary wealth of the 80 richest people in the world doubled between 2009 and 2014. In a deeply sobering and disturbing statistic, these 80 people – who could easily fit into an average-size village hall – own the same amount of money as 3.5 billion of the world's poorest citizens – just under half the population of our planet (Elliot and Pilkington 2015). The first implication for disaster recovery is that decreasing poverty would probably result in stronger dwellings and improved disaster protection. In 2010, within weeks, earthquakes occurred in Haiti (an exceedingly poor country) and Chile (a high-income economy) with inevitably differing consequences, as indicated in the stark comparison shown in Table 8.1.

The second implication is that wide inequality can produce a highly dangerous situation. It must be regarded as a sign of susceptibility to disaster in which conflict can break out as poor, highly disadvantaged people migrate to wealthier countries in large numbers or use violence to secure the assets of the wealthy.

The third implication for disaster recovery is the need for wealthy elites in any country recovering from disaster to use their vast resources to support the

TABLE 8.1 Comparison between the Haiti and Chile earthquakes of 2010

	Haiti earthquake 12 January 2010	Chile earthquake 27 February 2010
Magnitude	7.0	8.2
Gross National Income (GNI) (2010)	1,490	17,010
Deaths	222,000–316,000	526
Damage value	US$12 billion	US$30 billion

recovery enterprise. This requires them to pay fair taxes to strengthen weak government treasuries, and to support philanthropic efforts without paternalism. For wealthy national and international corporate bodies, this redistribution of wealth could be provided through corporate social responsibility (CSR) measures.

Third comparison between 1978 and 2015: evolving issues

The next chapter contains a discussion of options that relate to shelter and housing. In *Shelter after Disaster* (1978), these were all covered with the exception of the following aspects:

Relevant to option 2 – the use of evacuation buildings before *impending storms, flooding and volcanic threat:* This has no mention in the 1978 book.

Relevant to option 5a – support for the spontaneous provision of shelter by survivors: There is no mention of the use of cellular telephones by survivors to make contact with their distant relatives to request financial assistance for the obvious reasons that such technology did not exist in 1978 and rapid money exchange facilities had yet to arrive on the scene.

Relevant to option 4 – repair options: This important aspect of effective recovery is inexplicably omitted from the text, although it certainly took place in all recovery situations in the 1970s.

Relevant to options 1 to 8 – the use of information technology: Perhaps this is the most obvious omission from *Shelter after Disaster*. There is, of course, no reference in the book to the information technology revolution that followed the invention of the Internet, and its widespread use from the mid 1990s onwards. It is impossible to determine the overall impact of IT in developing knowledge of the shelter and housing sectors, or whether we would have arrived at the present point without the advance in cheap or free access to knowledge. Every option in this chapter has been advanced due to IT developments, but as with most areas of rapid technological change, this may have been achieved at a cost. This could be the decline in direct human contacts or public discourse and the need for officials personally to record their own experiences, file them and share them with their colleagues as they did in the pre-Internet era. For example, at the end of *Shelter after Disaster*, 12 pages are devoted to the detailed description of shelter and reconstruction policies and provision in six disaster case studies (Davis 1978: 100–11). These were reused in the UN guidelines (UNDRO 1982). Ian vividly recalls the effort that went into the assembly of these cases. It involved copious correspondence, expensive telephone calls, telex messaging, detailed library searches, interviews with officials and foreign travel to gather information.

Now in 2015, it is probable that all six case studies could be assembled in a single morning using the magic of Internet search engines and Skype communications. But alas, as with all progress, there is a price tag. The lengthy investigation process in the late 1970s yielded numerous by-products in the form of establishing some lasting relationships and professional networks, as well as developing research skills. For the authors, these gains were probably far more important than were the actual data collected. This raises the question for all involved in education and training concerning the urgent need in an age of rapid change and instant communication to recover the *unchanged* need for direct observation, personal documentation and human interaction.

Fourth comparison between 1978 and 2015: who is concerned?

Ian expressed an ethical concern in 1978, noting that the 'seedier' side of the problem of shelter is a battleground between conflicting priorities:

> I have found, in making detailed investigations of certain products, that their existence as disaster relief is not a response to the needs of the particular disaster. Rather it the result of an individual's whim, or sales pressure, or aid quotas from a donor agency. When one questions the logic with local or international officials the indignant shrug is a good indication of the level of concern for the recipients of misplaced aid.
>
> *(Davis 1978: 91)*

As disaster response in all its forms has become an exceedingly large business, one that involves millions of dollars, it is inevitable that the commercial world of finance, insurance, consultancies, contracting and manufacture will be heavily involved. But it is *not* inevitable that this opportunity will result in cynical exploitation of both disaster situations and a distressed population purely for commercial gain (US Chamber of Commerce 2012).

Fifth comparison between 1978 and 2015: the impact of early decisions

A further concern that Ian expressed in *Shelter after Disaster* considered the significance of immediate decisions in the 'long-haul of recovery' that we referred to in the preface to our book:

> In providing this protection (shelter or dwelling), any donor must be aware of the long-term consequences of his actions. The future form of a new

settlement and even the long-term economic development of a community is being determined in the early decisions that are made while flood waters are receding, or dust still hangs in the air.

(Davis 1978: 91)

This issue remains and has grown in scale. Typical examples include the siting of temporary housing that may well become permanent and have a major negative effect on future reconstruction plans. Another major change that can have a positive effect on communities concerns the way assistance is dispensed to survivors. This concerns the advantages of dispensing cash grants as compared to in-kind contributions. The former enable recipients to engage with formal banking, as discussed in Chapter 12 in the context of the Pakistan rural housing reconstruction programme.

Eight dilemmas in the shelter and housing sector

There are many dilemmas in disaster recovery management, and four of the most significant are discussed in Chapter 7. The following specifically relate to shelter and housing

First dilemma: planning vs plans

The most important observation one can make about planning, about sheltering, and about housing is that the process is usually more important than the physical outcome in the form of plans, shelters and houses. This is because planning is an approximate process that deals with developments in the future that cannot be known perfectly. Eisenhower's famous words, which may have first been stated by Napoleon, underline the issue: 'I tell this story to illustrate the truth of the statement I heard long ago in the Army: plans are worthless, but planning is everything' (Eisenhower 1957: 818).

The planning process enables the planner to find out many things about the activities to be planned in order to regulate them. Planning needs to be a participatory process that is considered to be the common property of all individuals and institutions that have a stake in it. It is usually a social process as well, in which consultation is the only way to ensure that plans are accepted, utilised and adhered to. A good plan thus represents a consensus among the interested parties, who are all aware of its provisions and their roles in its enactment.

Shelter planning relies on estimates of the residual number of homeless people after a disaster, and after those who can find alternative accommodation have been subtracted from the total. In theory, this can be carried out before disaster strikes by developing a clear understanding of the vulnerability of housing and businesses in relation to a given hazard and a particular level of impact. To meet the need to pre-plan recovery actions, the American Planning Association has

FIGURE 8.3 Temporary housing in Sendai, Japan for survivors of the 2011 Tōhoku earthquake and tsunami. The design and siting of this housing were pre-planned three years before the disaster (photograph by David Alexander).

developed an important publication: *Planning for Post-Disaster Recovery: Next Generation* (Schwab 2014).

In practice, few authorities have the foresight and resources to do this. Nevertheless, an example was cited as long ago as 1980 by Harold Foster, writing in his pioneering book *Disaster Planning* about the foresight of the city of Victoria, British Columbia in preparing recovery plans before disaster (Foster 1980).

More recently, the emergency planners in Sendai, Japan developed shelter and temporary housing plans (including the siting of temporary settlements and preselection of prefabricated house types) in anticipation of an earthquake a full eight years before the 2011 Tōhoku earthquake and tsunami (Figure 8.3). This foresight enabled the authorities to act swiftly in providing accommodation to survivors who had lost their dwellings.

Second dilemma: process vs product

At the conclusion of *Shelter after Disaster*, the process and product aim of emergency shelter was restated: 'to provide protection for a vulnerable family. It may take the form of a product, or it may be a process. It could start as a sheet of corrugated iron, which could ultimately become the roof of a house' (Davis 1978: 91). In 2008, Jim Kennedy and his colleagues examined post-tsunami reconstruction policies in Aceh and Sri Lanka. They referred to the 'product' vs 'process' dilemma that was raised in *Shelter after Disaster:*

> Of particular importance is Davis' . . . comment that 'shelter must be considered as a process, not as an object'. . . . Especially in a post-disaster context, shelter must be considered to be a series of actions for fulfilling certain needs rather than as objects only such as tents or buildings.
>
> *(Kennedy* et al. *2008: 25)*

The authors listed four needs: health, including protection from the elements, privacy and dignity for families and for the community, physical and psychological security and livelihood support. But they drew attention to the reality that 30 years after Ian originally espoused the notion of shelter being an integrated process, it is rarely implemented in the field. Instead, transitional settlement and shelter are often considered to be 'part of non-food item distribution rather than an ongoing exercise in supporting livelihoods, health and security needs' (Kennedy *et al.* 2008: 25). Rather than a lesson learned, it is still one to be relearned and implemented.

Third dilemma: supply vs demand

In any major disaster there is a convergence of interested parties. Some are local, some national, while others arrive from international directions depending on the scale of the event and the geographical proximity to wealthy industrialised countries such as the USA, or European countries. Some come with a genuine desire to provide support, others to file news reports or conduct research, and many other consultants, professionals and private contractors arrive to secure work, gain experience or sell their products or services. Although 'needs assessments' are conducted by agencies and government officials, thus opening the way for the demands of survivors to be expressed, the reality is that almost all commercial bodies and many NGOs have strong predetermined ideas about what the survivors and the situation need. These assumptions of what to deliver may be based on acquired past experience or may be in accordance with guidelines, manuals or humanitarian standards.

Does effective shelter and housing require a *supply-* or a *demand-*driven approach? As professionally trained designers have no monopoly on designing or creating, what can designers (or facilitators) usefully do to support the shelter, safe building and reconstruction needs of exceedingly poor, marginalised and vulnerable families? How can their latent design and construction skills and capacities be unlocked for the common good? We hope that as we explore the varied options, the material in this chapter will provide some guidance on this critical question (Davis 2015).

The *demand-driven approach* will not be applicable to middle-class families in industrialised countries where there is no possibility of owners building their own homes. But the underlying philosophy can be applied with concentrated desire by assisting groups to support survivors' expressed needs rather than make paternalistic choices on their behalf.

Fourth dilemma: universal shelter standards vs national standards

The most significant standards for shelter are described in the Sphere Standards. The minimum standards for shelter, settlement and non-food items are 'a practical expression of the shared beliefs and commitments of humanitarian agencies and the common principles, rights and duties governing humanitarian action that are set out in the Humanitarian Charter' (Sphere Project 2011: 243). These are 'qualitative in nature and specify the minimum levels to be attained in humanitarian response regarding the provision of shelter' (Sphere Project 2011: 240).

In the early 1990s, an international disaster management course took place at Cranfield University in the UK (both Ian and David later participated in these courses as lecturers). In one session of the course, the organisers invited a speaker from Sphere to visit and describe to the participants how the minimum standards would apply. The presentation resulted in a heated exchange between the speaker and participants who were national directors of disaster management in various countries. They wanted to know whether the standards were to apply in their countries. The affirmative answer then provoked the question as to who, within their respective organisations or within other parts of their governments, had approved the standards? The answer was that Sphere had not consulted with national governments as these standards were applicable to the international humanitarian community and they were not directed towards national governments. The national directors were not satisfied with this answer, insisting that it was the height of arrogance for Sphere to propose standards for disaster assistance without first undertaking detailed consultations with the host countries where they might be adopted.

This exchange raises two issues: first, the right (or not) to question the independent manner in which some international NGOs operate with disregard for the responsibility and authority of national governments. Second, given cultural and economic factors, should minimum standards be set internationally or nationally?

There are five standards for shelter and settlement, covering strategic planning, settlement planning, living space, construction and environmental impact (Sphere Project 2011). The undoubted value of the Sphere initiative has been the benchmark it has provided in order to establish minimum standards in the areas of shelter and housing, as well as other sectors of humanitarian assistance. Peter Walker has been one of the strongest advocates for standards. In 1996 he wrote of the need for universal standards for: 'quality of work, . . . needs assessment and distribution systems, management, relations with local authorities, and standards of accountability to donors and victims' (Walker 1996). Standards have had a vital educational value for agencies unfamiliar with shelter or housing, and if Sphere did not exist, there would have been much substandard work that would have resulted in hardship for survivors.

Fifth dilemma: development vs welfare approaches

One of the noticeable improvements in the fields of shelter and housing recovery since 1978 has been the growing awareness by decision-makers that relief approaches can have dismal consequences in terms of the creation of long-term dependency. This change has come about through the decisive influence of bodies such as the IFRC, Oxfam, CARE, UN-Habitat, UNDP and the World Bank, which have attempted to apply a strong developmental approach to all aspects of disaster assistance. There are still agencies and officials with an embedded relief philosophy, particularly those who come at the problem from a civil defence, police and military, or logistical background, who would probably subscribe to the views on the left-hand column of Table 8.2, favouring a welfare approach over a developmental approach.

An Irish architect, Maggie Stephenson has worked with UN-Habitat on a number of disaster recovery programmes: in Sri Lanka after the 2004 tsunami, in Kashmir (Pakistan) after the 2005 earthquake, and in Haiti after the 2010 earthquake. Ian had the privilege of visiting Pakistan and Haiti to conduct evaluations of these recovery programmes (Davis 2011a).

Maggie reflected as follows on the limits of those who seek to provide assistance. She asked: when does relief damage individual dignity and decision-making?

> [N]ot everyone designs or builds hospital, but many people design houses, think about their design, have preferences, dreams, nightmares, probably even build, fix or improve their houses. It is a challenge not only to know what we can *add* to these processes, but *how* to add [to] them and *how* to work with the people involved, deciding *what* the relationships would be.
>
> *(cited in Charlesworth 2014: 221)*

During one of our field study visits, she reminded me that the idea of *choices* facing disaster survivors involve far more than design choices, as they concern how to prioritise the use of money and resolve dilemmas about how extended families subdivide or stay together. Such choices may not be about design criteria but they will significantly affect the outcome. This became a recurring issue that we discussed during these visits to Pakistan and Haiti – how to ensure that disaster survivors, and not just the assisting groups, were able to make choices.

From these discussions, three imperatives emerged. The first is not to experiment on people who have no choice. The second is not to take away the right to choose, which is the last vestige of dignity for people who have very little. Third, the freedom to make choices, to have the opportunity to succeed or

TABLE 8.2 Comparison between a welfare and a developmental approach to shelter and housing provision (Davis 2011b: 210)

Characteristics of a welfare approach to the provision of shelter and housing	Characteristics of a developmental approach to the provision of shelter and housing
Beneficiaries are regarded as *victims*	Beneficiaries are regarded as *survivors*
Projects conceived by relief agencies and government officials who *direct* programmes	Projects conceived by relief agencies and governments who *facilitate* programmes
Key choices concerning shelter and housing are made on behalf of survivors	Key choices concerning shelter and housing are made by survivors
External assessments are made of shelter needs	Survivors assess their own shelter needs
Beneficiaries are regarded as *passive victims* needing support	Beneficiaries are regarded as *active survivors*, able to organize their own sheltering or reconstruction
Assistance is provided in kind, given a lack of trust that survivors will act responsibly if cash support is provided	Cash grants are provided to enable survivors with resources to purchase their own shelter materials, hire builders, etc. (and to secure other essentials), thus demonstrating trust by assisting groups
The overall stance is one of free handouts, or free services and free shelter materials	Wherever possible, survivors pay in cash for their own shelter, thus preserving their dignity
Minimal emphasis on training or capacity development	Maximum emphasis on training and capacity development
Shelter and housing seen in isolation from livelihood generation/recovery	Shelter and housing provision considered in close relation to livelihood generation/ recovery
Provision of shelter and housing largely placed in the hands of contractors	Provision of shelter and housing with user-build approaches
Shelter, transitional shelter, and permanent housing regarded as fragmented products under separate providers	Sheltering and housing regarded as a seamless process that recognises close relationships
Minimal participation of survivors in the shelter, housing reconstruction and recovery processes	The fullest possible participation of survivors in the shelter, housing reconstruction and recovery processes

even to make mistakes is fundamental to human development (Stephenson cited in Charlesworth 2014).

As noted in discussion of the previous dilemma, some of the most serious failures in the provision of shelter and housing derive from a *supply-driven approach* by governments, NGOs, international agencies and the private sector. This could be renamed a 'welfare model'. In this pattern, shelter requirements are assessed by expert 'needs assessors', the local government allocates communities and villages to assisting groups in accordance with their losses, standards of safety and sizes of unit may be issued by the authorities, and then shelter and housing units are provided. The assumption is that passive disaster survivors will be delighted with this 'handout process' and the associated provisions.

As described throughout this chapter, there are also examples of a *demand-driven approach*, and this could be renamed a 'development model' in which the surviving community is empowered so that it is active and plays a leading role in the process. In this pattern, survivors may assess their own shelter needs and some may well decide that they have more important priorities, for which they need financial assistance. They are allocated cash grants or rental subsidies by the authorities and possibly assisted in getting into the local banking system so that their incoming grants pass directly to involved families and thus avoid the financial middlemen, the point at which corruption often blossoms. Then, in close collaboration with assisting bodies in local government, local institutions or NGOs, discussions take place about the design and siting of shelter and housing. The government may play vital roles here in defining quality, space and safety standards. This may be an excellent opportunity for user-build approaches to be adopted, particularly in developing countries. These programmes provide training of craftspeople and builders, project managers, financial bookkeepers and others.

In this manner, there can be excellent integration of the sectors described in model 2 (Chapter 3) as the survivors become active in all aspects of shelter and housing. This can greatly help their psychosocial recovery. They can develop new skills and practise them, thus strengthening livelihoods, and they can plan and build shelters and dwellings as part of physical recovery.

Sixth dilemma: two-stage vs three-stage sheltering

This dilemma is fully discussed in model 16 (Chapter 4) and in Chapter 9 where options 6a and 6b, which look at 'transitional shelter', are discussed.

Seventh dilemma: house vs home

Most languages differentiate between the words 'house' and 'home'. And it is significant that in the language of disaster recovery, reference is often made in loss statistics to the destruction of houses or dwellings but never, in our recollections, to the loss of homes. However, the same literature does refer to 'homeless families' rather than 'houseless families'. Paul Oliver is an anthropologist who has written extensively on the anthropology of shelter (e.g. Aysan and Oliver 1987; Oliver 2006). Perceptively, he commented as follows on the distinction between house and home:

> A town is made of buildings, but a community is made of people; a house is a structure but a home is much more. The distinctions are not trivial, nor

are they sentimental or romantic: they are fundamental to the understanding of the difference between the provision of shelter which serves to protect and the creation of domestic environments that express the deep structures of society.

(Oliver 2006: 192–3)

The loss of a 'home', with all its associations and symbolism, may be irreparable. In 1887 Mark Twain wrote evocatively of such a loss, in a passage that was published posthumously as his autobiography:

A man's house burns down. The smoking wreckage represents only a ruined home that was dear through years of use and pleasant associations. By and by, as the days and weeks go on, first he misses this, then that, then the other thing. And when he casts about for it he finds that it was in that house. Always it is an essential—there was but one of its kind. It cannot be replaced. It was in that house. It is irrevocably lost. It will be years before the tale of lost essentials is complete, and not till then can he truly know the magnitude of his disaster.

(Twain 1924 [1887]: 180)

There is much to be learned about the way that the occupants of temporary shelter personalise their homes. For example, Figure 8.4 shows the personalisation of a prefabricated house in Romagnano al Monte, Southern Italy. It has survived for decades after the 1980 earthquake which provided the reason for supplying it. A different situation is illustrated in Figure 8.5, which shows 'barrack-style' transitional housing as used widely in Asian disasters, in this case the September 2009 earthquake in the Padang area of Sumatra. Here, families were given 10 square metres of internal space and very little furniture. Hence they carried out many of their activities outside their dwellings. Cooking and washing facilities were separate, and some communal space was donated along with the housing. Little scope existed for personalising this, but it was nevertheless done. Figure 8.6 shows a transitional shelter that was donated following the 2010 Haiti earthquake, in which a family of six sleep in a space of 18 square metres. They complained about the lack of space, the lack of security from theft, and the heat from the corrugated iron roofing which they have tried to reduce by hanging sheets below the metal covering. They also found the shelter very noisy when there is heavy rainfall.

Eighth dilemma: professional vs lay knowledge and expertise

In his 1906 play *The Doctor's Dilemma*, George Bernard Shaw inserted a widely quoted and typically provocative observation that 'all professions are conspiracies against the laity'. That quotation relates to a common dilemma that faces authorities in recovery operations. Ian recalls attending a side meeting in 2005 at the

FIGURE 8.4 Personalisation of a prefabricated house in Romagnano al Monte, Campania Region, Southern Italy, photographed in 2011, more than three decades after the 1980 earthquake (photograph by Ian Davis).

FIGURE 8.5 Transitional shelter in Sumatra, Indonesia (photograph by David Alexander).

FIGURE 8.6 Transitional shelter donated by the IFRC in Delmas 9 in Port-au-Prince, Haiti. The photograph was taken in November 2011, 21 months after the earthquake (photograph by Ian Davis).

World Conference on Disaster Reduction concerning ways to assess seismic risks and the creation of safe vernacular buildings: My observation about the chronic shortage of qualified engineers with a commitment to the design of earthquake-resistant vernacular buildings resulted in a comment from Amod Dixit, Executive Director of the National Society for Earthquake Technology (NSET, Nepal). Amod and his colleagues have developed a programme to train local builders, masons and carpenters in constructing safe, well-built structures. He strongly challenged my statement about the shortage of professionals and he stated that 'given the scale of the problem of unsafe settlements and dwellings around the world, the *only* way forward is to de-professionalise the process by training at all levels'.

In the previous chapter, we discuss the lack of accountability in the humanitarian sector, and earlier in the present chapter we discuss the lack of professionalism in the management of shelter and housing reconstruction projects – a context in which the same dilemma surfaces. In the following chapter, we examine eight options for sheltering and housing. Two of them are all about how survivors manage their own recovery and, in the process, create shelters and dwellings that relate to their own cultural values, their own safety and their own job creation ('Option 5a: spontaneous shelter' and 'Option 7a: user-build permanent dwellings'). So the dilemma is posed: how to balance the need for professional direction of shelter and housing projects, with the need for untrained, non-professional builders to take responsibility for the creation of their own shelter, dwellings and settlements?

An excellent example of the effective resolution of this dilemma – where use was made of professional engineering, planning and architectural skills in total harmony with strong participation of an active community and a self-build approach – was shown in the case study of the reconstruction of Malkondji (Chapter 1).

So this is not an 'either/or' dilemma: both approaches are required. There is *always* a vital need for skilled architects, planners and engineers with expertise in safe construction in all categories of building, including low-income vernacular construction. In disaster recovery situations, they have a particularly vital role in training or offering guidance in 'advice clinics' in order to create a multiplier effect. But there is a parallel need for an army of local builders and building craft workers who understand the principles and practice of safe, well-built structures, particularly dwellings.

Shelter preferences and functions

From 1975 to 1982, the Office of the United Nations Disaster Relief Co-ordinator (UNDRO) asked Ian to form and lead a small team of international consultants (Fred Cuny, Frederick Krimgold, Paul Thompson and Aloysius Fernandez) to develop the first set of shelter guidelines for the United Nations (UNDRO 1982). The team conducted research at ten disaster sites in different continents, covering floods, earthquakes and cyclones that had occurred between 1963 and 1978. In order to establish a benchmark for the study, questions were asked in the case studies about the preferences of disaster survivors as well as the functions of shelter. We were told by various officials that these studies were unnecessary as both were obvious and could be assumed, but we thought otherwise. From the answers to our questions, we were able to place the preferences in a ranked order and to list the many shelter functions. These findings are presented below.

Preferences of survivors for their shelter

Survivors show certain distinct preferences for their shelter in the aftermath of disasters. The evidence suggests that their priorities are ordered as follows (UNDRO 1982: 6 and Davis 2015: 46):

1   to remain as close as possible to their damaged or ruined homes and their means of livelihood;

2   to move temporarily into the homes of families or friends;

3   to improvise temporary shelters as close as possible to the site of their ruined homes (these shelters frequently evolve into rebuilt houses);

4   to occupy buildings which have been temporarily requisitioned;
5   to occupy tents erected in, or next to, their ruined homes;
6   to occupy emergency shelters provided by external agencies;
7   to occupy tents on campsites;
8   to be evacuated to distant locations (compulsory evacuation).

Functions of shelter

Emergency shelter has several vital functions. It protects against cold, heat, wind and precipitation. It is a place of storage and protection for belongings. It can help establish territorial claims associated with ownership and occupancy rights. It helps the establishment of a staging point for future action (including salvage, reconstruction and social reorganisation). It provides emotional security and answers the need for privacy. It acts as an address for the receipt of services (medical aid, food distribution, etc.). It provides a base for commuting to sources of employment. Last, it may be accommodation for families who have temporarily evacuated their homes for fear of subsequent damage (UNDRO 1982: 8 and Davis 2015: 47).

Despite these functions, the continued use of emergency shelter may be indicative of a lack of thorough analysis and applied research on shelter. During all the years since 1982, when the guidelines were published, we do not recall a single mention of the critically important issue of survivors' shelter preferences, or research findings or informed discussion concerning the varied functions of shelter other than protection from the elements and future hazard forces. But we make no claim to have scrutinised all the vast literature and guidelines on the subject, so perhaps these topics have eluded our reading.

The identification of sheltering and housing preferences by surviving communities and the awareness of shelter functions will inevitably vary place by place, context by context; and thus understanding needs to begin from micro-level studies that can be built into post-disaster damage and needs assessments. The collection of this information certainly does not require trained assessors with clipboards and standard questions. Disaster survivors, through their community leaders or representatives, can provide it.

Voices of survivors

In 2004, Ian was asked to provide a keynote address in a conference in Mumbai, India. At the conclusion of his presentation, the chairman requested any questions – 'for clarification', and Ian remembers that: instead of the usual points that crop up in such gatherings, a question that I had never heard before, or since, came from a person at the back of the auditorium who was clearly very angry: 'How does it feel to have built your entire career on the basis of other people's misery?'

I fumbled a rather unsatisfactory answer, but I have often reflected on that piercing question. It is all too easy for academics and officials working in the disaster or development field to become hardened and cynical when faced every day with pressing accounts of need, deprivation and anguish. I have found that one antidote is to make an effort to meet disaster survivors on any field mission and to read their stories or see films of interviews with them. Thus, we describe the varied options for sheltering and housing in the following chapter where we include some of these encounters.

Some accounts of the experience of people in disaster are heartbreaking, and often disturbing, as they describe the absence of safety, a lack of security or vanishing prospects of recovery. We are indebted to various individuals who are determined to document and share such poignant experiences. One of these is Hidetomi Oi, a Japanese water engineer, who has spent his working career involved in disaster assistance and risk reduction. He has now retired, but his last role was as Senior Advisor on Disaster Risk Reduction to the Japan International Cooperation Agency (JICA). He has made an outstanding contribution to this subject through his passion to document and learn from the voices of the vast mass of ordinary people caught up in disasters. He pursued this quest by working with colleagues to publish and create documentary films of these interviews. These include studies of the voices of flood survivors in Nepal two years after the devastating 1993 flood disaster (Oi et al. 1998) and in Central America five years after Hurricane Mitch of 1998 (Smith Wiltshire 2004).

In the Nepal study, various farmers or farm labourers describe the effect on their farms of the soil being washed away through landslides caused by flood impact. A 23-year-old woman, Nara Maya Lopchan, who used to break up stones from the river to form construction materials, lost her home and land in the flood. She recounted:

> Many people became *Sukumbasi* (landless) after the flood, but in the case of my family, we were so for generations before the flood and the same now too. Quite a few persons and groups came to our village to ask many questions. They compelled us to hold high expectations for assistance, but to our dismay nothing happened thereafter.
>
> *(Oi et al. 1998: 31)*

Her desperation was complete:

> I am waiting for almighty God to allow me to die. There is not a rule that the older people should die earlier than the younger. I am prepared to die any moment. I would rather die than lead a life with such hardship.
>
> *(Oi et al. 1998: 31)*

The authors of this collection of interviews concluded that:

> The most deplorable issue that emerged from the series of interviews is the fact that the number of people who managed to survive the disaster

have to further endure the aftermath of the disaster over years, throughout their lives, or even for generations. . . . Whereas the prevailing tendency among most donors seems to be the reluctance in providing assistance during the rehabilitation phase, a serious gap exists between the donor's assistance and the realities of people and communities in the aftermath of a disaster.

(Oi et al. *1998: 161–3*)

The interviews conducted five years after Hurricane Mitch provided insights concerning the survivors' *preferences* for shelter as well as the functions of sheltering, settlement and livelihood security:

A constant feature in these projects is the fact that all efforts concentrated on building and delivering homes in a more secure setting with fewer risks as compared to the original location of settlers affected by floods and landslides. But these projects did not consider granting secure livelihoods, which obviously includes economic reproductive options and the possibility of renewing their conventional everyday activities. . . . [T]he resettlement projects have placed them within urban contexts and dynamics, located distant from their farmlands or with serious limitations for them to re-engage in these activities in their new settlements.

(Smith Wiltshire *2004: 31*)

A more recent study of the voices of survivors in relation to reducing the risks they face has been vividly produced in text and three films by BOND, a consortium of British-based NGOs (Moss 2008). IRIDeS, the International Research Institute of Disaster Science at Tōhoku University, has built up a substantial documentation centre to record survivors' testimonies and experiences after the 2011 earthquake and tsunami in north-east Japan.

David recalls a particularly poignant episode during his investigations of the impact of the 1980 earthquake in Southern Italy: Days after the event, I conducted a house-to-house survey in Tricarico, Province of Matera, an ancient town that had been severely damaged by the tremors. I was particularly interested in an old stone building that I knew had been the local cathedral from time immemorial until supplanted by a Norman building in AD 1061. It had then been converted into a house and in 1980 was occupied by an elderly man, who lived very much in the traditional manner. His tenacious attachment to his belongings and way of life was touching and a strong illustration of the importance of house as home. The following is an extract from my diary for that day:

The kitchen contained a rough table, chairs, a dresser, cupboards and the blackened, rusty range, above which were racks of battered old copper saucepans. Everywhere there were vegetables: buckets full of chickpeas, strings of onions, peppers, chilies, tomatoes, rows of marrows, all drying or already shrivelled to desiccation. The only other room was the bedroom, which was also full of dried produce, apricots, nuts, brown bananas, figs. There, with the light shining weakly through the stone lancet, we could make out only three pieces of furniture: a tall wardrobe carved with an urn surmounting a broken pediment, a marble-topped washstand, and, most characteristic of all, the capacious and superelevated bed, where once whole families must have slept. It was an authentic peasant cottage, left at the half-past-six of its development during the landslide of time. Had he left home during the earthquake, I asked him. 'I couldn't do that', he answered in a thickly antique dialect, 'there's all my wealth to be protected. *Ho delle robe* [I have things . . .]'.

Last, David remembers a case in which teaching appears to have paid off: We advocate teaching theory, of the kind whose usefulness can more or less be guaranteed, as a means of understanding complex situations, including those that arise during the chaotic aftermath of a disaster. I was standing in the middle of an emergency operations centre that was in full function managing the emergency caused by a very recent and catastrophic earthquake. A man came through the door. He was dressed in a fireman's uniform and was completely covered in dust, from head to foot. It was the end of the day and his face, what could be seen of it beneath the layer of dust, bore the signs of exhaustion. He had just finished a turn of duty searching for people trapped under the rubble of collapsed buildings. With surprise, I realised that he was one of my former students on a Master of Civil Protection course that I had helped design and teach. He recognised me and said, 'You know, what you taught us was true – it worked'. Amid so many failures to connect, so much doubt about the ability of an academic education to contribute anything useful to the solution of severely practical problems, here was a case in which it all came together. He remembered my lessons about the dynamics of disaster and the relationships between the contributory elements, and they helped him to orientate himself in the difficult process of searching for survivors. Rarely does one have such satisfactions so they are all the more welcome when they happen, especially because in such cases they demonstrate that the gap between theoreticians and practitioners can really be bridged. What is true for the short term of emergency management can be equally true for the longer term of recovery and reconstruction.

The foundations of sheltering and housing

To introduce the eight options of sheltering and housing in the next chapter, we wish to create a pair of foundation blocks that support all the options.

There are two fundamental questions that recovery officials must *always* consider when formulating a strategy for shelter and housing in *any* recovery situation. The first is to determine the preferences of survivors, and the second is to establish and satisfy the *varied* functions of shelter. Without detailed knowledge of both issues, recovery officials will be placed in a dangerous situation that can closely resemble driving blindfold down an unknown, winding road filled with potholes.

This issue poses a challenge for inexperienced agencies and inexperienced officials venturing into the shelter and housing fields for the first time. They need to recognise that while their medical or nutritional colleagues spend their lives dealing with people with the same body functions requiring similar patterns of treatment, shelter is a fundamentally diverse issue. It is a subject with complex variables in terms of site, settlement, climate, size, shape, materials, culture, symbolism, resources, traditions, construction methods and political dynamics. Thus prescribed, generic answers or approaches that may be found in handbooks or Sphere guidelines *never* remove the need to understand specific preferences and functions (see the discussions herein about the dilemma of 'house' versus 'home').

In this chapter, we have seen that shelter and housing provision have changed dramatically in the past four decades on account of radical external changes. But the needs of survivors for safe homes, whether in the short or long term, is constant as they always need to re-establish their lives. The need for officials to face complex dilemmas in a positive manner remains a major challenge to anyone who fondly imagines that shelter is a simple problem that demands simple solutions.

In the next chapter, we reflect on our experiences of shelter and housing reconstruction. We explore the series of options that was introduced in model 17, 'modes of shelter and housing' (Chapter 4).

Notes

1   Communication between Ian Davis and Paul Sherlock, Oxfam, in February 2015.
2   Communication between Ian Davis and Graham Saunders in May 2011.

9

SHELTERING AND
HOUSING OPTIONS

Architects love designing houses. We love deciding what is possible with budgets, space planning for future extension, imagining a life, deciding on what shade of blue. Why would we want to take that away from others? Why would we not want to enable everyone to be architects themselves?

A good shelter programme should facilitate the range of choices, including less hardware-focused solutions (such as renting and staying with relatives), and accelerate and improve the long-term recovery or development of the housing sector.

(Maggie Stephenson cited in Charlesworth 2014: 193)

Having considered the wider context of sheltering and housing in Chapter 8, throughout this chapter we refer to the range of options for sheltering and housing that are depicted in model 17 (see Chapter 4; the model is also shown in Figure 9.20). We make no apology for the extended length of this chapter for it needs to cover much ground as we gradually assemble this model. We describe these options in detail, where possible using our experiences of the shelter and housing options. The eight options are a mixture of finite, visible products that can be delivered to beneficiaries, as well as social processes that are experienced.

In the quotation that opens this chapter, Maggie Stephenson describes the process of offering choices to survivors. Her points raise a question that needs to be addressed to the various actors noted in Table 9.1: 'Are people participating in our projects, or are we participating in their processes?' Tentative answers to this question may be found in this chapter.

Writing about these various modes of shelter is a relatively straightforward task – we describe the wide range of options that apply in both pre- and post-disaster contexts. However, the operational agencies in governments are likely to be radically different for any given option, and we have yet to come across a situation in which a single agency of government has responsibility for all the options we list. Within the NGO sector, there are certainly favoured options where organisations take on roles and other options in which organisations rarely get involved. The dilemma of whether to adopt a supply approach or a demand approach, which we highlight in Chapter 8, is evident throughout all sheltering and housing options.

The broad heading 'national governments' in Table 9.1 includes central and local government and the wide range of government bodies that cover the varied options, ranging from emergency management (civil defence) departments, housing and urban development departments, planning departments, recovery organisations, regulatory bodies, etc. Thus, no single government agency or department is likely to be responsible for all options. As we have already noted, overall coordinators with responsibility for all the pre- and post-disaster shelter or housing options do not exist. The implication is obvious that there are likely to be gaps, duplications, rivalries and missed opportunities.

However, there may be good arguments for retaining the status quo, in recognition that shelter and housing are a massive problem and that the best that can be hoped for is to devolve decision-making to local governments that understand local concerns. The one constant factor in relation to the options is present in any surviving family, which may experience the full range of options – they may be evacuated before disaster, survive the destruction of their dwelling, seek to repair it, improvise shelter, be offered a transitional shelter or become involved in a user-build construction of their permanent house on a relocated site. This sequence of phases and actions underlines the importance of their active involvement and participation at every stage and in relation to all options (Schilderman 2010).

Table 9.1 suggests possible roles for the actors involved according to the various options (*) and indicates where there can be a major role with extensive commitment (shaded cells); but there are always exceptions to crude generalisations of this nature.

When considering the wide range of shelter options that is the theme of this chapter, one important point needs to be emphasised. The following model suggests a progressive, step-by-step sequence. That may be the actual experience of certain survivors in some cases but, for others, the sequence from disaster to permanent dwelling may be an untidy process. As Henry Quarantelli (1982: 280) notes:

> Sheltering and housing phases do not usually develop in a neat linear fashion. In a given situation, some disaster victims may be entering the permanent housing phase while others are still in the emergency sheltering phase. Furthermore, in any given phase there may be several moves as a family goes from one temporary housing situation to another.

TABLE 9.1 Roles for key actors in sheltering and housing

Options	Survivors	National governments	International agencies (UN and IFRC)	National and international NGOs	Private sector
		Before disaster:			
1 Safe dwellings	★	★	★		★
2 Evacuation to safe shelter to escape impending hazards	★	★			
3 Unsafe dwellings on unsafe site	★	★			★
		After disaster (provisional shelter):			
4 Repair of dwellings (improvised or contractor repairs)	★	★	★		★
5a Spontaneous shelter	★	★			
5b Provided shelter		★	★	★	★
		After disaster (transitional shelter):			
6a Transitional shelter (temporary)	★	★	★	★	★
6b Transitional shelter (to evolve into permanence)	★	★	★	★	★
		Permanent dwelling:			
7a User-build permanent dwellings	★	★	★	★	
7b Contractor-build permanent dwellings		★			★
8 Relocated dwellings in relocated settlement	★	★			★

Note: ★ indicates possible role in a given option; shaded cell indicates potential for the actor to have a major role.

An erratic sequence has the consequence that:

> governmental organizations and relief groups may concurrently be dealing
> with segments of the population at different stages in the sheltering and
> housing activities after a major disaster. Sheltering activities may overlap
> with housing activities and some permanent housing may occur before
> some emergency sheltering is finished.
>
> *(Quarantelli 1982: 280)*

In some cases, a single family can obtain and use the multiple forms of shelter as
described by Quarantelli. For example, the enterprising family in Galle, Sri Lanka
shown in Figure 9.1 appear to be experts in knowing how to operate the aid
system to secure *three* forms of officially provided shelter! Following the
destruction of their home in the 2004 South Asian tsunami, they were allocated
a tent (on the left of the photograph), then a transitional shelter (set behind the
family) and finally (on the right of the photograph) they secured a permanent
dwelling. When Ian visited the area four months after the tsunami, all three forms
of shelter were in use for storage, keeping livestock or extended family living.
Note the pile of roofing tiles in front of the tent salvaged from their destroyed
house and awaiting future use.

FIGURE 9.1 Securing multiple sheltering options in Sri Lanka in 2005
(photograph by Ian Davis).

Throughout the following options, after noting situations that are relevant to given options, we highlight some lessons, largely drawn from our own experiences.

Options before disaster

Option 1: safe dwellings

In Chapter 2, we describe the different views of disaster recovery that result from diverse standpoints, and in Chapter 12, we consider the way the subject can be considered in a radically different manner when wearing a different hat. This was exactly Ian's experience in the following example.

Ian recollects: my initial academic teaching was a course in architectural history of the twentieth century, where I enthusiastically described the classic Fallingwater house by the world-famous architect Frank Lloyd Wright as one of the most famous houses in the world (Figure 9.2). It was built between 1936 and 1938 as a mountain retreat for Edgar J. Kauffman, a wealthy manager of a department store in Pittsburg, USA, and it has received numerous accolades including 'best building of the twentieth century' by the American Institute of Architects (AIA). I recall 'waxing poetical' in appreciative lectures over its architectural virtues of 'organic architecture' and 'interpenetrating space' and its

FIGURE 9.2 'Fallingwater', designed by Frank Lloyd Wright in 1935. The spectacular siting of the summer house came at a cost as the building and its contents have suffered extensive damage and have had to be restored as a result of the impact of flash floods and inadequate structural design (photograph by Ian Davis).

dramatic siting, perched on top of a waterfall in the ravine of the Bear Run River in Pennsylvania. And I still recall my excitement when visiting this heroic structure and noting the way it gradually appears as one approaches it down the river valley through the woods in a steep ravine.

But a few years later, wearing a different hat, my lectures had changed direction to ways to reduce disaster risk through prudent land use planning controls, which would prohibit development in unsafe steep slopes or areas subject to flooding. Now my descriptions of the house included reference to its reckless siting, against well-informed technical advice. During the planning stage, Wright arrogantly rode roughshod over objections, from two sources, about the dangerous siting of the house. His client hired a leading firm of Pittsburgh engineers to review the plans and they questioned the siting and the security of the foundations on account of the risks of undercutting and erosion by the stream and waterfall. In April 1936, they wrote: 'we recommend that the proposed site not be used for any important structure' (Hoffman 1993: 28–30). The second objection came from the engineers working for the Pennsylvania Water and Power Resources Board, which opposed the siting of the house due to the severe flood risk. Wright summarily dismissed the objections.

But he was wrong as it appears that floods did damage the structure. Exactly as the engineers had warned, floods occurred in 1938, 1954 and 1956. The terrified owner of the house was living there in 1956 when:

> water rose far above the living room floor, and although the terrace doors kept most of it out, the bridge to the guest wing was ... leaky. The stairs became a cascade. ... The house was hung with pendant scaffoldings of heavy timber, as we had begun to repaint. The scaffolding was caught in the wind and shook the whole house like a terrier shakes a rat. ... The damage to the property was enormous: to the house, nil; only much cleaning of mud and sand was required.
>
> *(Hoffman 1993: 110)*

In 2002, engineering work was required to strengthen the house on account of insufficient steel reinforcement to the concrete cantilevers. James Loper, the engineer in charge, wrote: 'we believe that some localised damage may have occurred during a flood' (Loper 2002: 24). In the memorable words of one commentator:

> I remember seeing a rhyme in a cafe on Pennsylvania's Route 381, which passes close by Wright's domestic masterpiece. It went something along the lines of: *'Frank Lloyd Wright built a house over falling water/which he really shouldn't have oughta.'* But nobody was going to tell Wright that back in 1935, when he turned up at Bear Run and sketched out a home for Edgar and Liliane Kaufmann in a matter of moments. He was never less than impetuous.
>
> *(Glancy 2001)*

Fallingwater was created by the vested interests of a powerful client and his architect against technical objections on safety grounds. During a lunch in the British Embassy in Managua, the capital of Nicaragua, in April 2012, Ian was introduced to further vested interests that surround building safety. He was able to observe at first hand the role of an ambassador as a salesman and the nature of cosy alliances between diplomats and commercial pressures. The lunch also raised the issue of securing safe construction of critical facilities.

Ian remembers: This was my first visit to the scene of a disaster – the embarkation point of PhD research into *Shelter after Disaster*. The field visit took place 12 weeks after the earthquake of 22 December 1972 that destroyed large sections of the city and killed 5,800 persons (Davis 1975a). I arrived in the city singularly unprepared: not speaking Spanish, not knowing anyone and knowing precious little about shelter needs or disaster recovery management. So in order to make contact with senior government officials, I visited the British Embassy who proved to be most helpful. They offered to arrange interviews for me with ministers and key officials involved in disaster recovery, as well as providing me with a vehicle and translator. However, before I was able to conduct a single interview, I received a surprising telephone call from the secretary of the British Ambassador, who invited me to lunch. I was collected by the Ambassador's wife who brought me to the official residence after a strange car journey. We travelled via Masaya, an outlying town where thousands of displaced families were living. She then demonstrated her personal disaster relief programme that consisted of stuffing banknotes through her car window. I observed in the mirror that there were hordes of children fighting in the dust for the disaster relief banknotes as they fluttered from the window of her Jaguar as we proceeded through the town. Her random approach to disaster relief ranks high in my list of 'most bizarre relief programmes' – a memorable lesson that the distribution of disaster relief can be an idiosyncratic and chaotic experience.

We arrived at the Ambassador's residence where I was duly introduced, but for some reason not to a fellow lunch guest; that was to come later. As we were served coffee at the end of the meal, the Ambassador introduced me to this English guest who was a salesman from a well-known British firm of building contractors. He told me that he was in Managua to seek to negotiate the sale of prefabricated hospitals since four hospitals had been destroyed in the earthquake. At this point, the Ambassador recited the list of my forthcoming meetings with Nicaraguan ministers that they had arranged on my behalf. He explained that in his view it would be most helpful for me to mention the possibility of purchasing the hospitals during my meetings since (I recall his exact words with some precision) 'a good recommendation will carry more weight from your lips as a detached British academic architect, than from myself'.

The salesman's description of this building system's recent construction in Nottingham, a city in the UK that is devoid of any earthquake risk, prompted me to ask whether these hospitals were designed to resist earthquake forces. His candid response was jaw-dropping: 'God knows!' After a rather uncomfortable pause, I then said that there was no way that I could recommend the hospitals to anyone in Managua who was managing the reconstruction programme and that I was astonished that this leading company would seriously propose their product without verifying such a basic and obvious requirement. The lunch ended abruptly and the Ambassador took me for an educational stroll into his garden where he asked me what I thought his function was. I told him that I assumed it was to represent the UK to Nicaragua and vice versa. He then added: 'my job, and yours, is to sell British products'. This came as something of a surprise as my proposed university research programme had never included the job of commercial traveller!

This lunch encounter was an early reminder of the pathetically low importance that a contractor and a diplomat placed on public safety back in 1973, both of them demonstrating scant regard for the fundamental need to ensure that all new hospitals must to be built to high standards in order to resist hazard forces.

Key lessons for option 1

These are the main lessons of experiences such as the ones recounted above. First, one of the primary aims of disaster risk management is the creation of a physical, built environment in which buildings are sited, planned, constructed and maintained to withstand extreme hazard forces, whether they are flash floods in Pennsylvania or earthquake forces in Nicaragua. Second, it is unwise to assume that powerful people in public service, key professions and building contracting understand, recognise or accept the principles and practice of safe design and siting. Third, the building contracting industry, in alliance with donor governments, may regard disasters as vital sales opportunities (see 'Third dilemma: supply vs demand' in Chapter 8). Finally, especially for critical facilities such as hospitals, building safety is a vital requirement of reconstruction planning (see 'Model 12: disaster risk reduction measures' in Chapter 4).

Option 2: evacuation to safe shelter to escape impending hazards

For many kinds of natural hazard (with the exception of earthquakes), warnings can be issued by the authorities for local residents to evacuate. They may come via multiple channels: mobile telephones, pagers, television, radio, sirens, flags being raised, police cars with loudspeakers or visits from community evacuation wardens. Evacuation transportation is normally part of these warning systems as well as the selection and management of temporary evacuation sites in cyclone shelters, schools or community buildings.

In 1992, while making a study of cyclone warning systems in the South Pacific, Ian visited the island shown in Figure 9.3, which is about 12 kilometres from Port Moresby in Papua New Guinea. He asked the village leader what happened to the community of fishermen and their families when a cyclone was impending. He explained that they were collected in government boats and evacuated to the

FIGURE 9.3   Island community near Port Morseby, Papua New Guinea
(photograph by Ian Davis).

mainland for the duration of the storm. However, he explained that when he was
a child, the island residents had no chance to evacuate, so all the village men would
stand around the base of the roof of the large village community hall and lift it
off its posts and then lower it over the entire village community like an upturned
boat so that they would remain protected inside the roof throughout the storm.
His fear was that one day they would not be evacuated and nobody would know
about this traditional form of cyclone protection.

In April 1986, Ian was invited by close friend Gustavo Wilches–Chaux to
Colombia to examine the devastation of the town of Armero and surrounding
areas, where a total of 23,000 people had died as a result of the eruption of the
Nevado del Ruiz volcano in November 1985 (Figure 9.4).

Ian recalls being taken, in an army helicopter, on a tour of the area affected,
accompanied by a local director of civil defence. The town of Armero was
covered by a sea of mud from the nocturnal lahar produced by Ruiz. The result
was the second worst volcanic disaster of the twentieth century, and more than
20,000 of the deaths occurred in Armero, which had a population of 29,000.
The volcano had been dormant for 69 years and the lahars caught everyone
unawares even though, weeks earlier, the government had received warnings
from multiple volcanological organisations to evacuate the area after volcanic
activity began to increase significantly. The site is now a cemetery, and the white
crosses visible in the photograph indicate the location of destroyed houses where
families are buried.

FIGURE 9.4 Armero, Colombia covered by a sea of mud from a volcanic lahar (photograph by Ian Davis).

As we flew back to the town of Ibagué, in Tolima District, the official shouted into the microphone of the headsets we were using to cut out the deafening helicopter noise that his next visit was to a 'slum settlement' called Barrio Industrial, where the local civil defence office had been asked by a local shopkeeper to provide signs to indicate an evacuation route. He was perplexed with the request as he had no knowledge of any evacuation plans for the area. He invited me to join him for the visit. We arrived at a settlement beside a fast-flowing river, the Rio Combeima. Evidently this river flowed from a nearby mountainous area subject to frequent thunderstorms that caused flash floods. The Barrio Industrial was an illegal settlement in acute danger of flooding; but the residents ignored the law and settled there on account of its proximity to the river, which they regarded as a potentially lucrative source of income. Gold fragments were carried down the river from the mountains and local prospectors sieved through the silt to find them.

The authorities had installed flash flood warning sirens, but they provided a maximum of just 12 minutes' warning time ahead of the arrival of a wall of water. The settlement was composed of rather flimsy improvised dwellings that were mixed in with some temporary housing provided by a leading international NGO for families displaced by the volcanic lahar. In a past flash flood, several of these houses had been destroyed and their occupants killed. While there may good reasons why local families were putting themselves at risk to live in the area, there can be no excuse whatever for an NGO to build temporary houses for disaster survivors, only to put the occupants at risk of another tragedy.

We found the only shop that sold tobacco and sweets and met the woman who had asked for help. She told us that there had been a flash flood in their barrio three months earlier that had destroyed several houses, including some of the recently built houses for disaster survivors. Two people had died in that flood. She was concerned that there had been no evacuation plan and that everyone ran in different directions in a chaotic manner when they heard the sirens. Therefore, entirely on her own initiative, she had organised a *comite de emergencia* of local residents in order to run community evacuation drills. She called them '*simulacro*'. One exercise had taken place in daylight hours; the other, in the middle of the night. As a result of these simulation exercises, the community decided on the best routes to escape to high ground (Figure 9.5); and two families had swopped houses – elderly people living near the river who had found it hard to move fast enough exchanged house with younger occupants who wanted to live nearer the river.

We asked her why she had taken this initiative. She responded that there was no other option as the government was not interested in the residents' plight, so they had to take matters into their own hands. All the shopkeeper wanted from the authorities were some signs to fix to telegraph poles to indicate the escape route for future evacuations.

This is an example of successful and totally unofficial evacuation planning at a local level; however, examples can also be found of successful regional evacuations on a massive scale. An outstanding example is in relation to Cyclone

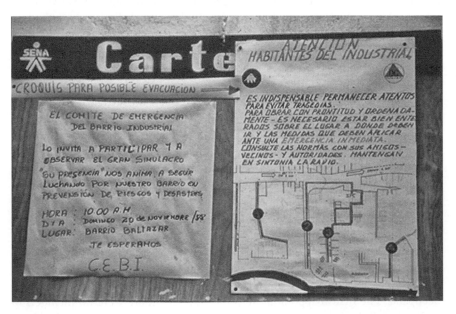

FIGURE 9.5 Map of the simulation evacuation route pinned to the door of the shop (photograph by Ian Davis).

Phailin in Odisha, India in October 2013, when almost 1 million people were evacuated from the vulnerable coastal belt prior to the arrival of a category 4 storm. In the end, it caused only 38 deaths (World Bank 2013).

Key lessons for option 2

First, the effectiveness of warnings is closely related to the frequency of hazards. Thus in Armero, the population and their authorities were not vigilant to the volcanic risk while nearby, in Ibagué, the regularity of flash floods created a response from a concerned community. Second, the advent of good early warning information on impending floods, landslides, tsunamis and volcanic eruptions is having a decisive impact in reducing loss of life, providing that good evacuation plans are in place. Third, in order to protect themselves from the everyday risk of poverty, low-income families are prepared to take risks associated with hazards. Finally, community initiatives to reduce risks may cost nothing but committed leadership is always needed, as it was in the case of Ibagué.

Option 3: unsafe dwellings on unsafe site

In 1986, Ian was leading a disaster management training course for NGO officials in Vijayawada, Andhra Pradesh, India. He recounts: As part of the course, the local participants reluctantly agreed to visit a housing project that their agency had built as part of cyclone reconstruction after the 1977 cyclone. Their hesitation may have been on account of this being the first time the agency had ventured back to the site in the nine years since the dwellings were built.

As our vehicles arrived in the village with the agency's name and logo, we were besieged by a large crowd of excited and angry residents. They led us at high speed to their settlement and immediately the explanation for their discontent became blatantly obvious. The concrete block houses were in varied states of collapse, with walls at strange angles on account of failed foundations set within a site with a very high water table. Many roofs had also failed, and we heard of the impossibility of living in the houses during monsoon rains. The one saving grace was that when the houses had been built, each family was given a coconut palm seedling that was planted next to each house to provide shade and a source of nutrition and possible income. These were now fully grown and were providing good livelihood support.

As we left the village, leaders pleaded with the NGO officials to return and rectify the problems of their house failures. So back in the class, as the course facilitator, I raised the issue with participants from the agency, asking them whether they would take any remedial action, especially on account of the houses being unsafe and in danger of collapse with the possibility of injury to their occupants. The first disappointing response was that there were no members of staff currently in the organisation who were present when the houses had been built – so why should they take responsibility for their predecessors' failures? The second reaction centred on the technical shortcomings of the original engineers and architects as well as the shoddy builders that they had employed – the responsibility must be 'placed at their door', certainly not at the door of the original funding agency. Thus, no responsibility was taken for the failures, demonstrating no accountability to the distressed occupants.

A heated discussion followed in which further support was manifest for the agencies' desire not to be involved in the problem, with other participants challenging this stance. The arguments ran as follows.

In support of the agency's position: First, why had the occupants not taken action to repair or rebuild their own collapsing dwellings? Why were they so passive? Second, how could we, as the agency, be held responsible for our actions since the approach had not been agreed when construction began? It is totally unreasonable to assume responsibility for our actions indefinitely.

Against the agency's position: First, how could a Christian agency ignore the manifest needs of a stricken community, given their mandate to 'love their neighbours as themselves' and with their commitment to both justice and mercy? Second, why could the agency not now consider the restoration of the village as a new development opportunity and raise funds for this?

Later enquiries revealed that no remedial action was taken by the agency. This depressing example relates closely to the discussion of 'downward accountability' in Chapter 6 where Table 6.1, 'The interveners in disaster', systematically describes this kind of situation.

In considering the safety of dwellings, there are two vital requirements for the occupants. First, occupants need to be made fully aware of the actions *they* need to take in order to ensure their own safety. Second, they also need to be aware of the actions they need to take to ensure that *their own dwellings* remain safe. A typical example is to install protective covers over all window openings when a cyclone warning has been issued. Then there are actions not to take, such as demolishing internal dividing walls – technically called 'shear walls' – which provide the structure with vital stiffening in earthquake-prone areas.

In recent years, a wealth of practical guidelines has been developed to provide advice on how to construct low-income vernacular buildings so that they are safe. Guidance on creating and maintaining a safe dwelling can be found in a series of documents (Agarwal 2007; Desai and Desai 2008; ERRA 2008; Szakats 2006).

Key lessons for option 3

First, as noted in the example of Orléansville – described in Chapter 4, 'What's in a name?' – shoddy disaster reconstruction can lead straight to a new disaster. Second, when projects are undertaken, quality controls are vital to ensure safety and good construction. Third, whether NGO or governmental, the extent of accountability of a given agency needs to be defined in relation to their long-term obligations to the beneficiaries. The issue of assisting groups' accountability

FIGURE 9.6 Model 17: options before disaster.

to survivors is explored in detail in Chapter 6 – Table 6.1 chronicles the reality of patterns of accountability to survivors. Fourth, the rapid turnover of staff has a decidedly negative effect on accountability. Finally, in disaster recovery, it is possible for secondary elements to eventually become the primary benefits, as in the planting of a coconut palm tree for each family following the cyclone in Andhra Pradesh, India.

Figure 9.6 refers to options 1, 2 and 3 ('before disaster').

Options after disaster: provisional shelter

Option 4: repair of dwellings (improvised or contractor repairs)

Two years after the 1995 Kobe earthquake in Japan, which resulted in the destruction of over 150,000 buildings, Ian met the Director of the Kobe Housing Reconstruction Programme and asked him whether, with the luxury of hindsight, he would have made different decisions about recovery planning. He replied that the authorities had seriously neglected the housing repair option. Some weeks after the disaster, in order to clear the streets and sites of earthquake debris, the government announced that they would not collect debris free of charge after a specified date. The result, in his words, was 'an orgy of wasteful demolition of repairable dwellings by their owners'. The Director believed that this was a mistake for three reasons: it resulted in the destruction of reuseable building materials, particularly from wooden houses; it delayed survivors in returning to their homes as repairing dwellings would have been a faster option than full reconstruction; and crucially, had the government promoted house repairs, this could have generated valuable work for small local builders and thus have helped revive the damaged economy. The Director believed that the fundamental reason for the neglect was the strong desire by politicians and officials for a 'clean slate', with empty sites awaiting the arrival of contractors rather than having to consider the 'messier situation' in which, in any given area, some buildings could be repaired, others could be made safe using retrofitting (strengthening) measures, and others could be demolished and substituted with new buildings.

The distinction between 'repair' and 'retrofit' is that the former applies to all aspects of a damaged building, whether structural or non-structural, while the latter refers to specific structural measures to make a building safe against future hazards.

Any repair or retrofitting operation for a severely damaged dwelling needs to follow a detailed damage assessment by an experienced structural engineer. Only superficial repairs can be undertaken by building users who do not possess technical skills relating to safe construction. The cost of major repairs can be similar to the cost of total reconstruction. When officials inspect buildings to decide whether to repair or demolish, it is vital to avoid painting any code on the outside of the building to guide contractors on the fate of the structure. We have seen evidence in several recovery sites where building owners have deliberately modified such markings to promote either demolition or repair in order to qualify for any compensation payments that may be forthcoming.

The World Bank's *Handbook for Reconstructing after Natural Disasters* provides wise advice on the repair option, with guidance on whether to repair or retrofit buildings or to demolish damaged structures:

> In reconstruction efforts, repairing and retrofitting a house may make more sense than demolishing and rebuilding it. Many practitioners and policy makers think that programs designed to repair and/or retrofit housing are difficult to design and implement. However, such programs can save many partially damaged houses, often with excellent results.
>
> *(Jha et al. 2010: 166–7)*

The guidance makes the important point that the repair option can have a decisive impact on improving the reconstruction process in terms of 'cost, environmental impact, speed, supply of resources, community participation and satisfaction' (Jha *et al.* 2010: 166).

After the Bam earthquake in Iran in 2003, technical staff devised the diagram shown in Figure 9.7 to assist officials in rating buildings for repair, retrofit or total demolition.

Key lessons for option 4

In summarising the key lessons from the experiences recounted above, the first point is that the repair option is generally neglected in recovery planning – this needs to change to avoid the waste involved in demolishing repairable buildings. Second, prior to decisions on repair, retrofit or demolition, it is essential for damage assessments to be undertaken by competent trained officials.

Option 5a: spontaneous shelter

The title of this option covers a wide range of ways that disaster survivors manage their shelter when confronted by the crisis of a sudden loss of their homes in a disaster, or by the gradual realisation that they have to vacate in the case of events such as rising floodwater or drought.

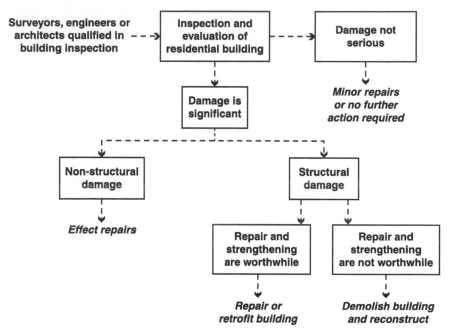

FIGURE 9.7 Decision-making on whether to repair, retrofit or demolish unsafe buildings (after ATC 2005 and Jha *et al.* 2010).

When Ian was *en route* to the city of Managua, Nicaragua to work amid the devastation caused by the 1972 earthquake, he recalls visiting the Disaster Research Centre (DRC) at Ohio State University in Columbus, Ohio: The co-directors of the Centre, the world's leading experts with specialist knowledge of the sociology of disasters, Professors Russell Dynes and Henry Quarantelli, were most generous with their time in guiding me through relevant literature in their library, which was the leading global resource on the sociology of disasters (see Chapter 2, 'Evolution of recovery studies').

As I departed, Henry Quarantelli asked me if I wanted some advice in relation to my forthcoming fieldwork on shelter issues. I told him I was still ignorant of the subject and needed any advice he could offer. He then made a memorable statement that I have often repeated to my own students: 'When you get to the disaster-affected area, you will find that there has been a gross exaggeration of damage and dislocation and a massive underestimate of local resources in tackling the problems'. In every disaster I have visited since then, this observation has proved to be accurate – with the exception of the 1985 Mexico City earthquake where the authorities played down the severity of the earthquake as they feared that the forthcoming football World Cup might be relocated.

In Ian's subsequent 42 years' work and David's 35 years' work in disaster recovery, we have not seen a *single* agency funding request following a major disaster that makes reference to the capacity of the surviving community to fend for itself or to sort out its own problems, including its shelter needs. Thus, there is a constant tendency to underestimate the local capacities of the surviving community, something that any experienced humanitarian aid worker will have observed in the field.

This neglect is obviously designed to maximise donations but remains contrary to the Red Cross Code of Conduct, which emphasises the need to recognise local coping strengths. In the 20 years between 1994 and 2014, 546 international NGOs from over 65 countries have become signatories of the Red Cross Code of Conduct (IFRC 1974). Paragraph 6 of this runs as follows:

> All people and communities—even in disaster—possess capacities as well as vulnerabilities. Where possible, we will strengthen these capacities by employing local staff, purchasing local materials and trading with local companies. Where possible, we will work through local NGHAs [non-governmental humanitarian agencies] as partners in planning and implementation, and co-operate with local government structures where appropriate. We will place a high priority on the proper co-ordination of our emergency responses. This is best done within the countries concerned by those most directly involved in the relief operations, and should include representatives of the relevant UN bodies.

Thus the signatories have all agreed to abide by this clause and also paragraph 10 which states:

> **In our information, publicity and advertising activities, we shall recognize disaster victims as dignified humans, not hopeless objects.** Respect for the disaster victim as an equal partner in action should never be lost. In our public information we shall portray an objective image of the disaster situation where the capacities and aspirations of disaster victims are highlighted, and not just their vulnerabilities and fears.

While agencies need money urgently after disasters, this is to support the powerful resources of survivors, who play vital roles in meeting their own recovery needs. Clearly, the staff in NGO fundraising departments must believe that if this resource was stated, it would reduce or eliminate donations. Their approach is understandable, but the effect is to distort reality and portray disaster survivors as passive, rather helpless victims who need assistance to sustain their lives. This omission appears to contravene what the agencies have pledged.

This neglect of the capacity of survivors is not confined to international NGOs. It also applies to governments, sections of the media and international agencies. In recent years, there has been recognition of the role of local communities in risk reduction measures. With strong support from the Red Cross movement, there has been a wide acceptance of the need for vulnerability and capacity assessments (VCA) in disaster risk assessment exercises (IFRC 2006).

But as noted, this recognition has yet to extend to the post-disaster context.

In the next sections, we describe situations in which each of us coped 'spontaneously' with disasters.

Coping with having a home bombed and with evacuation: Ian Davis, 1941

I was 4 years old when my family's house in Barrow-in-Furness, one of Britain's major shipbuilding towns in the north of England, was bombed by German bombers in 1941. Throughout April and May, the Luftwaffe's air raids sought to destroy the shipyard that was constructing submarines and warships. At about 10 p.m. on the Sunday night of 4 May, our family, as well as relatives from London who can come to stay in our house to escape the London Blitz, were all sheltering in our front room in the dark with lights out following the warning siren. We crowded together with much fear as we heard bombs dropping around us with terrifying explosions and great flashes of light that penetrated the blackout blinds. My father found a Bible and a torch and read the 23rd Psalm to us as we huddled together under our dining table as, unlike our neighbours, we had no shelter in our house or garden. For some unknown reason, our family had chosen not to go to the semi-underground concrete air-raid shelters, designed to provide some protection from bomb blast.

Then it happened. A landmine caused a massive explosion that was very frightening. A large chunk of concrete and other debris from our road crashed through the roof of our house, destroying my sister's bed and bedroom (she had been brought downstairs just five minutes earlier), a section of the top floor of our house was missing, and the entire house was filled with dust and soot from smashed chimneys.

At this point, we all left our shattered home, black with soot that covered us head to toe, to go to the official air-raid shelters which were about 100 metres away. While being carried on my father's shoulders, I heard anti-aircraft gunfire and saw searchlights sweeping across the sky as they sought to illuminate German aircraft for the gun crews. We spent the night in the crowded shelter, sleeping in tiered bunks, and I recall asking my mother why we should be given hot soup to drink in the middle of the night!

On Monday 5 May, the day after the raid, my father must have moved very rapidly in seeking a place for us to move to and organising a vehicle to take our family as well as a few clothes he had rescued from the debris of our ruined home. As we left, the roads were crowded with families leaving Barrow to escape a further night of bombardment. We went to Newby Bridge, a village about 29 kilometres away at the southern tip of Lake Windermere in the Lake District. I am sure that if we had had relatives in the region then that would have been our destination, but as we didn't, a kind friend arranged for us to stay in the house of strangers. I recall looking out of the window at the flashes of a further bombing raid in the night after we left. In our rush to leave Barrow, my parents were unable to do anything to protect our belongings within our ruined house and, as a result, most were looted.

Our family remained evacuated for nine months in a heavily overcrowded house with grumpy and very reluctant hosts who did nothing to make us welcome. The government required rural houseowners to accept evacuees seeking refuge from large towns and cities. I learned later that our stay caused my mother endless stress, while my father continued to work each day in the Vickers-Armstrong shipyard in Barrow as an engineer designing submarine engines.

Our own family's complex 'process of sheltering' from the threat of being bombed had many components: a driver with a truck to evacuate us, a willing friend to organise our temporary home, an official requirement for a rural family to share their home with strangers, a bicycle and a bus so my father could commute to work to support our family and, finally, reliable information, perhaps through a local newspaper or telephone, to advise us when it was safe to return.

But while this was an anxious time for my parents, myself as a 4-year-old, and my 7-year-old sister, the bombing and evacuation were certainly the most memorable experiences of childhood. Apart from my parents not having a shelter for our home, all their other decisions seem to have been eminently sensible.

The German bombing campaign of 11 air raids, intended to destroy the Barrow shipyard, failed as the bombs mainly hit houses or landed in the sea, beaches or fields rather than on their targets. So for the next five years or so, one of the risky joys of childhood was to play in overgrown bomb craters, conducting 'treasure hunts' with my friends as we searched for unexploded bombs as well as conducting exciting hunts for German spies that we felt might be lurking in local woods and sand dunes. We brought the attention of bombs to our schoolteacher or local policeman, neither of whom shared our enthusiasm as we excitedly reported our latest discoveries! Our elder cousin came to stay and we found a bomb washed up by the tide, and I recall her bright idea to place it on a rock where we pelted it with pebbles to try and explode it! A worried police officer came to our school and said very forcefully at one of our morning assemblies: 'Now children, this is a very important notice; do *not* bring any more bombs to the Police Station!'

The damage toll from the raids on Barrow was massive. The official record states that 10,000 homes were damaged or destroyed, representing a quarter of the town's housing stock. Six hundred houses were totally destroyed and major repairs were necessary to 1,400, one of which was our home. Rather astonishingly, only 83 people were killed and 330 injured out of a population of 75,000, which testifies to ineffective Luftwaffe bombing and an effective warning and shelter system (Trescatheric 1989).

One reason for including these personal memories in a book on disaster recovery relates to the high likelihood that children experience disaster very differently to adults, as my sister and I did. And what may be a dreadfully stressful experience for a parent may be a great adventure for a child, bringing a host of stimulating new opportunities.

Coping with sudden homelessness from an earthquake: David Alexander, 1980

My introduction to disaster studies came abruptly in November 1980 when I was made homeless by the Southern Italy earthquake, which took the roof off my rented accommodation in the Basilicata Region – one of the perils of living in an unreinforced masonry building of some antiquity that was located in a seismic area.

One month after the earthquake, on 23 December 1980, I made a tour of the affected area, which was large, mostly rural and geographically very diverse. Many of the 630 settlements damaged by the earthquake were located in highlands, and here the suffering of survivors was amplified by hardship and isolation. In mid afternoon, I drove along a mountain road that had been made topsy-turvy by liquefaction damage. On the other side of a farm gate I saw an extraordinary scene.

The farmhouse, a capacious, traditional stone building, had been knocked down by the tremors and was reduced to a pile of rubble about knee height. Evidently, its occupants had survived for they were outside and grouped in a circle around an enormous bonfire, on which one could see the burning remnants of a wardrobe and other pieces of furniture. A green canvas army tent was pitched close by. Perfectly immobile, the farmer, his wife, their children and an elderly couple were looking silently and intently into the flames. Next to them were the farm animals, cows, goats, chickens, geese and sheep, also mesmerised by the fire. The air was freezing cold and it had begun to snow.

I did not have the nerve to interrupt this communal reverie. As a coping mechanism it represented an extreme form of stoicism, and one only hopes that help and comfort arrived soon, as at Christmas the snow fell heavily. The residents of mountain areas tend to be inured to physical hardship and deprivation, but this does not mean that they are well endowed with the means to ward off the psychological effects of sudden loss and major disruption of their lives. The shock is usually greatest for people in old age. In the mountain communities of Basilicata, many of the elderly survived the earthquake only to die during the early aftermath as they lost the will to live and the ability to cope with unwelcome, radical changes in their lives.

Approaches to spontaneous shelter

There are numerous examples of spontaneous shelter being organised by disaster survivors and their relatives and friends (Figure 9.8). They include the following three approaches. The first is to purchase building materials and services following the receipt of remittance gifts or loans from relatives or friends who have left a country to form a diaspora. This rapidly developing process may well transform the disaster recovery situation in countries with a large diaspora of their population living abroad. The second involves improvised shelters constructed by survivors. Throughout our many visits to disaster sites we have observed an astonishing array of examples of ingenuity by disaster survivors as they improvise shelters and reconstruct their dwellings, often using debris they have salvaged.

The levels of creative innovation shown are particularly likely to be high in low-income groups where these 'coping mechanisms' tend to be strong. However, middle-class survivors tend not to have such evident skills, perhaps because they have more shelter options. Third, there is voluntary evacuation from the affected area to stay with host families or in shelters. In the words of Setchell (2012: 17):

> hosting by family and friends, or even by strangers, is socially defined, self-selected, culturally appropriate and typically provided before humanitarian actors arrive and—importantly—long after they leave. Hosting is, in fact, an effort to help, be it for social, family or even altruistic or nationalistic reasons, so how could it not be considered humanitarian in nature.

The International Federation of Red Cross and Red Crescent Societies has developed an important study concerned with practical ways that host families can be supported in their vital role (IFRC 2012).

FIGURE 9.8 Six days after the Guatemala earthquake of 4 February 1976, this photograph shows a typical example of improvised sheltering in a street in Guatemala City. This shows a complex mix of rope, canvas, blankets, plastic sheeting, wooden posts, salvaged furniture, etc. that provided temporary protection for a displaced mother, child and grandmother (photograph by Ian Davis).

The value and scale of the host families' contribution to displaced families is a remarkable example of spontaneous sheltering support as shown by the examples provided in Table 9.2.

In the Kobe earthquake (Japan, 1995) there were approximately 350,000 displaced persons. Between 40,000 and 60,000 left the city to stay with host families in various Japanese towns and cities. However, no government agency sought to determine how long they stayed away, who they stayed with or where they went. Such information on patterns of survival would also have been useful

TABLE 9.2 The role of host families as they accommodate survivors

Disaster	Survivors displaced by disaster	Survivors who left the area to live with host families, etc.
Kobe earthquake Japan, 1995	350,000	Approx. 40,000
Katrina hurricane USA, August 2005	1 million people were initially displaced in the Gulf Region. This figure included 300,000 school-age children. One month later in September 2005, 400,000 had returned. Some elderly survivors had been initially displaced in 1964 to the Lower Ninth Ward of New Orleans by the US Government's National Park Service. They were then displaced by Hurricane Katrina in 2005 to Texas, where they were further displaced by Hurricane Rita in September 2005 (Jackson 2006).	Population of New Orleans in April 2000: 484, 674; in July 2006: 230, 172; and in July 2012: 369, 250. Thus approx. 115, 424 displaced persons, who travelled to 50 US states initially to stay with host families, had not returned seven years after the disaster – 76% of the population in 2000 (Plyer 2014).
Haiti earthquake January 2010	2.3 million	Initial outmigration: 570,000 – 22% of the urban population of Port-au-Prince (Bengtsson *et al.* 2010). Another estimate: 500,000–600,000 (Esnard and Sapat 2014: 182).
Tōhoku earthquake/ tsunami and nuclear accident Japan, March 2011	450,000 displaced by earthquake/tsunami; 170,000 displaced by nuclear accident; 620,000 in total.	321,000 evacuated and relocated throughout 1,224 municipalities by December 2012 (JRCS 2013).
Yolanda typhoon Philippines, November 2013	4 million	400,000 living in 1,552 evacuation centres, with an additional unknown number living with host families (IDMC 2013).

to the authorities in managing the sheltering process following the Tōhoku earthquake tsunami and nuclear accident in March 2011. Websites were set up throughout Japan for support and also international sources offered to coordinate families who needed host family accommodation.

It is estimated that in the Haiti earthquake disaster of January 2010, 570,000 (22 per cent of the population of Port-au-Prince) initially migrated away from the devastated city in order to receive some form of shelter support when it was most needed. This information comes from a landmark piece of research (Bengtsson *et al.* 2010), which provided the first recorded example of mobile phone call analysis following a natural disaster. Innovative agency programmes were undertaken in Haiti to provide support to host families as they accommodated displaced survivors, and there are records of families building new dwellings adjacent to the houses of host families (Davis 2012).

The outmigration that followed Hurricane Katrina (2005) has been described as the greatest population movement in the USA since the Dust Bowl of the 1930s when half a million people were displaced (Esnard and Sapat 2014: 48). In many cases, these survivors had to cope with multiple problems during evacuation. Two American researchers studied the progress of survivors displaced by Katrina and concluded as follows:

> It was a continuing journey. They made several moves before even leaving the city—to and from the Superdome or Convention Center or other places of temporary shelter—and then they endured several moves from place to place. Some for example, moved in with family and friends. But as days extended to weeks, months, and even years, people were forced to move because friends' and families' personal and material resources were often strained beyond their capacity to continue to support evacuees. Further, as federal government and local community programs and housing support changed over time, evacuees had to shift back and forth from shelters to hotels, to apartments, to public housing.
>
> *(Weber and Peek 2012: 14)*

The data in Table 9.2 indicate that more than 115,000 people who left New Orleans and the areas affected by Katrina had not returned seven years after the disaster. This raises a number of important questions and issues. To begin with, why have these families not returned? Is this because their relocated homes offer better opportunities than their original residences, because they lack the resources to move back, or because they have no secure home or livelihood back in New Orleans or in the Gulf States?

When Ian visited New Orleans in 2012 to consider the implications for disaster recovery, he recalls discussing this situation with various people: I suggested that there may be considerable value in people *not* returning to their original homes, as this would reduce the population of a highly vulnerable location. This resulted in repeated responses that the return of people with the capacity to work was absolutely essential to running the vital oil refineries and related industries in the regions, and that the outmigration and failure to recover the pre-disaster population was having serious negative economic consequences. However, it is possible that a high proportion of those who chose not to return were elderly and not employed.

Key lessons for option 5a

The previous section has demonstrated the diversity of ways that survivors manage their own shelter needs. The value of accommodating survivors in host families in the immediate aftermath of a disaster is insufficiently recognised by authorities as a key coping strategy. In fact, it has an important double value. First, it is an asset to the authorities in terms of reducing their need to provide emergency accommodation. Second, for the traumatised families, there is a clear benefit in being accommodated with friends and relatives as their temporary stay probably provides them with vital 'comfort and security' far beyond the scope of any other form of emergency accommodation.

Authorities have largely ignored this vital resource, perhaps on account of the political desire to support 'visible solutions' that show their levels of concern in the form of tents, transitional dwellings, and so on. However, governments can do much to strengthen the resource of host families by collecting data on where survivors have moved to. This has a dual purpose: first, it will enable them to provide support to host families to help them to fulfil a vital role, continuing to house displaced families until recovery gets underway. Second, it will enable them to provide essential information to surviving families concerning their rights and recovery plans.

Option 5b: provided shelter

Both shelter and reconstruction planning can develop as ongoing processes immediately after a disaster has struck. However, at that point, authorities are unlikely to have all the information that they need in order to be successful. The provision of shelter is often made on the basis of an overgenerous assessment of needs, one which is tied to the total number of homeless people and families, not the residual number. Such calculations can ignore the voluntary movement of families to stay with host families. As a result, shelters are left empty and unused, and in any case, as time wears on there will be a gradual abandonment of those that are used in the initial phase of the disaster.

The planning of shelter requires the selection of appropriate sites and units. Sites need to be well drained and preferably close and accessible to the areas they are intended to serve, usually the damaged settlements from which homeless survivors are drawn. Landownership and tenure questions also need to be resolved. In the Mount Amiata area of southern Tuscany (central Italy), a group of five municipalities has chosen its areas for transitional settlement in advance of any future disaster. This was accomplished by developing criteria and carrying out local surveys of available land. If and when disaster strikes, the municipalities will be well placed to react quickly to any need for shelter that the local impact has generated.

A common mistake has been to locate shelter sites far away from survivors' homes or workplaces. Due to the collapse of factories or commercial buildings, many survivors will have lost employment; but others will be able to continue to work, and this is vital in terms of the need to re-establish damaged local economies.

The degree to which sites need to be urbanised depends on the terrain, the nature of access and the type of shelter that is provided. One simple short-term solution is simply to pitch army-style tents in a grassy field. If this is not well drained and conditions are wet or wintry, the site can easily become a quagmire. Caravans (mobile trailers) and container homes generally require a level site that is well drained, compacted and covered with gravel. Light-walled prefabs need concrete base plates and these are usually associated with full urbanisation, including roads, parking spaces and retaining walls. All forms of shelter other than those utilised for the shortest periods need connections to basic utilities: electricity, water and sewerage.

Planning must also deal with the problem of finding suppliers of shelter. It is seldom stockpiled as this is a highly expensive approach – stored shelters or shelter materials tend to deteriorate over time in outdoor conditions or take up too much warehouse space. However, many forms of shelter can be manufactured quickly. There is now a significant international market in container homes and light-walled prefabs. One can guarantee that, given half a chance, the companies that manufacture these items will endeavour to market their products in disaster areas. A major disaster may cause a significant surge in demand for temporary and transitional shelter. A choice must be made between using a single design and using a variety of models. Standardisation may reduce unit costs and possibly simplify maintenance, but it may also restrict the adaptability of units to specific needs – for example, housing larger families.

There are many instances in the history of shelter in which it was built in the wrong place or consisted of the wrong kind of building. The result is almost always abandonment of the shelter or drastic modification in order to make it respond to local needs, whether these are cultural or practical. Hence, there should be an assessment of what the needs are in the planning process. Do they, for example, include needs associated with the livelihoods of farming or fishing, such as space for animals, tools and equipment or boats? Are there any cultural requirements – for example, is the affected population accustomed to having a kitchen that is detached from or outside the main residence? What are the climatic requirements for shelter? It should not be riven with damp, washed away by torrents of floodwater, baking in summer heat or freezing in winter.

In many cases of shelter planning, there is a choice between aggregating shelters in an extensive park and dispersing them to many locations. The former enables services to be provided efficiently but risks creating a 'ghetto' of disaster survivors – an area which sets them apart from people who did not lose their homes. By dispersing shelter, it can be constructed in the gardens of people's homes, on their own land; although more readily if it is not the kind of prefab that requires a concrete base plate. Alternatively, small enclaves of shelter units can be created that efficiently utilise spare ground in the vicinity of the damaged settlement.

If it is efficient, contingent planning of shelter after a disaster can fulfil a basic need rapidly. Shelter cannot wait while studies are done and consultation takes place. However, hasty and ill-thought-out solutions waste resources, slow down the recovery and lower the morale of the survivors.

Forms of provided shelter

There is a bewildering range of ways that authorities provide shelter to displaced families as this option is highly popular with governments, NGOs, manufacturers and inventors. The issues of sheltering are well covered in the second edition of *Shelter after Disaster* (Davis 2015).

Move to hotels, railway coaches, ships, etc.: If there is sufficient capacity, hotel accommodation may be very convenient, although it is an expensive option. After the 2009 L'Aquila earthquake in Central Italy, about one-third of the 67,500 homeless survivors were moved into hotels. As the hotel capacity of the city is small and was diminished by the collapse of one of the largest hotels, most of the 20,000 evacuees were moved to tourist hotels on the Adriatic Sea coast. Here, they managed to preserve their sense of community, but they were more than one hour's drive from their home city and sources of employment. Cabins in ships, compartments in railway coaches, and so on can be used; but they tend to have problems of sanitation and the provision of basic services.

Use of caravans (trailer homes), mobile homes and containers: In both Japan and New Zealand, shipping containers have been converted into accommodation. Figure 9.9 shows their use in Christchurch in the Restart project, which was designed to bring life to the devastated centre of the city, and to the local economy, by providing them with a temporary retail area. Figure 9.10 shows a prefabricated dwelling designed on the basis of the shipping container in order to make it easily transportable using standard equipment.

The award-winning Japanese architect Shigeru Ban designed blocks of three-storied 'piles' of temporary container housing in Onagawa in Miyagi Prefecture, Japan following the 2011 Tōhoku earthquake and tsunami. This project comprised 189 apartments, each created within a converted container. The fitting out of the containers was undertaken by a group called Voluntary Architects Network. The cost of the project was US$465,000, an average of US$2,460 per unit. The project offered three options: apartments of 19.8 square metres for one or two residents; 29.7 square metres for three or four residents; or 36.6 square metres for more than four residents (Charlesworth 2014: 27–30).

Tents on campsites:

> The survivors lacked nothing: they have medicine. They have hot food. They have shelter for the night. Of course, their current accommodation is a bit temporary, but they should view it like a camping weekend.
>
> *(Italian Prime Minister Silvio Berlusconi, 8 April 2009)*

FIGURE 9.9   Surplus shipping containers converted into retail space in Christchurch, New Zealand after the 2010 and 2011 earthquakes had devastated the commercial centre of the city (photograph by David Alexander).

FIGURE 9.10   Standard container home based on the dimensions and specifications of a shipping container but constructed specifically as accommodation (photograph by David Alexander).

This memorable statement, made two days after the L'Aquila earthquake to survivors who had lost their homes and many of their family members, would be a strong contender if there was a prize on offer for crass political insensitivity. Some 28,000 people were accommodated in 171 camps, mostly in eight-person tents and for periods of six to nine months. They were well looked after, but there were problems of waterlogging and damage to the tents during summer thunderstorms. This solution could not have been employed during the winter.

The provision of campsites for disaster survivors has never been as extensive as it was after the 2010 Haiti earthquake. One month after the disaster, it was estimated that there were 1.5 million survivors in tents located on campsites. The Associated Press reported on 6 November 2013 that the number still in camps was 200,000. Then a further report by the International Organization for Migration (IOM) in June 2014, four and a half years after the disaster, stated that 8 per cent of the original camp population was still there. This followed their programme to provide rent subsidies for displaced families (IOM 2014). The figure amounted to a total of 103,565 people, or 28,134 households, who remained in 172 camps, suffering the extreme hardship of remaining in tents for almost five years. There are reports of forced evictions of camp dwellers and of fires being ignited in some camps by armed gangs who may have been employed by landlords seeking to clear their land. There is a report by Grassroots Online that the IOM statistics are flawed as IOM was instructed by the government not to include the sites of Canaan, Jerusalem and Onaville in the totals remaining in camps. It was assumed that the Government of Haiti did not want reports of large camp populations to inhibit new investment in the country (Gilbert 2013).

Some useful research was conducted in the Haiti camps in the initial period of occupation. A research team led by Anna Versluis, a geographer from Gustavus Adolphus College in the USA, interviewed a carefully selected sample of 53 camp survivors in March 2010, just two months after the earthquake. This may be the first attempt to determine an average profile of camp occupants through a sample survey, so the results are of interest despite the tiny sample from a population of 1.5 million. This is the profile:

- average age: 39 years;
- sex ratio: 60 per cent women, 40 per cent men;
- marital status: 30 per cent single, widowed or divorced, 70 per cent married or partnered;
- average number of children: 3.2;
- average number of years in Port-au-Prince: 24;
- percentage born in Port-au-Prince: 23 per cent;
- average years of education: 8;
- percentage unemployed or working in the informal sector only: 69 per cent;
- percentage with family living outside of Haiti: 51 per cent;

- percentage whose house was destroyed or damaged in the earthquake: 46 per cent destroyed, 52 per cent damaged;
- percentage who lost a family member (immediate or extended) in the disaster: 64 per cent;
- percentage who received material assistance after the disaster: 72 per cent.

The conclusions drawn from the study were as follows. A sizable proportion of households in Port-au-Prince were displaced. Most people in the camps received no material aid during the first seven weeks after the earthquake. Of those who received material aid from relief agencies, the assistance was modest. It mostly consisted of no more than a tent or several tarpaulins. However, the informal aid sector played a major role in providing material assistance. There were key differences in the forms of aid between the formal and informal aid sectors. The latter relied predominantly on cash transfers while the former provided aid in kind. Beneficiaries generally considered assistance in the form of cash transfers to be timelier and more effective than aid in kind (Versluis 2014). The experience in the Philippines after Cyclone Haiyan (known locally as Yolanda, and referred to by the local name in the remainder of this chapter) in 2013 was remarkably similar, although the timelines were shorter (Figure 9.11).

FIGURE 9.11   United Nations High Commissioner for Refugees (UNHCR) tents to accommodate survivors in the Tacolban area of the Philippines after Hurricane Haiyan (Yolanda) (photograph by David Alexander).

Evacuation shelters provided by the government: The 'official' sheltering policy for the initial phase in some countries, such as Japan, is to absorb the displaced population in very large numbers into requisitioned buildings, such as schools, community buildings, sports halls, etc. The survivors are then supported by volunteers, coming from all parts of Japan, who provide assistance in these 'shelters'. Table 9.3 indicates the vast scale of this operation in Japan following the Kobe and Tōhoku earthquakes.

One major problem in such indoor shelters is the lack of space for personal belongings and the lack of family privacy resulting from communal sleeping arrangements in school assembly halls and gymnasia. The architect Shigeru Ban became aware of the privacy problem following the Niigata earthquake of 2004: 'They put the victims in the gymnasium and people stayed there for six months

TABLE 9.3 The quantity and occupation of evacuation shelters following the Kobe and Tōhoku earthquakes

Number of evacuation shelters in school assembly halls, gymnasia, sports halls, community centres, etc.

	After 1 month	After 2 months	After 3 months	After 4 months	After 5 months	After 6 months	After 7 months
Kobe earthquake (Hanshin-Awaji Earthquake) 17 January 1995	1,138	1,035	1,003	961	789	789	639
Tōhoku earthquake and tsunami 11 March 2011	2,182	1,935	2,214	2,344	2,417	1,459	–

Number of people living in evacuation shelters

	Total displaced	After 1 month	After 2 months	After 3 months	After 4 months	After 5 months	After 6 months	After 7 months
Kobe Earthquake (Hanshin-Awaji Earthquake) 17 January 1995	350,000	307,022	264,141	230,651	209,828	77,497	50,466	35,280
Tōhoku earthquake and tsunami 11 March 2011	620,000	386,739	264,141	167,919	147,536	115,098	101,640	58,922

Source: data provided by the Cabinet Office, Government of Japan.[1]

FIGURE 9.12 Partitions used to provide privacy for communal living and sleeping in Japanese evacuation shelters (Paper Partition System 4 (c) Shigeru Ban Architects).

until temporary housing was ready. It was a terrible situation with no privacy. So I designed a partition system to give people privacy. It was very successful' (Ban, in Charlesworth 2014: 23). This system was made from cardboard sheets, paper tubes and cloth (Figure 9.12). The cardboard was used to cover the floor in crowded times at the beginning of the occupancy. After the population had decreased, it was used to create partitions for privacy at night (Pham 2011).

Renting accommodation in unoccupied buildings: In any society, especially those that are industrialised, there is always unoccupied accommodation and this can be requisitioned and used to accommodate displaced families. After the Mexico City earthquake of 1985, 20,000 families received temporary rental assistance in this manner (Comerio 1998: 134).

Provision of shelter materials and tools: In 1977, three weeks after a massive cyclone, Ian was sent to Andhra Pradesh to advise an agency on the shelter projects they could support. The casualties were enormous, with estimates of between 14,000 and 50,000 deaths and 3.4 million homeless. However, despite such massive needs, opportunities for external agencies to provide shelter seemed very limited as the surviving communities appeared to have met their own shelter needs. Gradually a possible explanation emerged as we observed large dumps, or stockpiles, of shelter materials – such as thatch for roofing, bamboo poles, etc. – that had been provided at the sides of roads for families to collect free of charge by local governments. The result was repair and improvised shelter activity on a vast scale, which occurred throughout the affected area.

What was significantly missing was essential practical advice, to come with the free building materials, on cyclone- or flood-resistant building techniques. The same was true in the Philippines after Hurricane Yolanda (Faure Walker and Alexander 2014). In India, one isolated attempt was made to fill this gap by a newly created organisation called ARTIC (Appropriate Reconstruction Training and Information Centre). This was initially supported by Oxfam, which attempted to apply the lessons they had gained from the Guatemala earthquake housing education programme that was in progress at this time.

Provision of patent shelter units: Ask any person closely involved in shelter following disaster about 'donor shelters' and they may take you to a filing cabinet in their office and pull out a bulging file of all manner of patent designs for shelter that have been sent to them for endorsement or purchase. In the 42 years that Ian has worked in this field, these designs have come in a continuous stream that has never dried up. The stream turns into a veritable river after any well-publicised disaster. These designs come from architectural, civil engineering or industrial design students and professionals, intrepid inventors, relief agencies and product manufacturers. Often, to gain momentum, they combine their resources to form consortia.

Probably the most extraordinary design yet to emerge is the 'Moss airdrop shelter'. *Time* magazine reported on the invention in 1976:

> Moss the tentmaker will not be fully satisfied until someone buys his favourite idea—an already tested shelter that can be rushed to earthquake, or other disaster-stricken areas. Carried over the site by helicopter and released in mid-air, it opens like a parachute and drops softly to earth, ready for immediate occupancy.
>
> *(Time 1976: 60)*

The publicity that accompanied the design stated that the shelter is to be jettisoned from an aircraft and 'through the differing accelerations of the air-resistant membrane and fast descent payload, is opened to the stable position in the air, landing upright and ready for immediate use' (Cuny, cited in Davis 1978: 49). Needless to say, it did not catch on and was not popular with users. A drawing of the Moss airdrop shelter appears in Davis (1978).

Back in the 1970s and 1980s, many of the leading agencies pursued the elusive quest for a 'universal shelter'. Ian recalls being asked to visit the Director of UNDRO (the predecessor of UNOCHA) in 1975: He asked me to consider designing such a shelter on behalf of the UN, and he even took me over to the window of his office in the Palais des Nations in Geneva to show me the well-cut lawn where he thought the prototype shelter product could be placed for a forthcoming press launch. Fortunately, the project changed direction and resulted seven years later in the first UN guidelines on shelter provision (UNDRO 1982). These have been refreshed and published as a second edition in 2015 (Davis 2015).

FIGURE 9.13 Polyurethane igloos provided by the German Red Cross and the Bayer Chemical Company for survivors of the Managua earthquake in Masaya, Nicaragua in 1973 (photograph by Ian Davis).

At that time, the West German Red Cross and the Bayer Chemical Company had developed their polyurethane igloos. These were supplied after earthquakes in Turkey, Peru and Nicaragua where 500 units were constructed (Davis 1978: 103). At peak, they achieved 45 per cent occupancy (Figure 9.13).

In 1975, Oxfam developed pentagonal shelters and used them after the Lice earthquake in Turkey, where 463 units were created and used with approximately 10 per cent occupancy. However, in response to a critical report in *New Scientist* concerning the potential risk of the shelters catching fire and resulting in toxic fumes that could kill the occupants, they abandoned the programme (see Davis 1978: 107).

In 1977, the UK Government's Overseas Development Administration (ODA), the predecessor of the Department for International Development (DFID), decided to develop its own 'universal shelter'. It commissioned the Industrial Design Department of the Royal College of Art to design the product. Ian was a member of the expert panel convened to review their solution. It included the head of the UK Building Research Station. Ian remembers: The designers presented us with a model of their solution, a lightweight plastic object with no windows and a hatch door that resembled a space re-entry capsule. A family of about six 'victims' could passively sit in a row on a form of extended sofa, but without any space to stand. Thus the occupants would become permanent 'couch potatoes' with no possibility of cooking, moving about, visiting a toilet or washing themselves. The review panel were totally perplexed and when we questioned the rationale for the design, we were told that the unit would survive any future earthquake and it would certainly float if a flood recurred! Inevitably, this ill-conceived project received the 'thumbs down' from our panel, never to be heard of again.

However, the move away from simplistic donor shelters as the ultimate 'supply-driven' products is an area of significant progress by UN agencies and international NGOs as, in more recent 'enlightened' years, it would be *almost* unthinkable for agencies such as the Red Cross, Oxfam, any UN department or any donor body such as DFID even to contemplate designing such products. The first 'almost' is the delivery of ferro-cement dome structures as permanent dwellings after the Latur earthquake of 1993 (this is described in Chapter 12; see Figure 12.15). A second 'almost' relates to the situation in Haiti following the 2010 earthquake. Perhaps due to Haiti's proximity to the USA, the authorities were bombarded with numerous examples of donor shelters or dwellings. They hit on a sensible way to manage these 'offers' by creating a shelter exhibition space, or park, just outside Port-au-Prince where agencies could inspect prototypes of each temporary or permanent design (Figure 9.14). It is not known whether any of these prototypes materialised into actual projects (Davis 2012).

FIGURE 9.14 'Innovative' house for the Haiti reconstruction, designed by Design and Architecture Service. The cost is US$12,000 per unit, and it is constructed of *'catch, mortier et bouteil'*. It carries a manufacturer's guarantee of 50 to 70 years service (photograph by Ian Davis).

Key lessons for Option 5b

As we noted in Chapter 8 in 'Shelter preferences and functions', the option of being directed to emergency campsites was firmly at the bottom of the list of favoured options, yet it seems to remain as the 'default' response of governments and many agencies. We also noted the limitations of tents and plastic sheeting relative to protecting people against extremes of climate, storing belongings, accommodating large families, pursuing occupational needs, and keeping pets and livestock. This section has shown the rich diversity of ways in which authorities meet sheltering needs. As this is the favoured area of work for most assisting groups, there is always a danger that excessive sums of money will be spent and resources will be deployed too lavishly in this phase of recovery, to the detriment of support for long-term permanent housing.

Most patent designs remain as novel ideas and never reach a development stage or field production. Of the many reasons for this lack of enthusiasm by decision-makers, probably the main one is that they are devised by people who are unfamiliar with disaster contexts; who have minimal awareness of the economics of shelter and housing, local building traditions, local climate constraints, the need to generate livelihoods in tandem with shelter and housing; and who have made no attempt to calculate the cost of transportation of their inventions by air freight and no grasp of the fundamentals of development. Thus, the implicit message concerning who decides and who designs, set in the quotation by Maggie Stephenson at the beginning of this chapter, would be a foreign language to many of those who invent and sponsor shelter designs.

Figure 9.15 refers to options 4, 5a and 5b ('after disaster: provisional shelter').

Options after disaster: transitional shelter

The subject of transitional shelter is discussed in detail in model 16 (Chapter 4 and reproduced in Figure 9.16).

Option 6a: transitional shelter (temporary)

We now discuss the significance of transitional shelter with reference to Typhoon Haiyan (Yolanda) in the Philippines in 2013.

Tacloban, a city of 220,000 inhabitants, is the capital of Eastern Visayas Region on the Island of Leyte in the eastern Philippines. On 8 November 2013, it was one of the principal settlements to be struck by Typhoon Yolanda, a hurricane that reached intensity 5 on the Saffir-Simpson scale with sustained wind speeds of 230 kilometres per hour and maximum gusts of 315 kilometres per hour. The dead and missing in Eastern Visayas amounted to 6,875 people, 94 per cent of the mortality experienced by the Philippines during the storm. In the region, 26,186 people were injured, representing 91 per cent of the national total. In Tacloban, the storm surge reached 5.2 metres, and on low-lying ground the

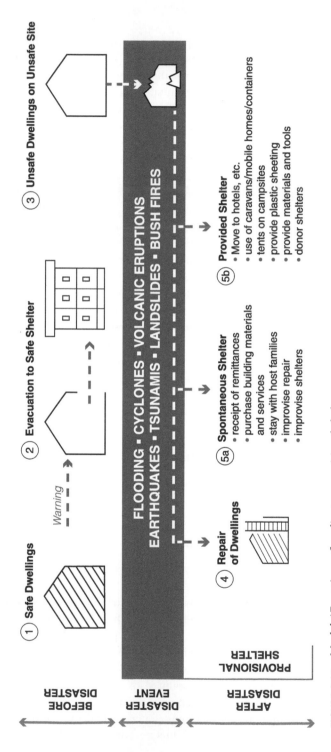

FIGURE 9.15 Model 17: options after disaster: provisional shelter.

SCENARIO 1: THREE-STAGE RECOVERY

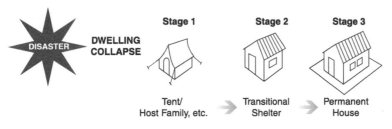

SCENARIO 2: TWO-STAGE RECOVERY

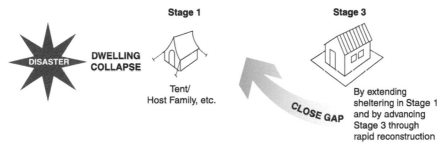

FIGURE 9.16 Model 16: scenarios for the shelter–housing continuum.

flooding extended for up to 1 kilometre inland. About nine-tenths of the city was severely damaged and many houses were completely washed away. Large numbers of people took shelter in the city's convention centre, located on the shore to the south of the urban centre, and they were drowned when the storm surge overwhelmed it.

Yolanda was one of the most powerful cyclones to make landfall since such phenomena were first recorded. To a certain extent the level of damage and destruction appears to have taken the Philippine Government by surprise but, in any case, damage to infrastructure was so profound that it proved very difficult to supply aid and assistance rapidly. Hence, according to contemporary estimates, barely one in five of the survivors had received any assistance after five days. Collectively, the Visayas Regions had 1.9 million homeless and 6 million displaced people.

Four months after the disaster, a team from University College London (UK) and the University of Tōhoku (Japan) interviewed many survivors in Tacloban and the neighbouring towns to the south, Palo and Tanauan. The aim of the survey was to examine the plight and survival strategies of survivors, particularly those at the bottom of the social scale.

There has been considerable debate in the literature on the question of what is the most appropriate strategy to manage the phase that links the immediate emergency and the onset of permanent reconstruction (Leon *et al.* 2009). Transitional shelter is widely used in this phase, but it has its opponents and detractors who argue that with adequate resources and organisation, it is possible to go straight from the improvised shelter that predominates in the immediate aftermath of a disaster to rapid, effective construction of permanent housing (Boano and García 2011; Clermont *et al.* 2011). At least, this is judged to be possible where vernacular housing is normally simple and does not take more than a few weeks to build.

The alternative view is that transitional housing is a means of bridging the gap that inevitably appears between improvised shelter and final reconstruction. Infrastructure may be too damaged to enable anything to be rebuilt rapidly, including houses; hazards may need to be assessed before planners can determine where it is safe to rebuild; and urban and regional plans may need to be drawn up or at least adapted to the post-disaster situation. In the case of Eastern Visayas, all of this is true.

The UCL-Tōhoku survey was restricted to a handful of *barangays* (local administrative districts), but these were carefully chosen to be representative of varying states of damage (Faure Walker and Alexander 2014). In many of them, there was relatively little social differentiation in that wealthier families tended to build bigger houses on larger plots in subdivisions that were also inhabited by the poor and needy. The house of the Barangay Captain acted as community centre and administrative headquarters and, in many cases, also as an evacuation centre during Typhoon Yolanda.

In the Tacloban area, there is a gradation of housing from the simplest one-room structures, built informally of wood and corrugated steel sheeting, to the more elaborate and substantial homes with two floors constructed from reinforced concrete. Such was the violence of the winds and storm surge in Yolanda that even examples of the latter were destroyed when concrete frames buckled, foundations were undermined by scour and infill walls were forced out of the frames and turned into fragments. However, there is abundant evidence that better building techniques could have saved more buildings in all categories: roofs fixed with 'hurricane straps', simple metal ties, improved concrete mix and design, better anchored posts, coverings for windows and doors, and so on.

The survey revealed that, as in so many other disasters, Yolanda has tended to increase social differentiation by disadvantaging the poor in the recovery process. The plight of many survivors of low social status amounted to outright destitution. At the bottom of the scale were those who had lost a breadwinner and sources of employment (e.g. in the destruction of fishing boats), were made homeless by the disaster, had no money to purchase relief goods, yet had to feed a large family. They ended up living in improvised shelter made of plastic sheeting provided by the international aid agencies and what materials (mainly wood and corrugated steel sheets) that they could forage from among the debris in the aftermath of the storm. In this respect, it is probably fortuitous that organised debris clearance and recycling was vastly slower in Eastern Visayas than it was in, for example, north-eastern Japan after the 2011 tsunami disaster.

The slowness of the response engendered widespread outbreaks of what the mass media described as 'looting' (Yap 2013). A better description would be 'foraging' or 'requisitioning' on the part of a population that was desperate for food, clean water and shelter. The looting, or whatever it was, led to some outbreaks of violence in the attempts to control it and some casualties in unregulated crowd movements.

In a disaster as large as that caused by Yolanda, and in an area with severe endemic poverty, there is bound to be a degree of dependency on outside aid. Many of the families interviewed in the survey endured the first four days of the aftermath with no food or water supplies. Government relief for the first four months was restricted to handouts of rice and canned foodstuffs. NGO assistance largely consisted of plastic sheeting, hygiene kits and cooking utensils. One NGO (Green Mindanao) provided traditional woven thatch roofing material, but only in certain areas. The day was saved by a single NGO, the Tzu Chi Foundation (Buddhist Compassion Relief) which was founded in Taiwan in 1966. Between November 2013 and January 2014, Tzu Chi handed out donations of 8,000, 12,000 and 16,000 Philippine pesos (US$175– US$350), calibrated with reference to size of family. Direct financial aid to survivors has been criticised as being a recipe for the misuse of funds (Peppiatt *et al.* 2001). However, in the Tacloban area, it appears to have been the single most important factor in stimulating the construction of transitional shelter for families whose only other options were to sleep rough or stay in overcrowded conditions with relatives. Among the respondents of the survey, there was a direct correlation between starting work in building a shelter and the date of receiving money from Tzu Chi. Those residents who, for one reason or another, missed the handout were visibly disadvantaged. Those who received it appeared to have used it wisely for activities connected with employment and shelter.

Unfortunately, there was a negative side. The most important raw materials for transitional shelter were wooden beams and laths, plywood and steel sheeting. With respect to the situation before the typhoon, the prices of all these materials increased by an average of between 20 and 50 per cent. In part, this was because of a lack of sawmill capacity that would have enabled a greater utilisation of the lumber derived from coconut trees felled by the storm; in part, it reflected a failure of government to control prices – never an easy task. Hence, the materials for building transitional shelter were beyond the economic reach of the poorest families who had to scavenge items of inferior quality.

The survey found that Yolanda had led to a severe contraction of the employment market. Fishing boats had been destroyed; so had vehicles, machinery and factories. Although some 'cash for work' schemes were implemented by the NGOs during the early aftermath, these were of short duration and the payments were meagre. Many of the poorest survivors were totally dependent on remittances or were reduced to scavenging. Some used their cash handouts to start small retail businesses, but the circulation of money in the severely affected *barangays* was greatly attenuated.

The Philippine Government passed an ordinance to establish a no-build zone within 40 metres of the coastlines most affected by Yolanda. Most residents surveyed were fully aware of this, but some were living in transitional or informal shelter in the designated zone (in north Tacloban, the shelters were also located on the seaward side of a clutch of beached ships). To remove people from the no-build zone, the government began a programme of constructing settlements of 'bunkhouses' – prefabricated shelters arranged in rows. These were built to a minimum standard of coconut lumber, plywood and steel sheet roofing materials. The size of each unit was about 10 square metres, although they could be amalgamated into dwellings of 20 square metres. Cooking areas, latrines and showers were accommodated in separate areas.

Concentrations of prefabs built on greenfield sites have been criticised as being liable to 'ghetto-ise' survivors by crowding them into enclaves. In our opinion, this is only a valid point if the prefab 'villages' remain occupied for a long time or if they markedly reinforce existing social differentiation. As was the case with similar constructions built after the 2009 earthquake in Padang, Indonesia, the bunkhouses erected in the Tacloban area were visibly lacking in the robustness needed for longevity. If they are evacuated in favour of permanent housing within a very few years, there will be no enduring issues. If, instead, their inmates are abandoned to live there without a permanent solution then there will be severe problems as the buildings deteriorate and discontent with living in crowded and cramped accommodation grows.

The population of the Philippines is about 100 million and is growing at a rate of 2.04 per cent per annum – one of the highest rates in Asia. Among the poorer classes of survivors in the Tacloban area, family size is large and women bear children from a relatively early age. Gender is an important issue in resilience during the transition phase. The UCL-Tōhoku survey found that there were many cases in which deaths were restricted to the adult male members of the family, who remained behind to defend their homes while the women and children evacuated in response to the warning of impending typhoon landfall. This meant that many widows had to assume sole responsibility for bringing up their children and recovering from disaster. The survey encountered other instances in which the husband or father of children had decamped long before the storm. At the same time, in the aftermath of the typhoon, employment opportunities for women were severely limited. Nevertheless, women coped, often heroically.

Gender is part of a wider problem of vulnerability. For many, lack of food security, assets and income led to a restricted diet, a situation not helped by the pernicious influence of soft drink manufacturers in the local area. The survey revealed a situation in which employment lost was very difficult to regain. The higher social strata were represented by people who had government employment or large remittances from abroad. They tended to live in much more substantial housing than that of the people without such attributes. The survey also found that virtually no one among the poor had received any technical assistance or advice in building shelter. Those who could afford it hired carpenters to build their prefabs, but this implies only a minimal level of expertise. The Philippines endures an average of 22 tropical cyclones per year. Should an intense one make landfall on Leyte Island during the lifetime of the shelters, it will devastate an already impoverished and enfeebled society and sweep away many of the prefabs.

Finally, the survey revealed a high rate of landownership. However, some of the land fell within the Philippine Government's no-build zone, which extends 40 metres perpendicular to the coast. Moreover, Western notions of landownership probably do not apply to situations in which boundaries are not defined in an adequate cadastre, where intense storms or tsunamis are liable to change the configuration of the coast, and where there is no guarantee that people will find that their land claims are respected in any compulsory relocation initiative.

In conclusion, the example of the Philippines demonstrates the importance of stability and continuity of employment and income in creating resilience to major disasters such as tropical cyclones. For the poor, simple, modest resources are the key to survival. In Tacloban, cash did not induce laziness or debauchery but, rather, enabled families to devise a survival strategy where otherwise none would have been possible. Alternatives to this would have needed to be complex, intensive and ubiquitous; no such solutions were applied.

In Eastern Visayas, a wholesale dependence on transitional housing could hardly have been avoided. Relief was slow in arriving in the disaster area and the resources to recover quickly do not exist. A relatively poor area afflicted by what some are calling the most powerful cyclone ever recorded could hardly spring forward into an advanced state of development. However, relatively simple measures could have been used to avoid the wholesale reinstatement of vulnerability; for example, by organising better employment conditions, reducing the price of construction materials and ensuring that technical assistance was widely available. As Kennedy *et al.* (2008: 24) note: 'Overall, "building back safer" might be a preferable tagline to "building back better" because "better" has multiple interpretations, many of which caused further problems, whereas "safer" provides a clearer goal on which to focus for post-disaster settlement and shelter.'

The Philippines ranks 94 out of 175 nations that are listed in the Corruption Perceptions Index (Transparency International 2013). Its score of 36/100 on the Index is well under the 50 mark, below which countries are deemed to have a serious corruption problem. At the same time, it ranks 3 in the World Risk Index (ADW 2012). This is evident in the fact that a third of the population had been affected by natural disaster over the last decade (IFRC 2013). The country ranks 97 in the Human Development Index, with a relatively low headcount of poverty but a high concentration where it occurs (UNDP 2010). The combination of a highly unequal society, inadequate transparency in governance and enormous vulnerability to natural disaster bode ill for the recovery of an area that is peripheral to national life, hence poorly connected in economic terms, and yet is at the front line for storms, tsunamis and earthquakes.

Option 6b: transitional shelter (to evolve into permanence)

This process is far removed from option 6a. While the transitional shelters in 6a are designed to provide a 'stopgap solution', they are expected to be replaced by permanent dwellings. In this category, transitional shelters are designed to evolve into permanent dwellings. This occurs as the majority of the building materials are recycled for reuse in the permanent dwelling. Where space is available, this form of transitional shelter (or housing) can expand by addition of rooms to the permanent dwelling.

Option 6c: move directly from provisional shelter to permanent dwelling

Option 6c is a way to cut out transitional shelter altogether by moving directly from some form of basic provisional shelter into a permanent dwelling. This process is presented visually as 'Scenario 2' in model 16 (Figure 9.16). In the discussion on model 16 in Chapter 4, we provide the example of this approach being adopted in Mexico City following the 1985 earthquake.

Key lessons for options 6a, 6b and 6c

The debate concerning the pros and cons of transitional shelter is ongoing and the discussion of model 16 in Chapter 4 highlights the pros and cons. Our position on the controversy is as follows. If it is possible to avoid erecting transitional shelters by extending the spontaneous sheltering process and by accelerating safe and effective reconstruction then this is certainly the preferred option. However, if there are severe climatic conditions from which survivors need protection, or if reconstruction is likely to be a protracted process due to the need to resolve legal titles to property, the need to revise building codes, and so on, then some form of transitional shelters will be needed. If transitional shelters are needed to satisfy the above reasons, they should: first, not block the sites of permanent housing; second, be capable of evolving into permanent dwellings to avoid waste; and third, be basic rather than expensive shelters that are of a similar price to fully reconstructed houses. The aim is to avoid the use of transitional shelters that consume large amounts of money and thus reduce the amount available for permanent reconstruction. This is shown in model 7 (Chapter 3), which concerns the escalating unit cost of reconstruction.

Figure 9.17 refers to Options 6a, 6b and 6c ('after disaster: transitional shelter')

Options after disaster: permanent dwelling

When considering the construction of permanent houses after disaster, there are two concerns. First, what is the typical level of destruction of dwellings in disasters and, therefore, how much reconstruction is required? Second, who

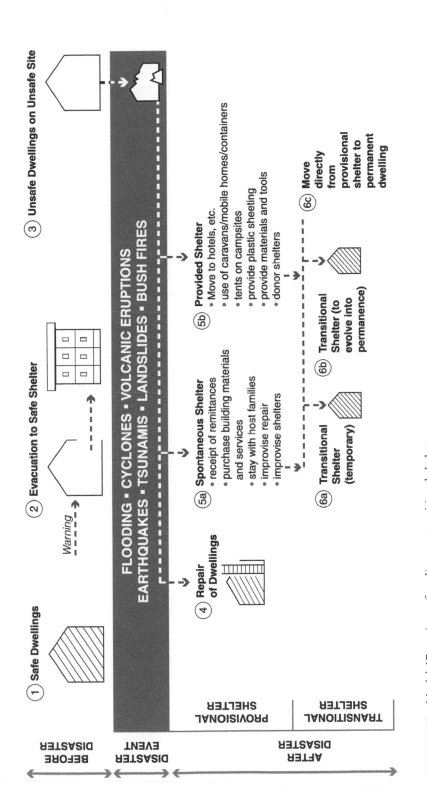

FIGURE 9.17   Model 17: options after disaster: transitional shelter.

reconstructs permanent dwellings after disasters? Regarding the typical level of destruction of dwellings in a major disaster, Bill Flinn (2013) graphically described the scale of the problem:

> Cyclone Nargis in Myanmar (2009) destroyed 450,000 houses, Cyclone Sidr in Bangladesh (2007) 400,000, the Pakistan earthquake (2005) 400,000 and the floods in Pakistan in 2010 an extraordinary 1.6 million homes. To put this into perspective, the number of houses built in England by the entire construction industry is typically in the order of 150,000 a year.

Concerning the question of who reconstructs dwellings, the Haiti earthquake of 2010 resulted in the destruction of 105,000 dwellings and damage to 100,000. In response, an astonishing 110,000 transitional shelters were built in the first two years after the disaster. In January 2012, Maggie Stephenson, who was working for UN-Habitat on behalf of the Government of Haiti, carried out a construction survey. This indicated that the government, international agencies and NGOs had together built a total of 5,189 houses while the residents themselves had rebuilt 50,000 without external assistance. The latter were unlikely to survive a future earthquake or hurricane.[2] These figures are probably typical of other disaster recovery situations in developing countries, where official bodies rarely succeed in building more than approximately 10 to 15 per cent of the total number of dwellings constructed. There are two sides to such massive 'self-build' activity. In positive terms, this is indicative of a high level of resourcefulness among the surviving population; but in a negative sense, it demonstrates the need to develop radically different strategies in order to avoid reconstructing vulnerability on a heroic scale.

There is an encouraging reminder that large volumes of houses *can* be constructed and repaired rapidly. In the Pakistan rural reconstruction programme, described in Chapter 12 under the title 'Yellow hat – optimism', 463,243 safe houses were constructed by adopting a user-build approach with untrained builders being taught to construct 'well' and 'safely'. Some 130,000 houses were repaired in a period of only three and a half years. Gauging the scale of this achievement by means of Bill Flinn's comparative figure, the rate of construction in this single housing project in rural Pakistan was the same as the entire housing construction programme throughout the whole of England.

Option 7a: user-build permanent dwellings

In *The Stones of Venice*, John Ruskin expressed a profoundly important concept, which provided the basic philosophy that lies behind the self-build movement.

> Understand this clearly: you can teach a man to draw a straight line, and to cut one; to strike a curved line, and to carve it; and to copy and carve any number of given lines or forms, with admirable speed and perfect precision; and you find his work perfect of its kind: but if you ask him to

think about any of those forms, to consider if he cannot find any better in his own head, he stops; his execution becomes hesitating; he thinks, and ten to one he thinks wrong; ten to one he makes a mistake in the first touch he gives to his work as a thinking being. But you have made a man of him for all that. He was only a machine before, an animated tool.

(Ruskin 1853: 161)

The advantages to the reconstruction of permanent dwellings of the user-build approach are extensive, with strong *internal* and *external* benefits. As an external benefit, it can lead to sustainable settlements and be an active expression of development. It fosters the active participation of surviving communities, mobilises communities and develops building skills, which in turn creates livelihoods. It helps develop leadership and reduce the risks of corruption. It can be used to build safety into reconstructed settlements. The approach lends itself to the provision of cash assistance directly to homeowners. Finally, as noted in Ruskin's famous and influential quotation, it has the internal benefit of endowing user-created dwellings with pride and identification. Opponents of the user-build approach often claim that it may be slower than contractor-build approaches, but the Pakistan experience provides contrary evidence.

Following the Guatemala earthquake of 1976, an alliance was formed between two international NGOs, Oxfam and World Neighbors, in order to initiate the groundbreaking Housing Education Program. Endowed with technical advice from Fred Cuny of INTERTECT, the aim was to teach rural labourers how to rebuild their dwellings 'better and safer'. It is now recognised as a pioneering reconstruction programme. The question that faced the organisers was who to select to lead the project. They were looking for a project leader who came from a rural background and who would command respect from the trainees. In an inspired move, they appointed a Christian pastor who was on the faculty of a Bible school in rural Guatemala.

The choice was made to make use of a remarkable individual who built his family a safe dwelling after a previous earthquake. As a young man, Pedro Guitz and his family had lost their house, and he decided to rebuild using earthquake-resistant construction. He did this by visiting a public library to determine the principles and practice of anti-seismic construction, and books were his only teacher. So in Tecpán, the town where he lived, his was the only house to survive the 1976 earthquake, and this excited much curiosity amongst the survivors who had lost their homes. Thus, Guitz became the main teacher, a person who was able to preach what he had personally practised (Figure 9.18). This user-build housing education programme has now been recognised as the forerunner of subsequent reconstruction initiatives in Afghanistan, India, Sri Lanka and Pakistan.

However, there was a grim and totally unexpected sequel to this enlightened programme. In the early 1980s, right-wing forces of the repressive Guatemalan Government, with strong support from the US Government of the time, sought to halt the spread of communism in Central America by cracking down on

FIGURE 9.18 Rev. Pedro Guitz explaining earthquake-resistant construction techniques to local builders in rural Guatemala, in the Oxfam-World Neighbors Housing Reconstruction Programme in 1976 (photograph courtesy of Oxfam).

anyone who they perceived to be left-wing social activists or local leaders. As some strong leaders had emerged during the Oxfam-World Neighbors Housing Education Programme, many were hunted down to be killed, and others had to flee the country to avoid the death squads.

The Guatemala earthquake was the first time that the user-build approach to housing – developed by John F. C. Turner in the 1960s and 1970s (Turner and Fitcher 1972; Turner 1976) – was applied in a disaster context with a specific focus on safe reconstruction. (Turner's vital contribution is noted in Chapter 2, 'Evolution of recovery studies'.)

The modest Guatemala earthquake reconstruction programme of 1976–9 is the ancestor of the ambitious Government of Pakistan rural housing education programme that took place almost 30 years later in Pakistan from 2005–8 (referred to above and described in detail under the heading 'Yellow hat – optimism' in Chapter 12). While the Pakistan programme was much larger than that in Guatemala, its basic aims and broad methodology were the same. Both programmes were attempts to apply developmental principles to housing reconstruction.

Option 7b: contractor-build permanent dwellings

The advantages of the contractor-build approach to the reconstruction of permanent dwellings lie in speed of construction, the ability to control quality and to scale up operations rapidly, and the ability to achieve transparency over disbursement (da Silva 2010: 75). One of the disadvantages is that although they acquire dwellings at the end of the process, surviving communities do not necessarily acquire the know-how, livelihoods, leadership, participation and pride in their created dwellings that come with the user-build approach. Moreover, there are much higher risks of corruption with the contractor-build approach. The tendering procedures can end in the award of contracts to contractors from outside the affected area, thus failing to generate essential local employment and demand for locally-based building materials when this is needed to support recovery.

An example of how survivors can exercise choice in the design of contractor-build housing comes from the construction market that arose after the Bam earthquake of 2003 in Iran. A few months after this event, the Government of Iran established an 'engineering and technical exhibition site' in Bam, where over 500 housing designs, construction techniques and building materials were exhibited to enable the surviving families to choose their preferred housing model. In addition, the Housing Foundation of Iran, working jointly with the UN Development Programme, organised a series of consultations to ensure that the views of female heads of households were heard and incorporated into house designs. This programme also offered builders training in safe earthquake-resilient traditional building techniques (Kianpour, cited in Jha *et al.* 2010).

Mahmood Hosseini and Yasamin Izadkhah from the International Institute for Earthquake Engineering and Seismology in Tehran observed that: 'The idea of convening a "construction market" worked better than building modular complexes and giving them to the homeless people. The liberty in choosing the type of materials and the architectural design was also much better for the beneficiaries' (Hosseini and Izadkhah 2004). However, there were two problems with this intended freedom of choice. First, a problem arose from the remote siting of the construction market and engineering offices, which was too far from the centre of Bam and the location of most of the residents who needed to visit the market at some point – and quite regularly in many cases (Hosseini *et al.* 2008). The second concern was noted by M. Mobasser, a researcher who studied the housing recovery situation in 2005–6 for her Oxford Brookes University master's thesis. She observed the lack of options provided to the residents, despite the exhibition showrooms of available designs. She further noted that the designs were determined by the simplest options that created the maximum profit (Mobasser 2006).

The second example concerns the Make it Right Foundation, set up after Hurricane Katrina struck New Orleans in 2005. The film star Brad Pitt, who had a home in New Orleans, visited the city's Lower Ninth Ward two years after Hurricane Katrina and was shocked by the lack of rebuilding progress in this historic, working-class community. There was a fear that the stricken neighbourhood would be forgotten. Therefore, he provided the initial resources and mobilised high-level support, including assistance from President Clinton and the Clinton Global Initiative, to build safe and sustainable dwellings in the area that had been the hardest hit in the most devastated part of the city. He commented as follows:

> We couldn't bring back the residents' heirlooms, photographs, or their lost
> loved ones, but maybe we could build a bridge home for those struggling
> to rebuild their lives before all opportunity atrophied and their lots fell prey
> to speculators. Maybe we could offer a more humane building standard.
> Maybe we could turn tragedy into victory. We would begin in the historical
> Lower Ninth Ward because of its iconic significance, and because it had
> the most difficult shot at coming back.
>
> *(Brad Pitt's 'Foreword' from Feireiss 2009)*

Therefore, the Make it Right Foundation convened 21 world-famous architects
to design climate-adapted, ecologically friendly homes. Then, over a period of
many months, Pitt and the architects met with Lower Ninth Ward homeowners
and community leaders to discuss their rebuilding needs and collaborate on
home designs. Surviving members of the community were able to choose from
21 innovative designs, including single-family and duplex houses, and to customise
their homes by choosing the paint colours, flooring, cabinets and countertops to
suit their styles and needs. The average single-family home that has been built
has a floor area of 130 square metres (Figure 9.19).

FIGURE 9.19 Frank Gehry and Partners, internationally known architects
based in Los Angeles, designed this duplex home (photograph by Ian Davis).

By 2015, the foundation had completed 100 homes, and these are now occupied by 350 people. The houses are built with advanced energy-saving credentials as well as with some safety measures. For example, they are elevated between 1.5 metres and 2.4 metres above ground level and are structurally engineered to withstand winds of 200 kilometres per hour and flood surges. The materials of which they are constructed must resist water damage and mould. Hurricane-resistant roofing, siding and window systems are utilised, and rooftops are designed to provide safe havens during catastrophic floods (Make it Right Foundation 2015).

In 2010, while conducting research for this book, Ian visited this project and left with conflicting reactions. The generous commitment of Brad Pitt, his colleagues and the architects involved in the Make it Right Foundation who wanted to reconstruct dwellings in this poor and highly vulnerable neighbourhood was highly commendable. The choice of house designs offered to the beneficiaries was an excellent feature of the programme, as were the high environmental and safety standards that were applied. Inevitably, Brad Pitt takes great pride from the achievements of the project:

> This neighbourhood which suffered such travesty, in fact the hardest-hit spot of the hardest-hit area, the icon of all that was wrong with the recovery effort, has now become the most ecologically performing and intelligently built community in all the United States. Someone in D.C. should come down here and seriously take a look at this.
>
> *(cited in Feireiss 2009: 471)*

He rightly noted that: 'the final paragraph of this story of "Make it Right" is unknown to me, because it is the community that will define it. As it should be.'

There is a negative side to the wide variety of house designs, some highly ostentatious and eye-catching, as they hardly contributed to the creation of a unified settlement or community. Instead, they resemble a trade exhibition of dwellings, competing for attention and buyers. This impression is enhanced by the procession of tour buses bringing a stream of curious spectators to see the houses as one of the latest tourist sights of New Orleans.

Owners, renters and squatters and reconstruction

In most industrialised societies, particularly in urban areas, the housing stock comprises a mixture of rented and owned housing. For example, in New Orleans in 2010, five years after the impact of Hurricane Katrina, there were 189,896 housing units (US Census Bureau 2015). In 2010, 47.3 per cent of the city's housing stock was owner-occupied. Damage surveys indicated that 61 per cent of the owner-occupied dwellings suffered major damage from Hurricane Katrina and 51 per cent of rented housing suffered major damage. According to the New Orleans Master Plan and Comprehensive Zoning Ordinance, 80 per cent of subsidised affordable housing suffered major damage.

Some approaches to reconstruction are more suitable to certain groups than others. The World Bank has provided guidance on the most suitable solutions for specific groups, emphasising:

> the importance of addressing the reconstruction requirements of owners who are landlords, since renters—a large proportion of the population in some countries, especially in urban areas—will be dependent on reconstruction by landlords. It is unlikely that a group of apartment dwellers (even if they were condominium or cooperative owners) would band together to reconstruct their units, particularly if reconstruction entailed relocation. However, this option is included here. More likely, they would liquidate their holdings and relocate elsewhere.
>
> *(Jha et al. 2010: 102)*

Table 9.4 lists some approaches for reconstruction in relation to the kinds of situation in which the beneficiaries find themselves.

There are examples of contractor-based reconstruction in Chapter 12. This includes a highly effective NGO-World Bank project in Banda Aceh, Indonesia

TABLE 9.4 Tenancy categories and appropriate kinds of permanent housing reconstruction

Tenancy categories of affected population	*Suitable reconstruction of permanent housing approaches*
1 House owner-occupant or house landlord	Any approach, whether 'user build' or 'contractor based' may apply.
2 House tenant	If a tenant can become a house owner-occupant during reconstruction, see 1. If the tenant becomes an apartment owner-occupant, see 3. Otherwise house tenants are dependent on landlords to rebuild.
3 Apartment owner-occupant or apartment building landlord	Owner- or community-driven reconstruction if owners as a group can function as a 'community'. Reconstruction of multifamily, engineered buildings will always involve contractors, but owners may not require help of agency.
4 Apartment tenant	If tenant can become a house owner-occupant during reconstruction, see 1. If tenant becomes apartment owner-occupant, see 3. Otherwise, apartment tenants are dependent on landlords to rebuild.
5 Land tenant (house owner)	With secure tenure, same as 1, house owner-occupant. Without secure tenure, same as squatter category (see 6).
6 Occupant with no legal status (squatter)	If squatter can become a house owner-occupant during reconstruction, see 1. If squatter becomes an apartment owner-occupant, see 3. Otherwise, squatters are dependent on landlords to rebuild, or they remain without legal status.

Source: adapted from Jha *et al.* (2010: 102).

in 2006 following the earthquake and tsunami (described under the heading 'Black hat – discernment'). This is an example in which the participation of survivors is paramount as all are able to exercise choice in the design of their dwellings and in the mode of construction. Some of these houses adopted a strong user-build approach while others were totally constructed by contractors.

Another contractor-based project described in Chapter 12 (under 'Red hat – emotions') is a post-cyclone project in Chinthayapalem in Andhra Pradesh, India in 1978. It was more problematic in its original design although it resulted in positive outcomes over time.

Option 8: relocated dwellings in relocated settlement

In Chapter 13, we conclude our book with a set of principles, the eighth of which relates to this option succinctly by stating: 'Relocation disrupts lives and is rarely effective. Thus, it should be used as little as possible' (p.310). The World Bank have provided some skin to fit on these bones as guiding principles for relocation:

- An effective relocation plan is one that the affected population helps develop and views positively.
- Relocation is not an 'either/or' decision; risk may be sufficiently reduced simply by reducing the population of a settlement, rather than by relocating it entirely.
- Relocation is not only about rehousing people, but also about reviving livelihoods and rebuilding the community, the environment, and social capital.
- It is better to create incentives that encourage people to relocate than to force them to leave.
- Relocation should take place as close to the original community as possible.
- The host community is part of the affected population and should be involved in planning.

(Jha et al. 2010: 77)

There are two important matters to note. The first concerns the micro scale of relocation where a given dwelling, or building, or road or element of infrastructure *must* be relocated, even if only a short distance away from some precipitous slope, river bank, ravine or earthquake fault. In such cases, it would be totally irresponsible to contemplate rebuilding *in situ*. It should be noted that, in response to the devastation of coastal communities following cyclonic surges or tsunamis, authorities in various coastal locations of Thailand, Sri Lanka and the Philippines have attempted to create coastal buffer zones in which fishermen are banned from building houses or work sheds on specific parts of beaches. There has been a pattern in which, in the political heat after disasters, such buffer zones were unrealistic and were designed before there had been any serious attempt to assess their impact on livelihoods. Later, following protests by the affected communities, the distances over which coastal building was banned inevitably became reduced (Ingram *et al.* 2006; Lloyd-Jones 2007; see the reference above to the buffer zone following Typhoon Yolanda in Option 6a transitional shelter – temporary).

The second concern relates to more macroscopic situations in which there has been ground displacement – where there is no alternative to relocating an entire settlement or community as the actual topography has been changed by disaster. This may involve the loss of land due to a tsunami, volcanic eruption, landslide, river flood or episode of coastal erosion, or where soil may liquefy due to seismic forces.

The relocation theme is a vital element in recovery planning and recurs in various places in our book. For example, Chapter 3 gives a description of the relocation of the city of Wenchuan in China after the 2008 Sichuan earthquake. We also discuss relocation in Chapter 7 as our 'Third dilemma: reconstructing existing unsafe settlements vs relocation to safer sites'. In this section ('Options after disaster: permanent dwelling') we discuss four examples of relocation and its outcomes in response to earthquake and hurricane risk.

Figure 9.20 represents options 1, 4, 7a, 7b and 8 ('after disaster: permanent dwellings').

Summary and conclusions

The essence of this chapter concerns the need to understand that effective recovery from disasters, sheltering and housing is best regarded as a seamless process rather than as a set of isolated options or the delivery of tangible products. However, many agencies and writers artificially divide the process up into tidy, well-defined phases and separate options. Seen from the standpoint of the survivors of a disaster, the actions that they or groups that assist them take at the outset of a disaster, or even before a disaster, will certainly have long-term consequences. Hence, this is a seamless journey from a crisis in their lives that we hope will eventually lead towards a safe and permanent dwelling.

Significantly, for operational reasons, for reasons of preference or to suit operational mandates, the vast majority of international humanitarian agencies 'cherry-pick' the aspects of sheltering that appeal to them, without acting in relation to the *totality* of this process. The result is that some are interested in emergency shelter options and others concern themselves with transitional shelter options, but few, perhaps for very good reasons, do not wish to enter into the demanding process of assisting survivors in the complexities of securing permanent safe construction. Thus, this is left to national governments, locally-based institutions, the private sector and the surviving community. This lack of integration opens the way for multiple failures to connect, which result from inability to recognise how a given option needs to flow naturally and sensibly into another.

The implication of this mismatch between the reality of a seamless sheltering process and the international community's selective focus are twofold. First,

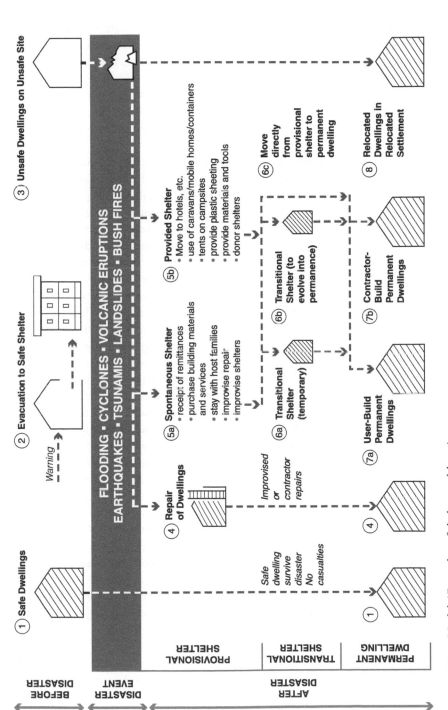

FIGURE 9.20   Model 17: modes of shelter and housing.

scarce financial resources are deployed in one phase, such as transitional shelter, while that money is desperately needed in permanent safe construction of dwellings and associated infrastructure. Second, actions taken in the immediate aftermath of a disaster need to be taken with regard to their long-term effects upon full reconstruction; but if the agency lacks vision, and perhaps is only interested in transitional shelter, it will inevitably fail to see this wider reality. Thus, either transitional shelters may be sited exactly where a permanent house needs to be or the agency fails to see the need for transitional shelter to evolve into a permanent, safe structure.

At the beginning of this chapter, Table 9.1 identifies the various roles of key actors in sheltering and housing. This highlights the challenge concerning the range of options that we have attempted to describe in the chapter. Table 9.1 shows that, aside from the role of national governments, assisting groups focus on the specific options that they favour and may take no role in others. For example, only a handful of international NGOs play any role in the creation of permanent dwellings whereas they are highly active in providing provisional or transitional shelters. While the reasons for this are understandable, because they relate to a given agency's mandate or interests, they nevertheless result in an obvious failure to look at the big recovery picture in terms of where resources are needed, or in failure to consider how a given option, such as the creation of transitional shelters, can inhibit the construction of permanent dwellings.

An example of this interaction between the various sheltering options was noted by researchers in Bam, Iran after the 2003 earthquake when they found that temporary shelters delayed the reconstruction of permanent housing (Hosseini and Izadkhah 2004). The main reason was that people became used to the temporary shelters and started to treat them as if they were permanent dwellings.

The artificiality of regarding the options or phases of recovery as watertight compartments is detrimental to the good idea of creating transitional shelters that evolve into permanence, rather than transitional shelters that are purely temporary. This view is far more likely to occur in agencies that play a role in the creation of permanent dwellings. This may explain why there is a neglect of our options 7a and 7b by international NGOs. This issue takes us straight back to Koenigsberger's penetrating quotation, which we cited in Chapter 2 (see p.32): 'Relief is the enemy of reconstruction. Therefore minimise relief' (quoted in Davis 1978: 66).

Hosseini *et al.* provided support for this wise assertion in their research after the Bam earthquake. They noted the gradual decline in the nation's concern in comparison to the outpouring of emotion in the immediate aftermath of the disaster:

> Thus the critical situation of the region was rapidly forgotten. This 'fall-off' of concern occurred just when the people needed more financial and emotional support. In many ways this was more necessary in later stages of the recovery process than in the early days after the disaster.
>
> *(Hosseini* et al. *2008)*

The only group of actors that experiences the full range of sheltering and housing options is, of course, the survivors.

If there is to be a positive change in coming years, it may be a belated awareness on the part of all assisting groups that they can play useful roles in supporting option 5a: 'spontaneous shelter'. Probably, there is a reluctance to become involved in this option as there is inevitably less visibility in supporting the accommodation of survivors by host families than in creating a well-defined set of new transitional shelters.

In reviewing the range of shelter options in this chapter, we have attempted to show how the subject relates to extremely important cultural and community aspects. Both rebuilt housing and transitional shelter must seek to preserve community and ensure that it functions in the aftermath of disaster. This includes social support networks and economic and logistical aspects of daily life, as well as the spiritual and psychological aspects of being in a community. As far as possible, shelter should not stigmatise or punish people for being disaster survivors. It should facilitate their recovery by allowing them to conduct or rebuild their normal economic activities and to engage in normal kinds of social interaction. Social support and solidarity should not be inhibited by the choices made in the provision of shelter, which should seek not to make divisions in society but to heal them. Shelter should not stigmatise or 'ghettoise' people as disaster survivors, and neither should it place them in situations of renewed vulnerability.

Shelter that is poorly designed from the functional, community and environmental points of view reduces the opportunities to recover from disaster. Well-designed shelter that is made with sensitivity to these issues can facilitate recovery, especially if it encourages social cohesion and the sort of participatory processes inherent in good governance. Shelter is not merely about housing; it is also about accommodating other activities, from workshops to worship, community gatherings, clinics, libraries, and so on. Absence of attention to these needs – and excessive concentration on providing *only* housing – will restrict the opportunities for recovery from disaster.

Unless vernacular housing in a particular area can be reconstructed with few materials and little investment of time and money, transitional shelter enters into the equation. There are unresolved issues about how 'permanent' transitional housing should be. More needs to be known about the balance between investing in transitional shelter and permanent reconstruction. As we have noted, with the usual shortage of funds, concentration on one of these issues would tend to restrict the other. If it is deemed necessary, transitional housing would be designed to provide sufficient support and to last long enough to tide people over until a vigorous programme of permanent reconstruction has been completed. This also requires thought to be given to how to achieve the progression from transitional accommodation to permanently rebuilt environments. With all of these questions, there is no single, clear-cut answer, but there is also currently a shortage of knowledge and expertise. (Alexander, in Davis 2011b: 201).

In the next chapter, we focus on 'resilient recovery' and seek to describe how recovery can become a sustainable process. We suggest ways to ensure that it will lead to a lasting recovery, producing stronger communities and safer living environments.

Notes

1 The statistics shown in Table 9.3 can be viewed in graph format at: www.cao.go.jp/shien/1-hisaisha/pdf/5-hikaku.pdf
2 Lecture by Maggie Stephenson at Oxford Brookes University, 10 February 2015.

10

RESILIENT RECOVERY

What is resilience, and what is resilient recovery?

If we advocate that recovery from disaster needs to be resilient, we must first define what we mean by resilience. This may seem an unnecessary step, but one publication (O'Brien and O'Keefe 2013) supplied a table of 28 different definitions of the term. The concept of resilience is more than 2,000 years old (Alexander 2013) and has acquired meaning in many different disciplines, including engineering, ecology, psychology, law, theology and statesmanship. Over the centuries, it has acquired a wide variety of overtones and interpretations. One is inclined to think that we have asked too much of it. Debates have raged around whether it should represent stability or change (Walker 1998; Manyena et al. 2011). To avoid the very common problem of becoming trapped in a definitional morass, we propose to use a single, broad definition of the term. Hence, we regard resilience as a combination of resistance and adaptability. This is inspired by nineteenth-century work in mechanics in which materials under load needed to have brittle and ductile characteristics so that they both resisted and adapted to the load by maintaining their structural integrity but deforming to absorb the some of the applied force. Society cannot eliminate hazard and disaster and so it must adapt, but it should also resist the impacts by 'hardening' everything from structural defences to institutions (Hyslop and Collins 2013).

For a society to be resilient, its healthy characteristics need to survive disaster without significant degradation or loss of integrity, and its unhealthy elements need to be contained and reduced. Essential elements include transparent administrative processes, accountable decision-making, participatory democracy, freedom from repression and corruption, peace and stability, guarantees on human rights, and adequate access to basic necessities such as food, shelter and employment. Generally, the less unequal societies (i.e. those without massive wealth gaps between rich and poor) are the healthier ones, as are those with the most active and fair democratic processes. Unfortunately, 69 per cent of countries score lower than 50/100 on Transparency International's Corruption Perceptions Index, indicating a serious corruption problem; and the proportion rises to 95 per cent in Eastern Europe and Central Asia (Transparency International 2013). This severely limits the opportunities for creating genuine resilience as it is undermined by corrupt practices. Work by Escaleras *et al.* (2007) and Ambraseys and Bilham (2011) has pointed to a strong correlation between earthquake disasters and corruption. Although correlation does not prove causality, the inference is a very strong one. Nonetheless, there has been some debate, largely unpublished, as to what corruption means (Soliman and Cable 2011). What matters here are large-scale attempts to illegally subvert public resources for private profit, whether that be pecuniary or related to the acquisition of power and influence.

In recovery from disaster, much emphasis is given to the concept of community. It has been argued (Davidson *et al.* 2007) that, except in specific circumstances, neither governments nor communities have much interest in disaster risk reduction.[1] Communities are often regarded as the ideal vehicle for both recovery and disaster mitigation. That may be true but in designing recovery strategies, we ignore the negative and infertile aspects of community-based recovery at our peril.

First of all, there is a need to define community. In some manner, it is a community of individuals who live, work or engage in recreation together and have some kind of common agenda, agreed goals or shared identity. The concept implies some degree of internal cohesion. However, communities can be riven by factional strife or rivalry. They can be dysfunctional, or at the very least they may be home to people with different and widely varying objectives. Moreover, there is a scale problem that is difficult to resolve. Although communities cannot be defined purely by their geographical characteristics, they may exist at the street, city block or neighbourhood scale, or across much wider areas. In a looser sense, community can be present at the level of towns or cities and, indeed, internationally as loose associations of people who communicate with each other with or without periodically meeting. Nevertheless, community implies a degree of interaction between people and perhaps also some common objectives, of which safety is a pertinent one here. The extent to which communities, at any geographical or conceptual scale, have cohesion and a therapeutic role for their members is variable and depends on circumstances. Time is an important factor, but communities which have existed for a very long time in close-knit form may either be a source of strength to their members or they may, possibly, be imbued with ancient rancour. One factor that does not usually determine the degree of community cohesion but certainly affects the status of a community is marginalisation (Polack 2008) – the state in which a community, or part of it, lacks political influence and economic power and, hence, the ability to determine its own affairs.

A resilient recovery may occur at the community level and, in any case, will certainly involve the community (Bretherton and Ride 2011). It may be determined, in part, by government policy or the actions of national or international bodies, but it must at least be accepted by local communities and, better still, these should feel some sense of ownership of it and active, autonomous participation in its activities.

Recovery from disaster can be described as resilient if it fulfils the following criteria. It must not stagnate. It must bring the affected area back to a level of economic functionality and positive social interaction. It must incorporate measures that not only repair the damage and make good the losses caused by the disaster but also go beyond replacement reconstruction to build a safer environment that is more capable of resisting calamity. At the same time, it must not recreate vulnerability. This will involve adapting to hazards. It will also involve ensuring that institutions are sustainable in the long term and that the funding stream, however modest it is, remains open. The sustainability of disaster reduction institutions needs to merge with the general sustainability agenda to create a society that is more resilient generally: not merely to disaster but also to climate change, resource scarcity, poverty and inequality, disease and accident, conflict, criminality and the many challenges that threaten our existence.

It is sobering to note that in the aftermath of the January 2010 Haiti earthquake, only a fraction of 1 per cent of the money pledged by foreign donors was destined to help the Haitian Government get back on its feet. Government in Haiti had been prostrated by the collapse of many key buildings, including the Presidential Palace, and the loss of many lives among the cadre of public administrators. Infrastructure was in a parlous state, where it existed at all, and taxation was hardly functional, such that government revenue was meagre or non-existent. Eventually, after a year had passed, 10 per cent of incoming funds were managed by the government, although not all of these had any impact on the need to strengthen public administration. Awareness of the Haitian Government's inefficiency, suspicion of corruption and desire to achieve results quickly meant that donors did almost nothing to help create the strong, functional institutions that would be needed to run the country over the long-term recovery period (Farmer 2011).

Resilient recovery is robust and enduring. It has mechanisms for solving problems, particularly about the apportionment of resources. It encourages a participatory approach and leaves many decisions to be made by the beneficiaries of recovery. It turns survivors into active protagonists, not passive recipients of aid. It has clear goals and specifies reasonable time limits within which to achieve the goals. It uses its resources wisely and does not recreate vulnerability through unwise decisions and lack of control over key activities. Transparency, accountability and participation are its bywords. Scientific information is respected and used in order to ensure safety and productivity.

Lessons from the past

Many publications have the phrase 'lessons learned' in their titles (e.g. Fallahi 2007; Mayunga 2012), but this is no guarantee that the lessons in question have indeed been learned. Lessons frequently present themselves for scrutiny, but they are routinely ignored – inadvertently or deliberately – or they are archived and forgotten. In our opinion, in order to be learned, a lesson has to be recognised, understood and incorporated into current practice such that there is measurable change for the better.

In recovery from disaster, there are common lessons that have been learned in some cases but not in others. With regard to transitional housing schemes, these are some of the lessons that planners have absorbed, often the 'hard way' by making mistakes and seeing their consequences:

- Units were not built to resist the dynamic pressures of future or ongoing hazards, such as wind, water or seismic forces, and hence would tend to collapse in renewed hazard impacts.
- Units that were inadequately anchored to the ground rolled over in high winds – with the occupants inside them.
- Units that lacked weatherproofing tended to decay rapidly in adverse weather, with frost damage, the spread of mould or rising damp.
- Units were designed or built with inadequate protection against the vagaries of climate, such as very hot or very cold weather.
- The design of units was countercultural; for instance, aggregating families in ways that were culturally unacceptable.
- Units were designed without reference to economic needs, such as shelter for animals or the need to maintain a workshop.
- Drainage was inadequate and the encampment flooded during torrential rain, or became a sea of mud.
- The transitional homes were situated in the wrong place with inadequate transportation links, such that the people who lived in them could not get to work. For example, housing was too far from the coast and hence fishermen could not easily mend nets and launch their boats.
- No effort was made to preserve the social fabric, and once people had moved into the transitional housing, they felt isolated and depressed as their networks of social support had disappeared.
- So much was spent on transitional housing that only limited funds became available for permanent reconstruction.
- Lack of connection between transitional housing and long-term permanent reconstruction has been a common problem. This is a question of ensuring that the two strategies connect, bearing in mind that it may be challenging to create the connection.

Many of the practical problems can be foreseen by using a simple risk analysis (see 'Model 9: probability/consequence risk assessment' in Chapter 4). This can be extended to include a site survey and an understanding of how items fit together.

The process of learning lessons from past experience is one of observation and reasoning. The Texan architect Fred Cuny was a luminary of his time – the 1970s to the 1990s. He spent much time on humanitarian missions observing how transitional shelter functions. In refugee settlements, he noted that lack of planning can increase the risk of transmission of fire or disease, reduce the opportunities to expand a camp, and diminish the role of social cohesion. A well-planned refugee camp can solve these problems. Cuny proposed the cross-axis plan in which the central functions of the camp (e.g. its clinic and administrative offices) are located at the axis and the collection of dwelling units grows outwards along and away from the axis. He also noted that siting the dwellings in enclaves with a central open space could increase social cohesion (Cuny 1977). Moreover, transitional shelter may start by being a mere donation of governments or NGOs – unless, that is, it is built by its occupants – but the longer it remains, the greater the opportunities for community to develop and shelter to be adapted to personal needs.

This book includes various examples and case histories, each of which constitutes a reconstructed scenario narrative in which there are lessons to learn. The scenarios give specific lessons derived from particular experiences of the reconstruction process. There are also general lessons. One is that the political, economic and social context of recovery from disaster is likely to change – sometimes radically – during the process. Inflation, the vagaries of markets, changing productive capacity, labour availability and skills, and investment trends all affect the economic outlook of recovery. Elections and demographic change are among the factors that influence the political and social conditions.

Readers of this book who are involved in recovery processes as managers, planners or field operatives may wish to design their own mechanisms for collecting and interpreting lessons to be learned. This involves looking for analogies in past events and translating experience to fit current and future circumstances.

Risk reduction strategies[2]

The United Nations' International Strategy for Disaster Reduction (UNISDR) oversaw the creation of the *Hyogo Framework for Action 2005–2015* (HFA), which is intended to convince nations to institute policies that will reduce disaster risk. The HFA has five priorities (UNISDR 2005: 6):

1   Ensure that disaster risk reduction is a national and a local priority with a strong institutional basis for implementation.
2   Identify, assess and monitor disaster risks and enhance early warning.
3   Use knowledge, innovation and education to build a culture of safety and resilience at all levels.
4   Reduce the underlying risk factors.
5   Strengthen disaster preparedness for effective response at all levels.

Hyogo is the prefecture of Japan affected by the 1995 Kobe earthquake disaster. A new strategy will be launched in Sendai, Japan in 2015 in the area struck by the 2011 earthquake and tsunami. This will urge nations to strengthen the monitoring and practical aspects of disaster risk reduction. It is expected that it will also make mention of vulnerability reduction for women and girls and the need to assist people with disabilities, as well as the importance of human rights in disaster risk reduction.

The HFA and its successor are instruments of international relations and also policy documents designed to be used at the national level with appropriate top-down influences to the local level. Local authorities can avail themselves of the assistance and tools for disaster risk reduction provided by UNISDR's *Making Cities Resilient* initiative (UNISDR 2013).

Structural resilience[3]

Resilience in a structural sense means designing buildings and other constructions to resist known or hypothesised forces. Overdesign – in other words, creating large margins of safety – is expensive, and often prohibitively so. In previous ages, there was much overdesign because forces could not adequately be calculated. That is not true of the modern era. Today, probabilistic calculations can be used to estimate magnitudes and frequencies, and structures can be designed on the basis of the probable maximum force that they will have to resist during their lifetimes or, in the case of hazards such as floods, the probable maximum magnitude of any inundation they will have to contain. However, all design calculations are made on the basis of a decision, or assumption, about the size of event that must be resisted.

On the north-east coast of Honshu Island, Japan, three nuclear reactor sites were affected by the March 2011 earthquake and tsunami. Two survived with limited damage, but the Fukushima site was overwhelmed, suffering catastrophic damage to four of the six reactors and disablement of backup systems designed to prevent and contain runaway nuclear reactions. It had been designed to resist a tsunami caused by a magnitude 6.8 undersea earthquake, whereas that which occurred had a magnitude of 9.0 and a millennial recurrence interval. With hindsight, the balance between estimation of risk and containment of construction costs was wrong (National Diet of Japan 2012).

Following the tsunami in north-eastern Japan in March 2011, sea walls and elevated roads and areas of housing are being constructed. The costs and environmental impacts rise dramatically if larger tsunamis with longer return periods are taken into account. The 2011 event may have been millennial in its return period. Does it make sense to design for an event with a very low probability of occurrence over a 1,000-year period? The answer is more a matter of societal preference – or political whim – than of scientific judgement. Hence, an area subject to a millennial tsunami offers the following rebuilding options.

- Build defences to resist a tsunami the size of the one that very recently occurred. This is likely to be uneconomical and will have substantial negative consequences for the environment, but it would offer very substantial protection.

- Build defences to resist a smaller tsunami, perhaps with a 100- or 200-year recurrence interval. There is a higher chance that such an event would occur during the life of any protected settlement, and of its defences. However, if a millennial event occurred again, the structures would be severely damaged or overtopped.
- Build minimal structural defences. This would give settlements a small degree of protection – perhaps against the 50-year tsunami – but in the long term, they would not be defended against medium-to-large events. Nevertheless, the minimal defences could be combined with other means of reducing or avoiding disaster, including evacuation.
- Build no structural defences. This implies that there would have to be alternatives. In the event of a tsunami, buildings and facilities would be sacrificed to the waves. However, taking into account the cost of building and maintaining structural defences, and the impact of the environmental modifications that they would entail, the losses may be comparable to the costs of protection by other means. Providing evacuation could be sufficient; lives would be saved and providing government did not abandon the damaged settlement, resources could be found for rebuilding.

These choices are valid for many types of hazard. However, there are additional considerations. There are many places around the world in which increased structural protection has engendered a false sense of security. No structural defence against natural hazards is 100 per cent proof against impacts – and likewise for anthropogenic hazards. The presence of good and excellent structural defences has often led to the overdevelopment of the protected area. When the defences eventually fail, the losses are greater than they would have been if the continuing presence of the hazard had been better recognised. For example, on the Mississippi and Missouri Rivers, 95 years of canalisation, dredging, and dam and levee building by the US Army Corps of Engineers ended in 1993 with the biggest, most devastating floods on record (Myers and White 1993). This was because the defences had encouraged urbanisation of the protected floodplains. The levees also encouraged the ponding of floodwater when they were breached or overtopped (Montz and Tobin 2008).

A reconstruction planner usually has a choice of methods to pursue. *Structural* measures involve building constructions that reduce the likelihood or effects of future hazard impacts. Anti-seismic buildings able to resist earthquakes, flood defences such as barriers and relief channels, retaining walls and drainage systems for unstable slopes, and physical barriers that contain or break the force of snow avalanches are all examples of the structural approach. To an extent, structural solutions have to be utilised, but experience suggests that over-reliance or exclusive reliance on them is unwise (Meyer *et al.* 2012). *Non-structural* solutions include land use control, emergency response measures (including warning and evacuation) and public education. *Semi-structural* responses include ecological measures (such as revegetating slopes) and portable defences (e.g. against floods). In most cases, it seems that the best strategy is the most diversified one, which includes a portfolio of measures from each of the three categories. This is reinforced by the fact that such measures tend to be complementary rather than conflicting with each other. Hence, resilient recovery can be achieved by adopting holistic strategies that spread the risk, and the response to hazard, over a broad range of measures.

Although the previous paragraph advocates a broad response, it is worth remembering that there may be weak points in any strategy and these may require special measures. Weaknesses in disaster risk reduction can lead to cascading disasters. These consist of two types: those that involve cascades of causes, or hazards, and those that produce cascades of effects (in certain cases the latter may be a subset of the former). The Japanese magnitude 9 earthquake, tsunami and nuclear radiation release of March 2011 was the cascading disaster *par excellence* of the last few decades. In this, the effects on manufacturing were spread around the world in line with the globalisation of such processes. Likewise, floods in Thailand in August 2011 led to shortages of components in the electronics and automotive industries in various countries.

The best way to understand concurrent, coincident and cascading disaster effects is to build scenarios of the processes that could lead to them. These are usually based on one or more 'reference events' from the past, in which the physical impact of the chosen event is adapted to current or future demographic, social, economic, environmental, institutional and governmental conditions. Scenarios of this kind are exploratory tools and should not be used with too much rigidity. There is usually a range of possible outcomes, conditioned by the input variables and the starting assumptions. Nevertheless, scenarios can be used to explore the probable robustness – i.e. resilience – of alternative recovery strategies (Franchina *et al.* 2011).

Governance and participation[4]

The term 'governance' originated in fourteenth-century English as a synonym for government; and thus it remained until it was briefly given the status of a euphemism for 'dictatorial decision-making' by international bodies which had to engage with dictators but did not know how to characterise their mode of government in open dialogue. However, under United Nations auspices, governance has since become a byword for responsible, transparent decision-making, usually with some form of participation by stakeholders. It follows from this that the best sort of recovery from disasters is one in which governance is upheld and strengthened in the name of fairness and good government (Shi 2012). The worst situations are those in which exploitation and marginalisation of sectors of society occur, with or without expropriation of the resources destined for recovery.

Marginalised groups are those that lack political and economic control over their destiny and are at the mercy of decisions made by other sources of power, which may appear to them to be arbitrary and uncaring, or exploitative. Since time immemorial, this has been the destiny of minorities and the inhabitants of high mountains and small islands. The solution, if there is one, is to form associations and work hard to draw attention to the conditions of the group and, crucially, to form alliances with external partners who can exert some leverage on the sources of oppression. Needless to say, in many instances the result has been even more brutal repression. Wherever they live, the marginalised are always the last to recover from disaster, and the people whose recovery is weakest.

Nevertheless, there are many cases in which recovery from disaster does not involve deep polarisation and rank exploitation, in which there is scope for social, as well as material, improvement. One response to potential marginalisation is the creation of what sociologists call emergent groups (Drabek and McEntire 2003). These are artefacts of disaster subcultures (Granot 1996) in which people from diverse backgrounds, and with diverse ages, educational levels, life experiences, and so on, come together on the basis of a common purpose. A typical form of group that emerges after disaster is an association of displaced people – those who have temporarily or permanently lost their homes in the impact. They will probably agitate for safe, rapid rebuilding of the local housing stock and designation of adequate funds to make this happen. In the aftermath of disaster, emergent groups have a chequered history of success and failure. Some become riven by internal quarrels; some are ineffective at advocating their cause; and some are deliberately marginalised by governments. Others have become highly effective political and social forces, and a select few have transformed themselves into permanent institutions, usually with charitable status, for the promotion of good in society. The success of emergent groups is usually a function of their cohesiveness, how articulate and determined they are in the public arena, and how they build networks of support. Obviously, they tend not to survive well where they are suppressed or harassed by government, or where the authorities refuse to recognise them as a legitimate source of advocacy and therefore refuse to listen to them. Nonetheless, emergent organisations may contain a wealth of human capital: organisers, technicians, publicists, experts such as engineers and architects, and so on. Good leadership and a reasonably democratic internal structure may carry them forward to substantial achievements. However, this assertion should be qualified by noting that much depends on the extent to which emergent organisations are regarded by the general population, and the authorities, as representative. If they are considered a mere faction, they may find opposition rather than a sympathetic ear.

Here is an example. In December 1982, the central Italian city of Ancona (population 104,000) suffered a major landslide on the northern fringes of the urban area. Some 280 buildings were destroyed, damaged or otherwise rendered unusable, and 3,661 citizens were made homeless. Initially, they dispersed to second homes, the houses of relatives, hotels and other forms of temporary lodging. Arguments began to rage in the scientific and planning communities about the causes of the landslide and the future state of the land that had moved (an area of 3.4 square kilometres). The displaced residents quickly formed an association and sent representatives to all official meetings to which they could gain entry. Much of the scientific debate took an esoteric turn – 'How deep was the shear plane upon which the landslide had moved?' – while the people from the residents' association stolidly demanded answers to more practical problems – 'Can we repair our houses, and if so, when can we start?' The landslide was unprecedented in size and controversial in terms of what had set it in motion. The controversy needed to be resolved in order to know whether it would move again and whether the areas could be resettled. This indeterminacy did not lead to a good relationship between the survivors and the scientists, and the politicians were caught in the middle. Suspicion of the other groups' motives further reduced the level of mutual trust. The situation eventually resolved itself, but it took 20 years to happen!

From these considerations, it should be clear that participation is not easy to assure, assuming that a government wishes to see it happen at all. Political support for disaster risk reduction waxes and wanes according to the salience of the issue on the political agenda. This tends to be high after a disaster and low in the intervening period between impacts. Social scientists have looked at various cases in the field in the hope of determining what frequency of disaster is optimal in terms of stimulating the public to adopt mitigation (e.g. Lindell and Prater 2002). They concluded that moderately large and frequent events are best at this. Very common impacts tend to be small and may inculcate a blasé attitude that fails to prepare the population for anomalously large events. Very large impacts tend to be infrequent such that preparedness lapses in the interval between them. This, of course, is a rather bald generalisation with plenty of exceptions determined by local culture, experience, expertise, the availability of data, foresightedness, social cohesion, and other such factors.

One problem with this is the way in which risk transfer is utilised in the community. Under ideal circumstances, risk can be apportioned fairly so that each member of the community faces up to it with the resources that he or she is able to devote to the problem and the knowledge that the collective response will provide solidarity to cover the rest of the needs. All too seldom is this the case. If a developer can, at a profit, build apartments on a river bank with no protection against flooding, or on an unstable slope, sell them all and escape the consequences when the river floods or the slope fails then this is the dark side of risk transfer. It is the nemesis of social solidarity and responsibility.

Under normal circumstances, urban planning should be the means by which development is guided to rational and productive ends – harm is avoided by ensuring that the fruits of development are not incompatible with each other and with the environment in which they are built. Planning is, at least potentially, an instrument of fairness in the community (the opposite of this is often termed 'ecological racism' – the placing of noxious land uses next to poor and marginalised communities). Planning has less room for manoeuvre when it comes to existing development, but sanitary laws, health and safety regulations, accessibility norms and redevelopment can be used as instruments for positive change. So can compulsory purchase by government agencies. Generally, planning is an underrated means of reducing disaster risk. However, the worst situations are those in which there is risk but planning is absent, planning fails to take the risk into account, or planning is merely the administrative expression of the will of the most powerful people in the community. In 1989, David toured the municipal planning offices of the 17 municipalities that lie on Mount Vesuvius in Southern Italy – a volcano that threatens the lives, homes and livelihoods of 650,000 people in these communities and 3 million in the area around them. In defiance of the law, most communities had neither an urban plan nor an emergency plan. The evidence on the ground, namely developments that encroached on recent lava flows, demonstrated that any planning that did exist was weak. In subsequent years, the situation improved (Rolandi 2010), but the team entrusted with the creation of the plan in one municipality found that the most blatantly illegal building, constructed without any of the requisite permissions, was the town hall!

Disasters are often described as a 'window of opportunity'. This is usually intended to mean that they lead to a period, usually the immediate aftermath, in which people of many different persuasions in the community can gain a consensus that certain actions are needed. In other words, the window swings open on a potential for positive change. However, it is also possible to think of a 'negative window of opportunity' that instead of creating resilience by increasing social solidarity, destroys it through opportunism. This can be political, economic, criminal or cultural. In some countries, the occurrence of a large disaster is the perfect opportunity for increased mafia activity, especially as the basis of power for mafia-style organisations is usually the construction industry, which moves into top gear during the recovery period. The death, deposition or merely distraction of politicians can give a free reign to unscrupulous political opportunists unless the electorate stands up to them.

In the context of resilience, there are various reasons why 'community' is a difficult and contentious word. The first is scale: communities have been perceived at every scale from that of neighbours on a street to that based on international teleconnections. It follows that the concept of community has no natural scale. Second, composition: communities can be extremely heterogeneous and thus be composed of people whose motives and expectations are highly varied. Third, power structure: communities may be dominated by certain people, clans or interests that subsume the expectations of other members. Fourth, harmony: there is no guarantee that the community is a therapeutic unit that exists to resolve conflict and promote the common good. Instead, it may be riven by factional strife and rivalry. Finally, there is no guarantee that disaster risk reduction is high on the community's agenda. Recovery from disaster may be, but this does not automatically mean that the community wishes to 'build a better world'. Nevertheless, as other parts of this book illustrate and as many commentators stress, the community is frequently taken to be the vehicle for disaster risk reduction. This is predicated on the fact that the impact of disasters is always local and therefore so should the response be, whatever the degree of harmonisation emanating from larger geographical entities. Responding to disaster widens community participation, and so it follows that the process of building resilience can have this effect as well.

Ensuring that, in seismic zones, tall buildings are constructed to have an appropriate degree of earthquake resistance is obviously a sophisticated and highly technical process. However, many aspects of resilience are not as complex and can be taught to a wide constituency of interested parties. How to recognise that a slope is potentially unstable, or that a river has a floodable area around its banks, can be taught to laypeople. Community-level courses can tackle hazard awareness, simple precautions, warning processes, improved building techniques, and so on. On the one hand, disaster risk reduction needs to be professionalised so that technical responses are adequate to needs, but on the other hand it also needs to be made a thing of the people.

Models of resilience[5]

As noted above, resilience involves a mixture of resistance and adaptation. The former is symbolised by the idea that an area affected by disaster should 'bounce back' and in so doing arrive quickly back at its former state of normality. As this evidently involves a degree of unmitigated vulnerability, theorists have modified the concept of resilience to one in which the community 'bounces forward', not back (Manyena *et al.* 2011). In other words, the disaster is used as an opportunity to rebuild systems and structures to a higher standard of resistance than before and, thus, to advance the cause of disaster risk reduction. This involves utilisation of the positive 'window of opportunity' in which the question of protection against disasters is high enough on the political agenda and fresh enough in the mind of the public that there is broad support for it.

Resilient recovery in Chile following the earthquake and tsunami of February 2010

The recovery following the Maule, Chile earthquake and tsunami of February 2010 provides a particularly good example of resilience that relates to model 11, 'resilient communities and settlements', in Chapter 4 (Comerio 2013, 2014a, 2014b). The earthquake is believed to be the fifth most powerful since seismic measurements began. The city of Santiago moved about 30 centimetres to the south-west, and NASA have calculated that the earth was knocked about 8 centimetres off its axis, thus shortening the day by one-millionth of a second (Platt 2012).

But of rather more immediate consequence for the citizens of Chile, 12 million people (or 75 per cent of the population) were affected, with 526 deaths in a disaster zone that extended over 500 kilometres of coastline and was 100 kilometres wide. The estimated loss was US$30 billion (or 18 per cent of the GNP); this sum was broken down into US$21 billion for physical assets and US$9 billion for business and indirect losses. A total of 370,000 housing units were damaged, of which the government were committed to rebuilding 222,000 units for low- and middle-income survivors, the rest being covered by insurance and private funds. The total for the government was further broken down into 113,000 needing rebuilding and 109,000 needing to be repaired (Comerio 2013: 1).

A resilient response was therefore needed, given the vast scale of the recovery challenge; and this was forthcoming to a remarkable degree, emphasising the strength of the four foundation blocks within Chilean society (as described in model 11): resourcefulness, rapidity, redundancy and robustness.

Resourcefulness: There was strong leadership that made good use of existing structures to avoid wasting time or resources (see 'Model 20: organisational frameworks of government for recovery management' in Chapter 4). One particularly wise decision was to use existing home sites, rather than remote sites, to rebuild houses. This kept people close to their jobs and family members and it enabled families to remain close to their house reconstruction, allowing them to monitor construction progress.

Rapidity: Within four months of the earthquake, 80,000 transitional houses were completed, and the reconstruction of permanent housing was accomplished by 2014 in an exceedingly rapid four-year period.

Redundancy: The dual pattern of strong top-down government authority in parallel to well-mobilised bottom-up recovery actions by citizens proved to be a vital 'fail-safe' approach. This is discussed in Chapter 13, principle 13 where the strength of governments and communities is represented on Mary Commerio's comparative model.

Robustness: Stephen Platt has noted a number of structural and non-structural risk reduction measures (as described in model 12 in Chapter 4) that were present in Chile. Together these elements comprised a robust safety strategy:

> Several factors contributed overall to the low casualty rate and rapid recovery. A major factor was the strong building code in Chile and its comprehensive enforcement. In particular, Chile has a law that holds building owners accountable for losses in a building they build for 10 years. A second factor was the limited number of fires after the earthquake, due to shutting down the electricity grid immediately. Third, in many areas, the local emergency response was very effective. . . . The fourth factor was the overall high level of knowledge about earthquakes and tsunamis by much of the population that helped them respond more appropriately after the event.
>
> *(Platt 2012: 32)*

Resilient international leadership of disaster recovery[6]

At present there is no specific UN agency or international agency with the mandate from governments to provide overall international leadership and coordination for disaster recovery. The closest candidates for such a role would be either the International Recovery Platform (IRP) based in Kobe, Japan or the

Global Facility for Disaster Reduction and Recovery (GFDRR), part of the World Bank in Washington, DC.[7]

The IRP is a 'thematic platform of the International Strategy for Disaster Reduction (ISDR) system' (IRP n.d.) and was established as a key pillar for the implementation of the HFA (UNISDR 2005). This was a global plan for disaster risk reduction for the decade that was adopted by 168 governments at the World Conference on Disaster Reduction in Kobe in 2005. Its key role has been to identify gaps and constraints experienced in post-disaster recovery and to serve as a catalyst for the development of tools, resources and capacity for resilient recovery. The IRP aims to be an international source of knowledge on good recovery practice.

The GFDRR has also been created within the World Bank to promote risk reduction within disaster recovery. In 2010, this facility made a significant contribution through the publication of the first set of comprehensive guidelines on recovery (Jha et al. 2010). In September 2014, GFDRR convened a conference in Washington, DC with the rather ambiguous and decidedly ambitious title of World Reconstruction Conference. The focus was on the same theme as this chapter: to 'strengthen resilient recovery and reconstruction in the post-2015 Framework for Disaster Risk Reduction' (GFDRR 2014a).

The conference statement was revealing as a snapshot of current viewpoints of the international recovery agenda of the participating delegates, which comprised 22 governments, NGOs and international networks. We reproduce the primary sections of the statement as follows:

1   Promote and ensure efficient, inclusive, and effective recovery and reconstruction interventions and measures through the institutionalization of post disaster needs assessments and recovery frameworks across regions and all levels of government. This would enhance risk governance, strengthen coordination, and empower communities and marginalized groups.

2   Provision for sufficient financial reserves and resources within government to manage and respond to disasters triggered by natural hazards, and formalized strategic and resource commitments towards equitable recovery planning, implementation and performance management; promoting more dependable and predictable international financial mechanisms for financing recovery.

3   Strengthening mechanisms for cooperation with services in areas of recovery and reconstruction that include standardized approaches for post-disaster needs assessments and recovery planning frameworks, and other support services such as sharing of information, data bases and rosters of experts, best practices, capacity building, tools, bilateral, regional and multilateral support to countries, and progress monitoring.

4   Strengthening readiness and capacity for recovery planning, implementation, and monitoring across regions and all levels of government, and

establishing clear roles and responsibilities for all actors in a recovery setting.

(GFDRR 2014a)

We value many of these suggestions, some being underlined in our book; but we question the high level of centralisation that is implicit in this statement as well in the grossly inflated title of the conference. Given the rich diversity of governments, cultures, levels of development and available resources of countries seeking to recover from 'natural' disasters, why is it so desirable to adopt standardised needs assessments and recovery frameworks? We have attempted throughout this book to encourage decision-makers in national governments to adapt any advice they receive in the light of their own needs and circumstances. In our view, resilient recovery is by definition locally defined and implemented recovery. We take the view that effective and culturally relevant recovery is bound to be better served by localised, devolved and adapted frameworks and responses, rather than by globally defined approaches created in the conference halls of Washington or at some world conference. Thus, we query the wisdom of the end of the fourth statement. Once again, given cultural, economic and political variables that exist and will continue, we do not believe that any international body such as the GFDRR, IRP or ISDR can establish globally applicable 'clear roles and responsibilities for *all* actors in a recovery setting' (GFDRR 2014a, emphasis added). The best they can hope for is to offer some carefully constructed guiding principles, tools and models for local adaptation and application.

This is an issue of accountability and trust by powerful international donor bodies rather than the exercise of control, and it takes us straight back to Charles Handy's insights that form the basis of model 15 and the 'trust-control dilemma' (in Chapter 4). On the 'trust' side of the pendulum, these words are written: 'The more trust you have, the less controls will be needed' (Figure 4.7).

The statement by the conference delegates is a reminder of the limitations of the *extent* to which the IRP or GFDRR, or the UN system generally, and the concerned international community can 'hold the ring' when it comes to leadership and coordination of disaster risk reduction and recovery. Ian attended the first World Conference on Disaster Reduction (WCDR) in Yokohama in 1994; 11 years later, he was present at the second global conference held in Kobe in 2005; and we have both been present at the Geneva Global Platforms. A strong impression comes across from the platforms and publications of these international gatherings that the organisers tend to believe their own propaganda which puts them firmly in the driving seat of policy and practice of global risk reduction and recovery. We suggest the reality is far different. Rather, they are in the driving seat for *certain aspects* of these complex subjects while a very considerable proportion of all risk reduction and recovery measures continues to take place well outside the orbit or sphere of influence of the GFDRR, IRP or UNISDR or of the World Conference.

The reality is that this work is vast in scale and is undertaken by related professions, institutions, government agencies, the private sector and individuals who may have never heard of international frameworks. Thus there is a need for long-overdue recognition of the wider dimensions of the risk reduction and disaster recovery discourse.

One of the key examples relates to safe building construction against high winds, the impact of floods and earthquake forces. This will be a cornerstone of any earthquake reconstruction programme as the business of government regulatory authorities, international bodies concerned with technical standards, and the work of the engineering, architectural, physical planning and construction management professions. Ian has attended several World Conferences on Earthquake Engineering, which occur every four years with up to 5,000 persons normally in attendance. There have been 15 of these gatherings dating back to 1956, 38 years before the first WCDR in 1994.

At such gatherings, he does not recall any reference to ISDR or to international frameworks, but this does not seem to matter. The participants collectively seek to reduce earthquake risks, often building them into recovery plans as well as into existing built environments since this is a central concern in their work. Since about 97 per cent of all earthquake deaths are due to the collapse of buildings, it becomes clear that the protection of buildings and infrastructure is largely outside the sphere of influence of ISDR and perhaps of any targets that may be formulated for the future. International agencies with a mandate for risk reduction and recovery, as well as NGOs, are unlikely to play key roles in building and infrastructure construction and physical planning.

Other areas of risk reduction and recovery planning that seem to proceed effectively without reference to international agencies include work on food security in drought-prone areas and work on river catchments and on environmental recovery with coastal protection measures by engineers, hydrologists, agronomists, environmentalists, etc. Here, flood or drought risk reduction is intermingled with land drainage, irrigation, erosion control, sewage, navigation, environmental protection, agricultural and urban development. Again, international agencies and NGOs are unlikely to play significant roles in such areas of work, which are normally the province of government departments.

Therefore we believe that disaster risk and recovery management is an infinitely wider concern than the agenda discussed at the GFDRR Washington conference in 2014 or at the Sendai WCDR in March 2015. Rather, disaster risk reduction and recovery in all its varied forms constitutes a major element in any society, far wider than any international frameworks; and a heavy dose of humility is needed by delegates and agency staff – i.e. recognising that they are only *a part* of the concerned international community and are certainly not in control of risk management, risk reduction or recovery.

So who has responsibility for resilient international leadership of disaster recovery? We suggest that the answer could be on the following lines.

International leadership is not *possible*: This is because the subject of recovery is far too great in terms of its scale and scope and due to the international donor community's need to defer to national government leaders who hold the full responsibility. However, the international community of donors, UN agencies, NGOs and professional associations can perform a useful service by facilitating vital research, documenting findings, sharing knowledge across frontiers and between continents, bringing national leaders together and becoming active partners 'on the ground' in recovery projects.

National leadership of recovery is the normal pattern: We discuss alternative patterns in model 20 in Chapter 4, with some guidance concerning which approach to adopt.

Local leadership of recovery, with close links to national policies, is essential: Local governments can play vital roles given their unrivalled knowledge of their own societies and their needs and resources.

Productive patterns of leadership are often dispersed, where all the above actors are closely integrated to link money, expertise, resources, authority and knowledge together. For example in Chapters 1 and 12, we describe successful recovery operations in India and Pakistan where productive partnerships were in place. In India, this was between an international donor government, DFID, a British NGO, Tearfund, a national Indian NGO, EFICOR, a sensitive architect, Laurie Baker, the Government of Maharashra, and a local community. In Pakistan, it was between the World Bank, the Government of Pakistan, the Pakistan Army, UN-Habitat, and local communities.

Summary

Most successful recovery will involve a certain amount of adaptation to risks. In Hilo, Hawaii after the devastating 1960 tsunami, the central business district was relocated inland and the waterfront was turned into a park. Other cities have created greenways around stretches of urban river in which rebuilding after floods has been severely restricted or banned altogether. Setback lines, also known as no-build zones, have been introduced in beach areas after storms and tsunamis in places as diverse as Florida, the Philippines and Sri Lanka. They are not without controversy as they restrict economic activity at the coast and are not necessarily well founded on hazard assessments.

Structural improvements tend also to follow the failure of structures or their components. Initially, most improvements in anti-seismic construction followed the failure of buildings in earthquakes. Information on this is still widely collected and utilised, but simulation, experimentation and model-building have taken over the process of assessing the seismic performance of structures.

Risks can be ignored, abated, transformed, transferred or shared. Many risks that are ignored tend to grow in importance and level of threat; for example, where a hazardous area is left to accumulate unplanned and unprotected urban development. Risk reduction needs to be a continuous process that adapts to changing vulnerabilities and probably also to both changes in hazards and improvements in knowledge about them. Hence, maps of hazard and vulnerability need to be refreshed at intervals or whenever substantial change has occurred. Prohibitions on land uses, building codes and other requirements need to follow these changes and the maps that depict them.

This chapter brings to a close our writing on varied aspects of disaster recovery. The next chapter gives our analysis of a survey of experts on the nature of effective recovery.

Notes

1   Dr Terry Cannon: personal communication and unpublished material.
2   See models 12 and 13, Chapter 4.
3   See model 12, Chapter 4.
4   See model 21, Chapter 13.
5   See model 3, Chapter 3 and model 11, Chapter 4.
6   See Chapter 4, model 20.
7   Websites of both bodies are listed among the selected websites in Appendix 3.

11

WHAT MAKES RECOVERY FROM DISASTER SUCCESSFUL?

A survey of expert opinion

Background

In September 2014 as we were nearing completion in writing this book, we decided at Ian's instigation to conduct a survey of expert opinion. We include the results in this chapter, near the end of our book, in order to draw together some threads that take this subject well beyond the scope of our limited experience. These insights from acknowledged experts in the field may indicate a coherent way forward in this bewilderingly complex subject.

We compiled a list of 56 leading experts who have specialised in recovery and reconstruction. The list included practitioners and academics from 19 countries and ten disciplines. As reconstruction is such a vital part of recovery, there was an inevitable preponderance of architects, planners and engineers. We are aware that as well as the five who did not respond to our invitation, we missed some key leaders in the field. We also recognise that too few national government leaders were consulted in the survey, and that simply reflects our own predilection as academics with a secondary interest in consultancy.

The general aim of the survey was to determine whether the focus we had adopted in this book is widely shared, as well as gathering fresh insights based on wide experience. Therefore, Ian wrote to the selected experts and asked if they would kindly respond to a single question, as follows:

> What in your view are the most important aspects of a successful recovery operation following a natural disaster?

Three contributors strongly objected to the use of the term 'natural disaster' in the phrasing of the question, and we fully agree with them. The word 'natural' was used in the question to guide the attention to disasters caused by 'hazards' rather than focusing on recovery following industrial or technological events, or terrorist bombs.

Fifty-one people responded, amounting to 91 per cent of those contacted. We were more interested in sounding out opinion than trying to achieve any kind of statistical generalisation, so the exercise involved a 'convenience sample' and was based on an open question. For the sake of brevity, Ian asked his correspondents to respond in 20 words, which is what many of them did while others felt less constrained by the target! One enthusiastic respondent expanded his contribution to just under 1,000 words, containing valuable insights but alas too long to include in full. The suggested word limit was designed to induce respondents to be synthetic and to give an answer of the 'in a nutshell' kind. The summary that follows concentrates on those aspects of the replies that were unexpected or different to *standard* answers.

In the following brief analysis, we have omitted the names of the participants. However, all contributors and their answers are listed in Appendix 2. It is likely that readers will reach different conclusions when they have seen the comments, according to their own questions, insights and perceptions. We place high value upon all the responses we have received, but with one notable exception (see 'Gaps in the survey'), we do not wish to identify individual contributions to this discussion. Moreover, a good many of the most interesting responses were echoed by other participants in this modest survey, as can be deduced from Appendix 2. We should note that, as none of the contributors was aware of any of the other comments, this 'blind approach' helped to avoid any mutual influence on their answers.

Long-term benefits of recovery

A common thread in the responses was the belief that the reconstruction process should provide lasting benefits in the long term. Whatever the status of the country affected by a disaster – rich or poor – there is a consensus that recovery and reconstruction need to be combined with development, and this in turn needs to focus on providing sustained and sustainable benefits to local society. One vital aspect of this is safe, well-established tenure of land that is relatively free from hazards.

The responses suggest that the community of disaster recovery specialists has well and truly espoused the idea that creating resilience should be a 'bounce forward' not a 'bounce back' process; in other words, one that focuses on eliminating future vulnerability. This is strongly linked to the ways in which disaster reveals *present* vulnerability in a *post hoc* manner and thus sets the agenda for vulnerability reduction. Past risks should not be recreated by recovery, which should be part of a constant process of risk reduction.

Local ownership of recovery

There is a further consensus that recovery and reconstruction are local processes and that there should be a high degree of local ownership of decision-making.

An appropriate byword is *self-determination*. Coupled with this, there needs to be strong, clear leadership. And coupled with this, local participation in the planning and decision-making processes is a key requisite, as is transparency and democracy in the processes involved in helping a community to recover. Aid and assistance need to be free from bias and corruption. As one respondent noted, democratic participation in the reconstruction process has a synergistic effect on democracy in general in a particular region or country. In addition to leadership and inclusive decision-making processes, expertise and technical assistance are essential to the process so that good use can be made of resources and resilience can be built into the process in matters such as the siting of reconstructed facilities in relation to hazardous areas and the quality of rebuilding. Various respondents noted that recovery is an open-ended process. It does not have a fixed ending but should flow into forward development. However, it should grow seamlessly out of the early humanitarian or civil protection response to disaster.

Holistic recovery

Respondents noted that reconstruction is not merely a question of rebuilding housing. Infrastructure and community are other elements that must be rebuilt. Some respondents pointed out that, for both ethical and practical reasons, priority should be given to providing women and children with a safe environment. Respondents emphasised that another vital factor of rebuilding is to recreate livelihoods and relaunch the local economy without allowing outside forces, by design or ineptitude, to undermine local attempts at economic regeneration. Public–private partnerships can be very useful to this process, but rapid privatisation of services and industries should be avoided until it can be ascertained that the potential consequences are not harmful to society and community.

From the responses to the question, it is clear that successful recovery from disaster is seen to be a holistic process, but one that has many facets and hence can be subtle and complicated. It must adapt to circumstances that change during the period in which it occurs. Clear, consistent leadership, with strong support from the local community and the beneficiaries of reconstruction, must create a 'joined-up' approach. This must avoid the fragmentation that occurs when different ministries and agencies fail adequately to communicate, liaise and work with each other, and this results in a mixture of duplicated efforts and tasks that fail to be completed.

Ethical recovery

Recovery must be ethically sustainable. This means that transparency and accountability must be allied with justice and equity. Human rights must be respected and the maintenance or restoration of safe, secure conditions must be a fundamental goal. The organisations and systems that drive the recovery process need to be identified, understood and perpetually held to account.

Recovery is an occasion in which something can be done to help local people achieve a better life than the one they had before the disaster. However, transparency and accountability are essential if people's aspirations are to be met, goals are to be realistic and public trust in the processes at work is to be maintained. The restoration of social networks is a key part of encouraging local capacity to grow. Moreover, efforts should be made to heal the rift that is likely to occur between survivors and others in the community. A requirement of resilience may be to diversify livelihoods so that plurality affords some protection against the destruction wrought by the next disaster to strike the recovering area.

Cultural recovery

One respondent noted very cogently that society and culture created the exposure of the past, manifest the vulnerability in the present and will assign and negotiate future risk. One important challenge of reconstruction is to preserve benign cultural values while helping those that foster vulnerability to change for the better. Recovery cannot take place successfully without full acknowledgement of cultural values, the constraints they pose and the opportunities they offer. However, if local cultural traits include failure to observe building codes, cultural change must be effected.

Respondents indicated that much of the suspicion that recovery experts once had of cash transfers has disappeared. They are now seen as a way of energising the recovery process and empowering local people. This is not to argue that cash handouts are the solution to all problems, but providing checks and balances can be instituted, cash is a powerful and flexible instrument for jump-starting recovery.

Defining a strategy for recovery

Faced with the need to allocate resources between shelter, infrastructure, productive employment, social services, markets and protection works, governments need to have a clear means of establishing priorities, and these need to be acceptable to the beneficiaries. Two overriding priorities should be to strengthen governance and to create a long-term policy for recovery. However, this needs to be a flexible action that can be adapted to changing circumstances. Hence, processes of mediation and negotiation need to be created or strengthened, and full weight needs to be given to them. Strategy formulation needs to benefit from a robust and verifiable theory of change, which involves monitoring trends and developments so that adaptation can take place. During recovery, it is important not to neglect the sources of knowledge, both external and indigenous, that can contribute to the wise design of strategies and processes. One example of the challenges faced by reconstruction planners is the need to site rebuilt facilities so that they are safe from hazards but at the same time productive and fully functional. For example, it would be wrong to site housing for those citizens who make their livelihood by farming or fishing too far from the fields or the sea, yet safe locations need to be found.

Gender and recovery

The literature on disaster recovery clearly indicates how important it is to consider gender, given the enhanced vulnerability of women to disaster impact, their vital function in keeping families together and the reality that their specific needs are often neglected in recovery operations. However, women are still under-represented within the disaster studies community, and only 9 of the 51 respondents to this survey were women and only 3 contributors specifically commented on the needs or roles of women in recovery.

Gaps in the survey

While most of the topics noted above secured multiple endorsements from survey contributors, there was a noticeable absence of comment on certain themes that we regard as vital elements in successful recovery operations. It took one of the eminent 'father figures' of disaster sociology, Russell Dynes, aged a youthful 90, to be the only respondent to propose that 'creativity' is an essential attribute of recovery (see the 'Green Hat – creativity' discussion in Chapter 12). This omission reminds one of another gap – no contributor commented on the need for a '*genius loci*' that gives meaning, or a 'sense of place', to localities or confers identity or cultural continuity to the recovery of an environment (see the discussion in Chapter 6, '*Genius loci* and preservation of the identity of places and human settlements'). A third significant gap concerns the absence of any reference to the need for governments to recover, both in their functions and their facilities, as a *sine qua non*, without which they will be unable to promote and lead the overall recovery process that is their primary responsibility. Finally, only two responses mentioned the need to create an efficient organisation to coordinate the recovery process (see model 20 in Chapter 4; Chapter 10, 'Governance and participation' and 'Resilient international leadership of disaster recovery'; and model 21 in Chapter 13).

The essence of effective recovery

Respondents were unanimous that recovery from disaster needs to be effected with realism, but that it should be an open process which reinforces democratic values and contributes towards greater fairness in society. Inefficiency, conflicting aims, discrimination and exploitation will never garner support for recovery processes, and without that support they are unlikely to be successful in the long term. In synthesis, one respondent summed up the requirements of successful recovery in 20 adjectives, as follows:

Respectful, responsive, resilient, informed, impactful, impartial, inclusive, collaborative, coordinated, consistent, targeted, timely, transparent, appropriate, accountable, legal, prioritised, efficient, developmental and humanitarian.

Summary

We believe that this is the first survey to be undertaken of 'expert opinion' regarding disaster recovery. We are encouraged by the respondents' high rate of response, enthusiasm and level of consensus concerning the ingredients of effective recovery. In conclusion, the responses to the question 'What makes things work well?' are bound to be somewhat utopian. They may represent a counsel of perfection, but in this case they are based upon many years of observation, trial and error, case histories and personal involvement. The insights are derived from highly varied professional backgrounds in recovery from many different kinds of disaster, in many diverse places and in widely varying circumstances. We recognise that the reality on the ground may fall lamentably short of the ideal situation, but we feel strongly that it is essential to have an idea of what the best answers to the problems should be. The challenge is to operationalise the precepts given here and to tie them to specific realities on the ground. The circumstances of disaster are usually to a greater or lesser extent hostile to innovation – but rarely without opportunities for positive change.

12

'THINKING HATS'

As we show repeatedly in Chapters 3 and 4, a model is a simplification of reality. It is intended to make a complex phenomenon understandable by leaving out the 'noise', or inessential detail, and concentrating on the 'signal', the essential elements. It follows that a good model is an elegant and robust simplification of reality, one that is adaptable to diverse circumstances and will endure when called into question. A good model does not explain all aspects of a problem or a phenomenon, but it does throw light on the matter. It is the starting point for new insights and the development of new, creative solutions.

The models, observations and experiences described in this book are many and varied. By way of conclusion, we feel that a coherent summary is needed. We set this out in two parts: in this chapter, there is a wide-ranging discussion concerning ideas, emotions, problems and opportunities; and in Chapter 13, a series of emerging principles are set out.

Edward de Bono's 'thinking hats'

For the discussion, we have adopted Edward de Bono's concept of the six thinking hats (de Bono 1985). This was described in model 19 (Chapter 4). Each of the 'hats' represents a different way of thinking about the basic challenge: recovery from disaster. Colours are associated with viewpoints and moods, which in some cases are negative or positive:

- black hat – discernment;
- white hat – information;
- red hat – emotions;
- yellow hat – optimism;

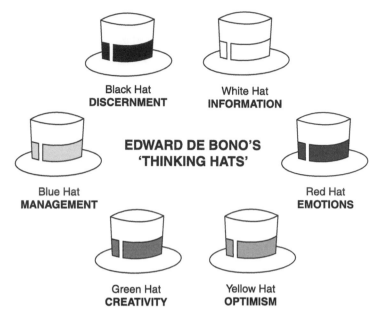

FIGURE 12.1 Model 19: Edward de Bono's 'thinking hats'.

- green hat – creativity;
- blue hat – management.

George Atkinson, a British architect who became the head of the UK's Building Research Station, wrote about reconstruction as far back as 1962. In his perceptive paper, all of de Bono's thinking hats are strongly in evidence. He reflected on past disaster reconstruction and the behaviour patterns he had observed among officials and planners (we have inserted the references to applicable hats):

> a wave of enthusiasm for planning [red, green and yellow hats], sometimes accompanied by the lobbying of planners; an attempt to determine the cost [white hat] and to seek aid far and wide; a period of great realism as resources are balanced against plans [black hat]; a waning of enthusiasm as the problems which reconstruction poses become more complex and less easy to solve [black hat], a sifting out of carpet-baggers as prospects of easy pickings recede; if fortune is kind the setting up of a reasonably effective reconstruction organisation [blue hat] ... or, if fortune is less kind, the wasting of resources on over-ambitious and ill-advised projects which enrich almost as many pockets as they rehouse homeless families [black hat].
>
> *(Atkinson, cited in Davis 1978: 89)*

The applications of the thinking hats model to disaster recovery are summarised in Table 12.1.

TABLE 12.1 'Thinking hats'

Thinking hat	General application	Application to disaster recovery
Black hat	Negatives/dangers/difficulties/ weaknesses/being a devil's advocate/making a judgement/ why a given approach will fail	• More complex/cascading disasters to recover from • Silo mentalities and how to change this narrow culture
White hat	Objective facts and figures/asking probing questions: • What do we know? • What do we need to know? • How do we get the information we need?	• Are we collecting the right sorts of data; for example, on qualitative assessment of recovery progress? • Where to store information
Red hat	Expressing feelings and emotions/ likes/dislikes/intuition/hunches	• Anger management over corruption and abuse of human rights
Yellow hat	Brightness/optimism/ opportunities/strengths/exploring the positives/probing for value and benefit	• Ways to celebrate/share the achievements in recovery management
Green hat	Being creative/new ideas, concepts and perceptions/ways to improve/ alternatives	• Where/what are the innovations in recovery management?
Blue hat	Focus and control/what we have learned/thinking about thinking/ what is next/where we are going	• Extracting the dominant lessons from disaster recovery studies

We now consider de Bono's model in more detail, as a means of clarifying the processes of recovery after disaster.

Black hat – discernment

De Bono's black hat is associated with negative factors, dangers, difficulties, weaknesses, the need to be a devil's advocate or make a judgement, and the explanation of why a given approach is bound to fail. De Bono termed this approach 'discernment'.

As population rises and the world gets ever more complex and interconnected, disasters are bound to become more challenging. They will probably be larger, more frequent and more widespread, and have impacts that will be more difficult to ameliorate. The degree of global interconnectedness is such that cascading disasters will be more common, and the effects of the cascades will touch more aspects of life in more widely dispersed places. For example, the Fukushima nuclear disaster, triggered by the 2011 Tōhoku earthquake and tsunami, resulted in the cancellation of Germany's nuclear power programme; and the flooding in

Thailand in 2011 had a negative global impact on industrial supply chain management. Climate change and rising sea levels look set to increase the intensity and magnitude of meteorological disasters. Each of these disasters will be followed by a recovery phase and these will place increasingly large burdens on governments and international support systems.

Cities such as Tehran, Tokyo, San Francisco, Istanbul and Kathmandu carry an extremely large earthquake risk. This means that they urgently need scenario formulation, in order to understand the effects of such disasters, and pre-planning in order to know how to tackle them when they occur. A good example of pre-planning took place in Miyagi Prefecture in Japan, where the location and form of transitional housing were planned well in advance of the Tōhoku earthquake and tsunami (see Chapter 8, section on 'First dilemma: planning vs plans'). This enabled disaster managers to provide accommodation for surviving families rapidly and efficiently after the event.

Disaster risk reduction is a transverse field that touches upon more than 40 different disciplines and professions. A major challenge exists in the need to consider problems holistically, not in terms of single disciplines. The latter tendency is often described as the 'silo mentality', in which a given field or sector is isolated and fiercely protected from intruders. The main challenge is to make and consolidate the connections between different systems through which people deal with recovery and reconstruction: natural, technical, organisational, political and social. We need to create education programmes, including education or training for established professionals, that widen the perspectives of people who deal with recovery from disaster and help ensure that their approach is more interdisciplinary, transdisciplinary – or perhaps 'non-disciplinary', which means that it should be conditioned by the exigencies of the problem, not the discipline through which they are viewed. This is relevant to model 2 (Chapter 3) in which we highlight the need for holistic integration of five key sectors.

Corruption, lack of transparency, failure to widen participation in decision-making and denial of human rights are all significant negative factors in recovery from disasters (Alexander and Davis 2012). Where human rights are infringed or restricted, people are likely to lack access to the information needed to create a healthy form of recovery and they may also lack the ability to act and make appropriate decisions. In many cases, lack of transparency in decision-making is associated with corruption, in which funds are syphoned away, codes and standards associated with public safety are deliberately ignored, and power structures are distorted for the benefit of the powerful and corrupt. The large sums of public finance that are disbursed during disaster recovery operations are always a temptation for corrupt officials. In July 2014, Ray Nagin – the mayor of New Orleans when Hurricane Katrina struck in 2005 – was sentenced to ten years in prison for corruption. His case related to bribes that were paid by city contractors looking for work, including rebuilding contracts after Katrina (*The Economist* 2014).

Besides these factors, lack of a participatory approach is usually the result of a 'top-down' process of decision-making where decisions that affect people's lives are made remotely and transmitted to them with little concern for their appropriateness, popularity, efficiency or ultimate efficacy. It is a valid principle of recovery that the beneficiaries should be fully involved in the process and anything that is done for them should involve decisions with which they can live and of which they, in general terms, approve. This issue is represented in Chapter 13 (see principle 13 and model 21) with a discussion of the strength of community participation in relation to the strength of government.

The Berlin-based international NGO Transparency International regularly produces an international Corruption Perceptions Index. The 2013 report indicates that of the 177 countries in the world, two-thirds score below 50 on a scale that extends from 0 (highly corrupt) to 100 (very clean). Thus, 69 per cent of all countries listed indicate serious corruption problems. The regional picture is particularly alarming, with 95 per cent of countries in Eastern Europe and central Asia and 90 per cent of countries in sub-Saharan Africa perceived as having serious corruption problems (Transparency International 2013). It should be noted that corruption is not measured directly in the report as it would be too difficult and dangerous to do so, and the endeavour would doubtlessly be thwarted by the corrupt.

A probing analysis of corruption in Sri Lanka conducted by the Humanitarian Policy Group in 2008 concluded that:

> The vast amount of humanitarian assistance that entered the country after the tsunami exacerbated corruption risks. . . . [and] many politicians at the national, provincial and local levels and other non-state actors used the large influx of resources as an opportunity to increase their political capital amongst their constituencies and for personal enrichment.
>
> *(Elhawary and Aheeyar 2008: 6; also see Davis 2014b)*

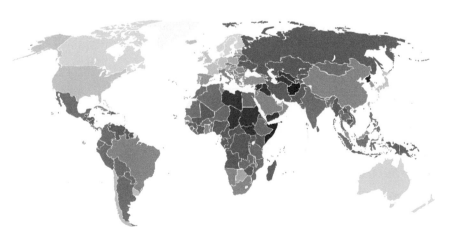

FIGURE 12.2   International Corruption Perceptions Index 2013. The deeper the grey colour the higher the perceived level of corruption; the lighter the grey colour the lower the level of corruption (Transparency International 2013).

It is worth remembering that in 2005, 193 countries signed the United Nations' *Hyogo Framework for Action 2005–2015* (UNISDR 2005) and thus formally indicated their commitment to risk reduction. Their concern is obviously rooted in a noble ethical stance. However, if the Transparency International report is to be believed, there is hypocrisy in the 'paper-thin' ethical values of corrupt governments who sign agreements that refer to governance and accountability but do not follow their precepts. The Transparency International report indicates that about two-thirds of these same countries are regularly corrupt in their dealings (Transparency International 2013).

Many experts in recovery like to concentrate on the positive aspects and the concept of progress. It is as well to have an optimistic view of the process, but not to the extent that this obscures one's ability to see things as they really are. Corruption is endemic in many countries. Power structures and the wealth that follows and motivates them are heavily unbalanced and decision-making is flawed because it represents the interests of the few and not the many. To pretend that the world is otherwise is to put one's head in the sand. Instead, recognising the limitations that this situation involves will allow the problems to be confronted and – at least in some cases – rectified. It is encouraging that international finance institutions (IFIs) are now proposing specific ways to minimise corruption (Davis 2014b; Jha *et al.* 2010).

Ian witnessed a housing reconstruction project in Banda Aceh (Sumatra, Indonesia) following the 2004 Indian Ocean earthquake and tsunami. Uplink, an Indonesian NGO consortium working in partnership with the World Bank, devised an effective way to combat corruption. Groups of survivors who had been neighbours before the disaster were brought together again to participate in the reconstruction of their homes. They were allocated individual cash grants. Each group carefully chose a scrupulously honest person from among themselves to act as group treasurer for the community project. Ian asked one group (see Figure 12.3) whether this process guaranteed that the finance for the project would be secure. A unanimous answer came back from all the widows: 'This is our money. Of course we are all keeping watch to make absolutely certain that it does not go anywhere it should not!'

We are convinced that a fairer world is possible but only if we recognise how unfair it often is and what needs to be done to make it fairer. The magnitude of the task should not be underestimated.

White hat – information

> There's a danger in the internet and social media. The notion that information is enough, that more and more information is enough, that you don't have to think, you just have to get more information – gets very dangerous.
>
> *(Edward de Bono 1985)*

FIGURE 12.3 A group of widows who lost their husbands, and also in many cases their children, when the tsunami devastated Banda Aceh. On account of her honesty and bookkeeping skills, the woman holding the microphone was selected by her colleagues to be the group treasurer for their housing reconstruction project (photograph by Ian Davis).

FIGURE 12.4 Houses built in Banda Aceh by the widows shown in Figure 12.3. The project was managed by Uplink, an Indonesian NGO consortium that worked in partnership with the World Bank, which financed the project. Occupants were given choices in the design and decoration of their houses, as can be seen in the colour scheme and the fourth house with its external access stair to the upper storey (photograph by Ian Davis).

De Bono's white hat refers to information in the form of objective facts and figures. The white-hat wearer asks probing questions about what we know, what we need to know, and how we can obtain the information we need.

Data are the raw material of information, which is in turn the raw material of knowledge and wisdom. As the tastiness of a cake is a function of the ingredients with which it is made, so knowledge and wisdom are only as good as the data on which they are based. In many fields of disaster risk reduction, including recovery and reconstruction, we lack an adequate basis of evidence. Assembling empirical data from past disasters runs the risk that future events will be quite different, and the lessons of the past will have only limited applicability. However, to date, most disasters have been partial reruns of previous events. There is still a lack of understanding of how these fit together and give us a complete picture of trends and tendencies, regularities and generalisations – the evidence base, upon which decisions should be made.

Quantitative assessments of the quality and efficiency of reconstruction are rare and so are longitudinal assessments. These are particularly important in the light of how social, economic and physical conditions change during the life of a disaster recovery programme. If longitudinal assessments were standard practice, we would have more information on how villages, towns, cities and regions either recover or fail to do so. One vital development is a groundbreaking series of long-term studies of the way various communities in India (including Malkondji) have reconstructed their settlements. The work was carried out by Jennifer Duyne Barenstein and her Indian colleagues (Duyne Barenstein *et al.* 2014).

We embarked on this book with a longitudinal case history of the Indian village of Malkondji (see Chapter 1) that Ian studied in a pair of linked evaluations in 1996 and 2011. If knowledge of this reconstruction process had been confined to the first evaluation, which is the normal pattern of case studies of recovery, perception would have been limited and lessons would have been only partial. The return study after 15 years not only enriched the original project evaluation but also significantly expanded our understanding of the underlying processes.

The challenge is not to collect information according to the rules of pure induction – i.e. indiscriminately – but to do so in a targeted way that helps devise solutions to problems. For example, as described in model 16, ('two- or three-stage shelter and housing recovery' in Chapter 4) and options 6a and 6b on 'transitional shelter' (Chapter 9), the role of transitional shelter in recovery remains generally controversial. Serious questions persist on the subject regarding whether their presence impedes permanent reconstruction. Specifically, was it prudent for agencies in Haiti to build 110,000 transitional shelters? Second, are transitional shelters adequately linked to the preceding and succeeding phases? Third, does their short-term value justify the effort and resources expended? These are the sort of questions that would benefit from more comprehensive evaluation on the basis of past disasters.

More effort needs to be devoted to identifying the areas in which data, and therefore knowledge, are scant. The qualitative assessment of progress in recovery and, likewise, assessment of the quality of reconstruction are examples of matters that currently need more attention. The views of beneficiaries, their prospective roles as participants in the recovery process and their perceptions of recovery from disaster all need to be examined more closely. Generally, research on recovery from disaster has been confined to particular disciplines, particularly architecture, engineering and economics, and has been less copious than research on the emergency phase. The balance needs to be redressed, but with a much more interdisciplinary approach.

It is a challenge to place information in accessible but safe places. Disaster researchers are well aware that, after disaster, information tends to be highly perishable. This means that it can easily be lost. Accessible banks of data on recovery from disaster are needed, and they should be resilient and protected so that diverse conditions can be compared in the interests of broadening knowledge of processes. Information can also be manipulated politically with false data, gagging orders, destruction of records, and so on. This is harmful to the process of developing an objective picture of events and the problems that they entail.

It would be valuable to monitor the use of what information and data we do have. Throughout the field of disaster risk reduction, there is a perennial gap between research and practice in which the first of these consistently fails to inform the second. More research is needed not only on the fundamentals of recovery but also on the utilisation of scientific knowledge in the process, with the objective of narrowing the gap between academics and practitioners.

We would like to strike a cautionary note: 'objective reality' is not such a clean and unassailable concept as one might think. All human realities are conditioned by experience and beliefs. Some of those realities are harmful, and we believe that fatalism is usually one of these. However, this should not be a pretext for automatically discounting different views of the world. Science has given us an objective philosophy and methodology of investigation, but the complexity of the world means that it is heavily interlaced with value judgements and we need to be cognisant of them. We need to be aware that our way of viewing the world – as academics and observers of disaster recovery – may differ from that of other interested parties. The challenge is both to accept some of these different views and, where that does not seem wise, to convince the people who hold them to adopt a different stance. Perhaps this is closer to de Bono's 'red hat' than his white one.

To summarise, there is an urgent need to remedy the lack of serious attention to recovery from disaster, which is a significantly neglected field. This applies to writers, academics, universities, professions, governments and international donors, including United Nations agencies, international financial institutions such as the World Bank, the Red Cross system and lateral, bilateral and multilateral organisations. A particularly good example of an agency seeking to focus attention on effective disaster recovery options is UN-Habitat, which regularly produces compilations of shelter projects (e.g. Ashmore 2010, 2013). These reports have become a valued inventory of shelter in its varied forms. The International

Recovery Platform has also become a valuable clearing house of reports of recovery from various agencies.

It is also vital to tackle the problem of isolated structures and narrow thinking that is the prevailing norm within the sectors and line ministries of most national administrations. Such 'silo' or 'stovepipe' mentalities in government line ministries, UN agencies and universities remain a staunch enemy of risk reduction and recovery. The reasons for such isolation and competition for attention and resources are often far stronger that the forces that promote collaboration and well-integrated structures and policies. Thus ways need to be found that result in 'joined-up approaches' across sectors and change to this narrow culture with all its dangers of unbalanced recovery.

Some member states of the European Union have sought to reorganise disaster risk reduction and response in such a way as to reduce the dependency on a single ministry. The model appears to have originated in Italy and has become a non-binding EU directive, adopted by, among others, Sweden and the UK. Competency in disaster risk reduction and disaster response is the preserve of the Cabinet Office under the nominal chairmanship of the Prime Minister and his or her technical delegate, the head of the service. Major decisions are made by the national cabinet, chaired by the Prime Minister or Deputy Prime Minister, and implemented by the service or the various ministries in collaboration with one another. In practice there remains a residual dependency on the Home Office or Ministry of the Interior, but this arrangement does help ensure that particular ministries are working together.

Red hat – emotions

> Human behaviour flows from three main sources: desire, emotion and knowledge.
>
> *Plato*

De Bono's red hat is one that expresses emotions, likes and dislikes, intuition, hunches and so on. In studying recovery after disaster, we have encountered negative aspects that make us angry: corruption, abuse of human rights, and exploitation of situations for personal gain. Disaster can be a 'negative window of opportunity' that fosters delinquency and unethical behaviour. These factors need to be studied, in as much as one can do so in safety and without one's efforts being thwarted by the interested parties. Moreover, emotions need to be confronted and evaluated coolly.

Red hats can easily be elusive as in Western cultures there is often a taboo against the expression of powerful emotions. In government offices, donor organisations and academic institutions, there is a strong bias towards the *white* hat environment of hard facts and figures, as if the pains and agonies, as well as triumphs, can be airbrushed out of the picture of disaster. The example that follows seeks to capture the intensity of emotions experienced in a disaster and its aftermath. We believe that the suppression of emotions, or failures to understand them as they affect ourselves and others, diminishes our understanding of this subject as well as our ability to empathise with survivors in their long struggles against adversity.

In November 1977, Ian was working as a consultant for the British NGO Tearfund, which had formed a partnership with EFICOR, a Christian NGO from India that had been working actively in Andhra Pradesh in cyclone reconstruction projects. The cyclone of 1976 resulted in 14,204 deaths, mainly in the area where the storm surge had engulfed low-lying coastal settlements (Davis 1981). The local government decided to allocate (perhaps more accurately 'auction') villages for reconstruction assistance to appropriate NGOs. Inevitably, the results were erratic, as some communities secured massive support while others received very little. Villages tended to follow religious affiliations. These were Hindu, Moslem, Sikh and a minority of Christian villages of varied denominations. Therefore, the government followed a pragmatic course and allocated the predominately Christian village of Chinthayapalem, in the Guntur District, to the Christian agency EFICOR for reconstruction assistance. This community was situated less than 5 kilometres from the sea, and the storm surge had killed many residents and destroyed livestock and buildings.

In 1977, many internationally-based officials were present in Andhra Pradesh to run cyclone relief and recovery programmes. They had virtually travelled directly from Guatemala, which had suffered a major earthquake in 1976. Here, an exciting new approach to reconstruction of rural dwellings had been pioneered by Oxfam and their US partner World Neighbors. This involved merging the immediate relief and reconstruction phases in a seamless manner with the introduction of training programmes in safe earthquake-resistant construction (see Chapter 9, 'Option 7a: user-build permanent dwellings'). Ian was present and gave some initial advice on the early development of this programme with Fred Cuny of INTERTECT. Fred and his colleague Everett Ressler, using the Guatemala model, formed a body that still exists, called the Appropriate Reconstruction Training and Information Centre (ARTIC).

Ian and his ARTIC colleagues attempted to persuade EFICOR to adopt this 'Building for Safety' housing education model for the construction of housing in Chinthayapalem. However, they were totally uninterested since their eyes were fixed on a 'showpiece' reconstruction project with efficient and fast delivery. This project was designed by Indian engineers and built by external (not local) contractors, all with close links to EFICOR. So a unique opportunity to build houses and also teach unskilled labourers how to build well and safely was missed and our emotions were full of sadness and anger at the short-sightedness of the decision. For Ian this was particularly disturbing since he was a board member of Tearfund at that time and they were the main project donor – his face was red, even if his hat did not match it!

Ian vividly recalls the noisy meeting with survivors in a hired marquee pitched among the ruins of the village, where ambitious plans were outlined while proposals and cheerful perspectives were pinned on boards around which the excited villagers crowded to see what they could expect. EFICOR, with Tearfund's financial and technical support, proposed raising the village two metres in order to provide protection from future storm surges. It also proposed to create fish ponds, purchase new livestock to replace those killed, build a new church and pay contractors from the nearest large city, Vijayawada, to construct substantial masonry dwellings. It was clear that the new Chinthayapalem was going to be vastly superior to the settlement it was going to replace. There were at least two reasons for this, both wearing red hats.

The first reason related to the plight of those affected by the cyclone, who stirred the emotions of Tearfund supporters enough to induce them to contribute a very large sum of money. On account of UK Charity Commission rules, this had to be spent in a limited period of time strictly within the area affected by the cyclone. In 1977, Tearfund was still in its infancy and progressive disaster response approaches were at an early stage of development. The second reason related to passionate sectarian emotions arising from the local religious context. The Christian minority in this part of Andhra Pradesh believed that they were always the last in line for government assistance, which always went first to the dominant Hindu communities. Hence, local Christian leaders reasoned that it would be a good example of how to correct past prejudice and injustice to provide a local Christian minority community with an exemplary model village embodying the best practices in the region. This spirit seemed to include a competitive desire to demonstrate superiority and generosity so that the government and other religious groups could see that 'Christian charity' meant strong assistance from Christians to Christians.

The meeting ended and the team from EFICOR and Tearfund walked back to their vehicles. Ian recalls: as I was about to climb into our Land Rover, a highly articulate and angry man came forward and grabbed me by the arm in order to talk – or rather shout – at me. He had a white headdress (even if he was metaphorically wearing one of de Bono's red hats!). It became clear that he was a well-educated community leader of some stature and he asked me what 'my agency' was going to provide for his village? I assumed he must be a member of the village in which we had been meeting and replied by summarising the offers that had been made in the meeting that had just ended. He stopped me in my tracks by saying he was not from Chinthayapalem; he was from the next village – Muthayapalem. I responded that it was likely that another agency would be looking after his reconstruction. This response raised his anger to fever pitch as he claimed that his village and numerous other settlements had totally fallen through the net, despite their losses being similar to those of the village that was to be rebuilt. He shouted at us as our cars moved away, and his final outburst is etched in my memory: 'How can you as a Christian man give these people all this help, far more than they need, while you give us nothing?'

The sting in his words demanded that I return to the area to see what had transpired. Two years later, I went back and found that the new dwellings on their raised mound were set next to ruined villages that had still to be rebuilt,

FIGURE 12.5 The tiled roofed house in the foreground is part of an original dwelling built in 1978, but all the other houses have been built subsequently (photograph by John Noble).

FIGURE 12.6 Reconstructed houses in Chinthayapalem – photograph taken in 2014, 36 years after construction. Note the damage to the roof, which is probably in need of reroofing. However, these houses have probably well exceeded their normal lifespan (photograph by John Noble).

exactly as the angry leader had anticipated. I have often reflected on this emotional encounter and I came to share this man's righteous anger: at injustice, at weak allocation policies, at the failure of Christian agencies to allocate their money according to need – in accordance with one of their central beliefs – rather than in response to creed.

But are such impressions, emotions and concerns that stretch back 36 years accurate in relation to the *present* situation in 2014 in Chinthayapalem and in neighbouring villages? Red hats of emotion always need the white hats of reliable information. Therefore, some detective work was needed to locate a person in this remote part of rural Andhra Pradesh able to conduct interviews with village leaders and take photographs to record the current state of the community and the status of the original houses built in 1978 as well as determining whether the adjacent village was ever rebuilt.

The following interview, using questions set by Ian, took place in July 2014 with Chinta Yesuratnam and Nakka Vijayendrarao who were village heads during the reconstruction period of 1977–8 and who still live within the village. Evidently they were delighted that we had remembered the village reconstruction, and were pleased to be questioned by John Noble, a local resident from the local Guntur region.[1]

JOHN: Are you satisfied with your village or not? If you are, what is good about it, and if you are not satisfied, what are the weaknesses and problems?

CHINTA AND NAKKA: We are well satisfied and very happy. Before the 1977 cyclone, we lived in mud houses, and even with light rain they could collapse. But after EFICOR built our new houses, we are happily living here with no trouble, even in times of cyclones.

JOHN: Is the village mainly composed of a Christian community or not? If it is mixed, can you say what communities now live together?

CHINTA AND NAKKA: The village is entirely composed of a Christian community.

JOHN: In your view, is your village superior to others in the area, or average, or less good than other villages? Any reasons for your answers?

CHINTA AND NAKKA: Chinthayapalem is superior to other villages for 20 years because we had very good houses built by EFICOR. But now the majority of the houses are collapsing since they are at the end of their life. Some people are repairing and rebuilding their houses, but some residents do not have the financial resources even to repair.

JOHN: Have you maintained any link with the agency that built the settlement: EFICOR? If yes, what form of link – and if no, is there any reason?

CHINTA AND NAKKA: We have not kept any link with EFICOR; they never came back to our village after they left when the construction was completed. The EFICOR office in the village was burnt in a fire accident in 1981. The leader of the reconstruction project, Albert Lilly, returned to the village once.

JOHN: How many houses are there and roughly what is the current population?

CHINTA AND NAKKA: There are 104 old houses that were built after the cyclone and 10 new houses. The current population is around 600.

JOHN: Have there been any major changes or new houses built since 1977?

CHINTA AND NAKKA: Two years ago, we built a new church in same place and 10 to 15 new houses were constructed beside the village.

JOHN: Has anyone attempted to explain to the community about cyclone risks and what they should do to protect themselves and their animals and dwellings?

CHINTA AND NAKKA: The village heads asked for the cyclone shelter after the houses were built. The EFICOR team helped us to get government funds and they constructed a cyclone shelter by adding some more money to the fund released by government in 1981.

JOHN: Were the neighbouring villages eventually rebuilt after the 1977 cyclone, and if so, by which agency?

FIGURE 12.7 Houses in Muthayapalem, the neigbouring village to Chinthayapalem – the photograph was taken in 2014, 36 years after construction. These houses are not the original dwellings built by CASA after the disaster; they collapsed and have all been replaced (photograph by John Noble).

CHINTA AND NAKKA: Muthayapalem, Matsyapuri, Anand Nagar and Hanuman Nagar were rebuilt after 1977 cyclone by Church's Auxiliary for Social Action (CASA).

JOHN: What is the position of those villages rebuilt by other agencies like CASA?

CHINTA AND NAKKA: Our houses that were built by EFICOR were far better than the houses in these villages that were constructed by CASA, and those houses eventually collapsed completely.

JOHN: Do the villagers evacuate to cyclone shelters when a storm is coming? If so, where is the nearest one and how much warning did they get?

CHINTA AND NAKKA: Yes, in the 1990 cyclone, villagers evacuated to the cyclone shelter beside the village that was built by EFICOR.

JOHN: Are you using the fish ponds, how does the cyclone shelter function in normal conditions, and what became of the training you were given by EFICOR for the welfare of your village?

CHINTA AND NAKKA: We are raising income from the cyclone shelter by using it for storage of grains, and the fish ponds are leased out to cultivate aquaculture. The EFICOR team trained some of our villagers in some professions like carpentry, tailoring, etc. to develop their livelihoods. They are still using these skills in their work.

This interview shows that the project appears to have succeeded and that Ian's fears and those of the angry village leader were never realised. The original superior-quality dwellings have lasted longer than inferior cheaper housing. The livelihood creation work in the late 1970s still yields positive fruit. And the neighbouring damaged villages were eventually reconstructed by a Christian agency for Hindu communities. Thus the overall assessment resulted in positive emotions on the part of the village occupants.

FIGURE 12.8 Fish ponds created for aquaculture and trees planted to form a cyclone shelter wind break in 1978 as part of the cyclone recovery programme (photograph by John Noble).

FIGURE 12.9 A new house being constructed in 2014 using traditional construction techniques and bamboo and thatch building materials. The structure does not appear to have any of the design features of a safe house that will resist cyclonic wind forces, such as a low pitch roof, or to contain any structural safety measures, such as triangulated cross-bracing (photograph by John Noble).

The recent interview and photographs are a reminder that all disaster reconstruction is a time-constrained process with an 'expiry date'. In this case, most of the dwellings needed to be replaced well within a 40-year period, and as Figure 12.9 indicates, without public awareness and the training of builders in safe construction, such replacement will rebuild vulnerability as well as a house.

The general success in rebuilding Chinthayapalem was to be followed a decade later in 1993–6 when EFICOR rebuilt Malkondji, which, as Chapter 1 describes, was also a highly successful reconstruction project. However, while Malkondji was built by local masons who were taught safe construction, that was not the case in Andhra Pradesh. Therefore, it is interesting to speculate whether the construction in Figure 12.9 would have built safety measures into the design if the original settlement had been built using the progressive, user-build training programme, which would have been passed down and replicated, as has happened in many other contexts.

This example from India contains elements of frustration, of missed opportunities, with red hats firmly in place. But as we have described, despite the mixed emotions, this is a story with positive value as expressed in the pride and satisfaction of residents in their village (definitely wearing yellow headgear) over a long span of three decades.

Recovery from disaster can be an inspiring process in which good practice warms the heart and reaffirms confidence in human ingenuity, social values, public service and charitable intentions. In October 2001, 118 people were killed in the collision of two aircraft at Milan's Linate Airport. Subsequently, an emergent group was formed by the relatives of the deceased and this later turned into a charitable foundation that has worked hard to promote safety in both aviation and surgery. Although not without problems and conflicts, the group has had a cathartic, healing effect on its members in their bereavement and has been a force for good in the world.

Yellow hat – optimism

The yellow hat symbolises an optimistic response to problems. The optimistic thinker explores positive elements with a sense of brightness. He or she searches for opportunities, emphasises strengths, and seeks positive value and benefit. We hope that the reader of this book has also gained a sense of the positive developments that have taken place. Recovery from major disasters in China and Japan has been efficient and extremely rapid. One may fault aspects of these recovery programmes (as noted in Chapters 3 and 10), but one has to admire the dedication and zeal with which the process has been carried out.

If one looks for them, there are plenty of examples of progress of the kind that begets optimism. For example, economists have drawn attention to the crucial role of local economic stimulus and the restoration of livelihoods in ensuring that recovery takes place adequately.

In the field of recovery from disaster, one of the main achievements of the past 30 years has been the realisation that 'top-down' solutions, dictated from outside the disaster area and its milieu, are not effective – rather, ample participation is needed from local citizens who are the interested parties or stakeholders in recovery.

This has led to a fascinating series of experiments in how best to involve local communities in determining their own future after disaster. The process begins with participatory research (Özerdem and Bowd 2010) and continues with transparent governance and participatory methodologies in recovery management (Gokhale and Mistry 2012). Focus groups, town meetings, community projects, and so on, are evidence of participatory strategies in action. However, it is important to note that there are no panaceas and the process is easily misconceived (Davidson et al. 2007).

The Pakistan earthquake of 8 October 2005, when over 73,000 people died and 3.5 million had damaged or destroyed homes, was followed by the most remarkable reconstruction project that Ian has witnessed in 42 years of visiting recovery sites. His optimism and encouragement relate to many successful elements in a definite 'yellow hat experience'. The sheer scale of the rural housing reconstruction operation was a daunting achievement by any standards, with the building of 463,243 houses and repair of 130,000 within three and half years.

FIGURE 12.10   Pakistani masons constructing an earthquake–resistant dwelling as part of the EERA/UN-Habitat/World Bank user-build rural housing reconstruction (photograph by Babar Tanwir, UN-Habitat).

FIGURE 12.11   The distribution of information materials used in the EERA/UN-Habitat/World Bank user-build rural housing reconstruction (photograph by Sheikh Ahsan Ahmed, UN-Habitat).

FIGURE 12.12 Timber structure of the earthquake-resistant rural housing with a lightweight roof of corrugated steel as part of the EERA/UN-Habitat/World Bank user-build rural housing reconstruction (photograph by Ian Davis).

But these dwellings were not delivered and constructed by an army of contractors for passive beneficiaries – rather they were built using locally available materials, with the addition of imported concrete blocks and corrugated iron sheeting. These dwellings were built in the rural housing tradition by earthquake survivors, well trained in seismic safety construction as well as in good building practice (ERRA 2008).

A creative innovation for Pakistan was that for the first time in their disaster assistance approach, cash grants were allocated directly to families as a subsidy to pay for their own dwellings and recovery. For poorer families, it was enough to cover two rooms; but the majority of people built more, mobilising their own resources in addition to the cash grants. A total of $US1.6 billion was provided through the government, but people mobilised a further US$2 billion, which they also used in compliance with the required building standards. The cash grant therefore acted as leverage. The banking system was strengthened to have more branches and more capacity. People accessed funding either through their bank accounts or through the post office where there were no banks. This provided a means to monitor dealings in a transparent manner. There were mechanisms put in place to mitigate the risk of any bank or post office officials charging commission. This was done by ensuring that all recipients knew the terms and conditions, that they knew their rights, and that they had access to a secure channel to report and act on complaints.

For many beneficiaries, this method of financing their reconstructed dwellings was their introduction to banking – an essential prelude to further development opportunities for each involved family. This was also a vital way to combat corruption as each family received their money directly, thus cutting out middlemen – a frequent source of past corruption.

An astonishing feature of this project was the positive role of the Pakistani Army in multiple roles from assessing damage to participating in the builder training programmes that were led by ERRA with support from UN-Habitat. Much of the credit for this must come from the inspired leadership of Lieutenant General Nadeem Ahmad, later to become the Chair of the National Disaster Management Authority (NDMA) and the Deputy Chairman of the Earthquake Reconstruction and Rehabilitation Authority (EERA).

Fortuitously, ERRA was well supported by the World Bank and UN-Habitat in the funding and technical development of this programme. Experienced Irish architect Maggie Stephenson developed a highly effective working relationship with General Nadeem and his colleagues, including Waqas Hanif who effectively managed the housing programme from start to finish. Maggie Stephenson was able to share the wide experiences of UN-Habitat from other Asian countries with her Pakistani colleagues in developing household- and community-driven programming (Mumtaz *et al.* 2008; Arshad and Ather 2013).[2]

In 2011, Ian nominated General Nadeem and ERRA for the UN Sasakawa Award for achievements in Disaster Reduction and they received a certificate of merit. He recalls: On the evening following the award ceremony, I was invited to meet him with Maggie Stephenson for a celebration meal. During the evening, I told General Nadeem that I was perplexed about this ambitious reconstruction programme since it was virtually unheard of for a national military establishment – with implicit 'top-down' command and control doctrines – to so enthusiastically embrace a 'bottom-up', fully developmental approach in rural reconstruction. This was expressed in community development with high levels of trust in adopting cash grants and a strong capacity-building training emphasis. He responded that he had always been interested in community development and had promoted such projects in his home community in rural Pakistan. Nadeem explained that:

> when the 'user-build' reconstruction approach was first explained to me by Maggie and her colleagues from the UN, then it was obvious that it could bring great benefits to our rural communities who had lost their homes, and in many cases their relatives as well as their livelihoods. I was also aware of the political importance of rapid recovery in housing, and the value of equitable coverage for social and political cohesion in Pakistan.

Thus, a valuable lesson from my visit to Pakistan and our dinner conversation was the reminder of the danger of stereotypes that can easily condition attitudes.

Military leadership does *not* necessarily result in rigid top-down command and control approaches. It was clear that General Nadeem's military background had prepared him with leadership skills, enabling him to motivate his team under difficult circumstances as well as to demonstrate the value of coherence and consistency among partners, planning for scale, etc.

The Pakistan success shows us that anything is possible in those rare yellow hat cases where power and resources, knowledge, organisation and vision come together. This is a reminder that leadership is essentially a quality, not a designation. But luckily for the rural housing recovery in Pakistan, something rare happened in General Nadeem in that the two converged.

Green hat – creativity

> There is no doubt that creativity is the most important human resource of all. Without creativity, there would be no progress, and we would be forever repeating the same patterns.
>
> *Edward de Bono, 1985*

The green hat represents thinking about creativity. It involves new ideas, fresh concepts, novel perceptions, new ways to achieve improvement and creative alternatives. It rarely grows out of rational, considered thought processes. It was Picasso, one of our greatest creative artists, who once declared that 'the chief enemy of creativity is good sense'.

Recovery from disaster is a process that strongly needs a creative approach that is capable of bringing innovation and redundancy – in the sense of different routes through the problem-solving maze. In an economic sense, disasters involve a process of accelerated consumption of resources. They thus create resource shortages, often instantaneously. The process of recovery is one of getting the best value from scarce available resources. Some of that problem is technical, but much of it is organisational. Creativity does not flourish in a vacuum. A broad culture of disaster risk reduction is the essential basis for creative work. Most of the scope lies in connecting disparate elements of a recovery system so that they function better in concert.

In a disaster risk management training workshop for mid-career officials held at Cranfield University in 1998, there was a discussion about the qualities needed to ensure effective outcomes. One group compiled a list that did not include creativity while the second group placed it highly on their list. This resulted in a fascinating debate. The leader of the group that saw no value in creativity, a senior police officer, argued that in his experience of leading disaster operations, this was the last quality he wanted in his teams. His concern was for operations. Staff should not invent solutions but should instead follow standard operating procedures (SOPs) to the letter. At this point, the leader of the other group (metaphorically wearing a green hat) strongly argued for creativity, and for highly creative staff, on the grounds that one often does not have the resources one needs in disasters and, in such situations, one needs to make a lot out of a little, which requires high levels of imaginative invention.

The lack of recognition of the importance of creativity in disaster recovery is also reflected in the findings of recovery survey described in Chapter 11. Only one of the 51 contributors highlighted the importance of creativity.

Another example comes from the recovery after the Southern Italian earthquake of 1980, which took between 17 and 20 years. During this time, local people lived in temporary accommodation and reconstruction was slowed down by perennial shortages of funds. The town of Sant'Angelo dei Lombardi, in the Province of Avellino, did not even begin reconstruction for four years after the earthquake. However, these were not years marked by inactivity. Geological and foundation surveys, detailed planning and public consultation took place. Historic monuments were assessed, and projects were designed to restore them. Expansion plans were made so that the town could grow rationally. The decision was taken not to demolish all damaged housing but to evaluate carefully what could be repaired and what could not. As a result, when the reconstruction was finally completed, a considerable amount of the town's historical character, its *genius loci*, had been preserved in a reconstruction that made it a safer, more rational settlement with greater potential to adapt to twenty-first-century needs. The reconstruction was prudent in that its leaders did not rush into it, but nor did they waste time. It was also creative in that the result was a more functional town, one that commemorated

FIGURE 12.13 Resettlement housing built after the 1688 landslide in Pisticci, Basilicata Region, Southern Italy and since modified (photograph by David Alexander).

FIGURE 12.14   Frauenkirch in Davos, Switzerland, destroyed by avalanche in 1602 but rebuilt in 1603 with the reinforced 'prow' to deflect the impact of avalanches, thus protecting the church and its occupants (photograph by Ian Davis).

the disaster but also provided better parking and traffic circulation, improved public open space and rational opportunities for further expansion.

Much earlier, the landslide of 1688 at Pisticci in the Province of Matera devastated three neighbourhoods of peasant dwellings and killed 400 people. Because level land was in short supply, the houses were rebuilt *in situ* on top of the now stabilised landslide deposit. In order to minimise the risk in case ground instability started up again (which indeed it did) or earthquakes were to occur (which they have done), the dwellings were built as one-storey units (Figure 12.13). The design and orientation along the contours of the landslide deposit show considerable creative insight. Unfortunately, in recent times, some of these buildings have had a superstructure of up to two storeys added to them, which is unwise in the light of natural risks. Most of Pisticci (population 17,250) suffers from landsliding as it is situated in the midst of one of the most unstable terrains in Europe.

The Swiss of Canton Graubünden are well aware of the devastating effects of avalanches on buildings and infrastructure. Many roads are now protected by expensive concrete avalanche roofs, which allow snow movements to pass over the road without damaging anything on it. An earlier strategy was to construct buildings with reinforced walls on the upslope side and to use V-shaped 'prows' to divide avalanches and direct them around it. The church at Davos Frauenkirch offers a historic example of this (Figure 12.14). It was built in 1350 but was largely destroyed by avalanche in 1602, being rebuilt with the reinforced 'prow' the year after. It has been emulated by new buildings further downslope, which have curved, reinforced walls and small windows on their upslope sides. This is an example of how adversity forced creativity in the search for a building design that would resist the force of avalanches without damage.

Paradoxically, there are places where creative solutions are showered upon disaster-stricken localities (see Chapter 9, Option 5b: 'Provided shelter – provision of patent shelter units'). They come in the form of instant shelters of novel design that arrive in a steady stream from opportunistic manufacturers, inventors, industrial designers, intrepid architects and, especially, architectural students. Most designs never make it to field application, but occasionally they are used and the results have not been encouraging or satisfactory for their occupants. In 1976, Ian wrote as follows about the various relief agencies he had visited in Geneva and Washington:

> I found that a familiar pattern was for officials to say 'emergency housing' and walk over to a filing cabinet which virtually overturned as it was opened. The drawers were bulging with '57 varieties' of shelter types. The vast majority of these concepts mercifully have never left the drawing board or filing cabinet, but this seems no deterrent to the ingenuity and persistence of designers.
>
> *(Davis 1978: 49)*

To design and create an untested innovative *temporary* shelter with a projected lifespan of a few weeks or months for survivors suffering from the trauma of a disaster would appear to be callous and reckless folly, so what words are appropriate when an experienced agency creates an untested, innovative *permanent* dwelling with a projected lifespan of 20 years or more for a poor family who require security and comfort?

One design of a permanent 'donor shelter' that did progress beyond the drawing board was built after the Latur earthquake of 1993 in India. In the village of Gubal, near Kilari, the Adventist Development and Relief Agency (ADRA) erected a series of ferro-cement domed permanent dwellings, the original intention being to provide structures that would withstand a future earthquake due to the robustness of the dome shape (Figure 12.15). However, in the second phase of the project, the domes were wisely changed into rectangular rooms. When Ian and colleagues visited the site in 2011, the distressed occupants wrongly assumed we were from ADRA and had come to assist them. They had a litany of complaints that included: concerns that the domed structures leaked and local builders did not know how to repair cracks in the ferro-cement on account of the unfamiliar technology; difficulty in dividing the domed internal spaces to create some privacy; the fact that the domed form did not fit rectangular tables, beds, shelves; the dwellings became intolerably hot in the summer, etc. They complained that throughout the past 18 years no one from ADRA had visited the site to see how they were coping with the dwellings.

FIGURE 12.15 Ferro-cement domed dwellings provided by ADRA in the village of Gubal following the 1993 Latur earthquake. This photograph was taken 18 years after construction (photograph by Ian Davis).

This is an example of an untested novel technology that was adopted by a well-established agency, which had the resources to secure and apply excellent technical advice and also the ability to maintain the structures when they failed. The result of this adoption of a totally inappropriate piece of creative design appears to have been 18 years of dissatisfaction as well as abject misery for the unfortunate occupants – and all this was provided by a *'relief'* agency.

Therefore, the problem with the failure of such invented shelters does not lie in any shortage of creativity. Designers have failed, and continue to fail, to understand the dynamics of disaster recovery and incorporate the local cultural, climatic, social, environmental and developmental requirements of sheltering into their thinking and designs.

There are also examples where a good creative idea, technique or technology is misapplied. Such was the case with the base-isolated transitional housing constructed after the L'Aquila earthquake of 2009. This was a bold idea to provide shelter for 15,500 people in six to nine months by constructing three-storey apartment blocks that were not designed to be anti-seismic as the concrete rafts on which they stood were given pendulum base isolation, hence reducing the need for anti-seismic reinforcements in the superstructure above (Figures 12.16 and 12.17). However, one can question whether the enormous expenditure on such refinements was necessary given that the buildings were intended as temporary accommodation and are not particularly suitable for

FIGURE 12.16 Transitional apartments in the L'Aquila area, Southern Italy, built after the earthquake of 2009, the *case antisismiche sostenibili e ecocompatibili*. Each building of this kind was designed to sit on 40 columns capped with pendulum base isolators (photograph by David Alexander).

FIGURE 12.17 Pendulum base isolator under a transitional apartment building at Bazzano, L'Aquila, Southern Italy. Seismic activity would displace the column, but the building on the concrete base plate at the top of the picture would undergo attenuated shaking due to the damping effect of the cup-and-ball isolator (photograph by David Alexander).

permanent occupation. The whole project assumes the character of a large experiment in architecture and urban planning, and one that is successful only from certain points of view (ECA 2012).

Of all de Bono's hats, perhaps the one least in evidence in disaster recovery has been the 'green hat' of creativity. One reason may be that that disaster recovery is complex and always seems to require a bewildering array of bureaucratic processes and an army of bureaucrats. Rather like the policeman described in the example above, many of these officials are most comfortable with their checklists and their obedience to standard operating procedures. Nobody with any experience of disaster recovery operations would imagine that it is possible to recover across all fields without well-developed plans, policies, strategies and tactics. But such patterns or procedures *all* require innovation, imagination and resourcefulness, and these creative skills need to be welcomed and valued wherever they are found. They also need to find their place in training and education for all involved parties.

Blue hat – management

Last, de Bono's blue hat is about strategic management. This involves determining the focus and objectives of strategies, assessing what has been learned, anticipating what comes next and considering where to take the process. This is a very broad activity and one that also involves 'thinking about thinking'.

Recovery from disaster needs adaptive management. This is something that keeps its fundamental objectives consistent but adapts its methods to changing circumstances. The best recovery from disaster has clear aims and objectives that are well known to the beneficiaries and fully supported by them. This also means that priorities are set and adhered to, but not without changes if they must be adjusted to changes in the controlling circumstances. The management consultant Peter Drucker has reminded us that: 'long-range planning does not deal with future decisions. It deals with the future of present decisions' (1974: 125). Good management of recovery is an inclusive process that coordinates all agencies and interests involved and makes sure that they are communicating with each other.

Our experience suggests that many things in the recovery process depend on individuals with talent – the talent to observe, interpret, communicate and coordinate – and we have highlighted examples throughout this book. It is necessary that such people hold responsible positions which enable them to have an impact on critically important processes. This in turn depends on having healthy organisations that can see the virtue of hiring and supporting such individuals. Sadly, it is common to find that institutions are 'mediocracies', run by and for mediocre people, to whom excellence is inimical and who would not dream of hiring people who are more competent than they are themselves. This is not what the complex and challenging process of recovery from disaster requires.

In this chapter, we have considered many aspects of strategic management, seen though different lenses, or by wearing varied hats:

- A good strategic recovery plan will recognise and anticipate the *black hat* of potential problems and continually explore ways to tackle such difficulties as they arise or, where possible, long before they even begin to appear.
- Effective planning and strategic thinking is dependent on the *white hat* of a flow of accurate information, but leaders and managers should not confine their attention to the knowledge of facts alone.
- This will be balanced by the *red hat* of passionate emotions that can be expected in the context of any disaster recovery operation as lives and futures are at stake. A strategy devoid of emotion is akin to a marriage without love.
- The *yellow hat* of cheerfulness, hope, good humour and optimism will always be needed to cope with the 'slings and arrows of outrageous fortune' that characterise most reconstruction efforts.
- For the obvious reason that environments, comprising dwellings, people, jobs and trees, require the work of an army of artists, designers and innovators, creativity, the *green hat* of invention or improvisation, is an essential attribute. Also needed is the work of creative managers and leaders who have the rare ability to maximise the value of limited resources. The magical quality of *genius loci*, which we discussed in Chapter 6, that gives surviving communities a sense of place and well being rarely comes by accident. Rather, it is the product of liberated creative minds.
- Finally, the *blue hat* is essential headgear as the complexity and inherent risks attached to any convincing strategy demand concentrated thinking, debate, planning, organisation and reviewing.

It will be exceedingly rare for any inspired leader of recovery to wear all these hats. They are more likely to be found in well-constructed teams composed of leaders and followers, optimists and realists, those who gather and process data and those who act with intuition, calm tacticians and passionate dreamers. And, of course, in political environments where apathy and bureaucracy prevail, one must ensure that there are 'movers' as well as 'shakers'.

Valediction

The 'six hats' thinking process is not an infallible model. For example, we see evidence of a sort of 'pink hat' that bridges the gap between information and emotions through the different views of what constitutes useful information and different value systems. We also catch a glimpse of a possible 'orange hat' in which positive emotions inspire an optimistic response. There is also a species of 'grey hat' that links objective facts about risks and threats to risk assessments. The reason why it is not quite a white hat (one symbolising information) is that, however rigorous one strives to be in risk assessment, the process involves trying to create objective facts about unknown circumstances. However, we feel that de Bono's classification of thinking modes helps clarify the ways in which the multifaceted problem of recovery from disaster can be viewed and helps bring a sense of order to the multiplicity of opinions about it.

One topic that is missing from this chapter, and often missing from the debates about recovery, is religious faith. Within the NGO sector, a high proportion of groups are motivated by their religious convictions. This is also a major strength, or 'coping mechanism', for many survivors during the trauma of disaster and dislocation. This was the case for thousands of people in Haiti following the 2010 earthquake (Mooney 2010). Moreover, the knowledge and authority of religious leaders can be a strong positive force that helps local recovery and resilience. Yet, in the context of disaster, there can also be a negative side to faith in the form of bigotry and prejudice as one faith is set against another or where recovery assistance is allocated on sectarian grounds rather than according to need, as we highlight in this chapter in the red hat description of an ill-advised project in Andhra Pradesh.

Thus, the Red Cross were wise to include this concern in their Code of Conduct for Disaster Relief, item 3:

> **Aid will not be used to further a particular political or religious standpoint.**
>
> Humanitarian aid will be given according to the need of individuals, families and communities. Notwithstanding the right of NGHAs to espouse particular political or religious opinions, we affirm that assistance will not be dependent on the adherence of the recipients to those opinions. We will not tie the promise, delivery or distribution of assistance to the embracing or acceptance of a particular political or religious creed.
>
> *(IFRC 1974: 3)*

For believers, faith sits under many hats and can be a multicoloured experience It can be a white hat, based on a belief in historical events – for example, as expressed in Christian worship and in the reading of the Creed. It can also be yellow – full of joy, hope and confidence. It can be under the red hat of deeply expressed positive or negative emotions. And in all disaster contexts there is the inevitable challenge to faith, the dark black cloud of suffering, anger and anguish. Faith should never be taken as an excuse for fatalism about recovery or disaster risk reduction. Indeed, interpretations of religious texts tend to err on the side of activism, not fatalism (e.g. Ghafory-Ashtiany 2009).

The epilogue to this book consists of quotations from two ancient sources that positively resound with a message of hope. The first, taken from the prophecy of Isaiah is profoundly yellow hat – 'a light rising from darkness'. The quotation captures the reality of faith with confidence in a God of justice and compassion, a poetic vision of triumph after disaster, the 'repairer of broken walls' (aptly portrayed on the cover of the paperback version of our book!).

Hence, there may be some additional coloured hats to add to de Bono's classification of thinking modes that help to clarify the ways in which the multifaceted problem of recovery from disaster can be viewed, helping to bring a sense of order to the multiplicity of opinions about it.

To conclude this book, we move from these varied perspectives on recovery to a series of emerging principles.

Notes

1 John Noble, a local resident in Andhra Pradesh, kindly undertook these interviews on behalf of the authors in July 2014.
2 We are grateful for the support of Maggie Stephenson in writing about the Pakistan recovery experience, for the invitation to Ian Davis to visit the reconstruction sites in 2008 and for providing details of the rural recovery programme.

13

EMERGING PRINCIPLES OF RECOVERY

Primum non nocerum. (First, do no harm.)

Hippocrates (c. 400 BC to 370 BC)

As we discuss in Chapter 2, we are aware that there is no shortage of principles for shelter, reconstruction and recovery; in fact, we have written or contributed to some of them over the years (Alexander 2002: Chapter 7; Jha *et al.* 2010: 1–2). One is reminded of Groucho Marx's famous dictum: 'Those are my principles: if you don't like them, I have others.' But, despite the value of such lists, we include below a set of conclusions that emerge from our joint study. We will end with a restatement of the most important principles by structuring them in accordance with model 14, 'project planning and implementation', which we introduced in Chapter 4.

As we have noted, many things in the recovery process depend on individuals with talent: the talent to observe, interpret, communicate, coordinate and lead. It is necessary that such people hold responsible positions which enable them to have an impact on critically important processes. This in turn depends on having healthy organisations that can see the virtue of hiring and supporting such individuals. Sadly, it is common to find that institutions are often 'mediocracies', run by and for those for whom excellence is inimical. This is not what the complex and challenging process of recovery from disaster requires.

The principles are grouped according to the scheme shown in model 14 (reproduced in Figure 13.1; see also Figure 4.6).

Underlying principles

1　While all disasters are different in scale, nature, impact and recovery actions, critical lessons can be deduced from past experience. They can usefully be shared widely and applied in such a way as to help ensure effective recovery.

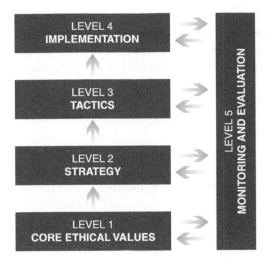

FIGURE 13.1 Model 14: project planning and implementation.

2 Authorities need to devise their own sets of guiding principles in order to support their recovery tasks. The principles should cover five levels that relate closely to the prevailing culture and system of governance, namely core ethical values, strategies, tactics, implementation and monitoring and evaluation. While core ethical values, strategic principles and evaluations can be shared widely across cultures and continents, tactical and implementation principles vary from case to case and are specific to local contexts. Therefore, at the outset of recovery, wise recovery managers will seek to develop appropriate principles to guide tactics and applications.

Level 1: core ethical principles

3 The equitable distribution of recovery resources should be based on the needs of beneficiaries rather than their status. For those affected by recovery, fundamental rights should be established and secure tenure of property should be guaranteed.
4 Anti-corruption measures should be devised and applied in order to ensure that resources flow to meet vital needs and do not corrupt those who handle them.
5 At the heart of all effective recovery operations lie the needs of survivors. A good recovery policy helps reactivate communities and empowers people to contribute to rebuilding their homes, lives, livelihoods and environment. The leaders and managers of recovery need to give account for their actions to the object of their concern – disaster survivors.

Level 2: strategic principles

6 Policies and plans for recovery should be financially realistic. It should be recognised that, as the political commitment to allocate financial resources

to recovery declines over time, budgets will also decline. Thus, there is a need for pragmatism to ensure rapid recovery by generating and maintaining a political consensus and by dedicating funds to the process.

7 Except where specialised coordination of complex, cross-disciplinary matters is needed, existing ministries and institutions should be used to facilitate and manage recovery.

8 Relocation disrupts lives and is rarely effective. Thus, it should be used as little as possible.

9 To ensure that recovery is effective, every effort must be made to strengthen government and governance.

10 To contribute to long-term development, recovery must be both safe from future hazards and sustainable.

11 Reconstruction is an opportunity to plan for the future and conserve the past.

12 Pre-disaster planning should be used to prepare for disaster events and subsequent recovery.

Level 3: tactical principles

13 Effective disaster recovery requires strong community participation. Therefore, management structures must empower local people but ensure harmonisation with higher levels of government. But strong community participation needs to be balanced with a strong governmental role, both being essential ingredients for effective recovery.

The various case studies cited throughout our book can be assessed and compared according to the following useful matrix devised by Mary Comerio to indicate graphically the levels of effective community participation and the government's role from weak to strong (Comerio 2013: 37).

14 In a major disaster, there must be central control of resource flows and international liaison, but municipal government must have sufficient autonomy to manage the recovery at the local level.

15 Due to the dynamic, rapidly evolving situation after a disaster, the process of planning recovery is more cyclical than linear. The tactical sequence involves assessing needs, planning, testing the plan, implementation, monitoring and evaluation, then reassessment, planning, testing, and so on.

Level 4: implementation principles

16 Material and human assistance imported into a disaster area should augment, complement and reinforce local initiatives, not supplant or duplicate them.

17 Successful implementation is based on the timely supply of accurate information. This must include quantifiable data, such as the numbers and types of dwellings that have to be constructed, and qualitative data, such as the social well being and psychosocial status of the surviving population.

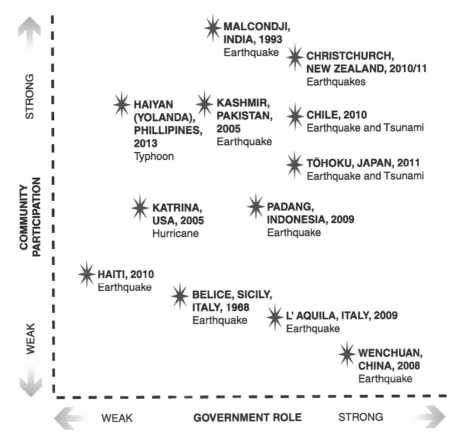

FIGURE 13.2 Model 21: the Mary Comerio comparison of recovery management approaches[1] (reproduced courtesy of Mary Comerio; see Comerio 2013, 2014a, 2014b).

Note: the model relates to the case studies of recovery cited throughout our book.

Both sets of information need to be merged and communicated to operational staff in government and civil society.

18 The key resources for implementation are organisation, leadership, authority, cash, flexible plans, overall commitment to the task at all levels, and a clear vision.

Level 5: monitoring and evaluation principles

19 The process of building resilience by evaluating progress against agreed benchmarks should be monitored and evaluated no less frequently than every six months. The following elements should be evaluated:

• reconstruction goals;
• dedicated budgets;

- accountability to avoid corruption;
- staff training;
- protection of critical facilities;
- how to build redundancy into disaster risk management systems;
- disaster risk reduction;
- how to strengthen emergency services.

20 It is necessary to devise ways to ensure that lessons about how to promote resilience in recovery operations have been learnt, documented, stored, disseminated and acted upon.

Concluding observations

To conclude, we summarise some of the most important lessons of this book.

Social dynamics of recovery

Recovery from disaster is a human social process as much as a technical one. Although international bodies such as the United Nations place great reliance on policies, it is people who, in reality, interpret them and act upon them. There is therefore a need for inspired leaders – and we have identified several in our text – who have a broad appreciation of its complexities, a recognised ability to communicate well, and talent in the process of coordination. Strategies need to be articulated well, but they should be adaptable to changing circumstances. In terms of their outcome, those people who have devised them and put them into effect need to be accountable to the public and its representatives. Participation of stakeholders is a paramount need, but it must not be allowed to descend into a free-for-all in which no real progress is made because competing interests block all developments. Conflicts between stakeholders need to be reconciled by firm but democratic processes.

Sharing information

Where recovery from disaster is prevented or subverted by vested interests, and where people who fight to make it accountable or effective are persecuted, there is room for public outrage. Protest can achieve change although sometimes at heavy cost. Good ethical principles should never be abandoned for reasons of expediency. At the other end of the emotional scale, good achievements in recovery need to be celebrated and shared so that good practice is disseminated and lessons are learned in a multilateral sense.

Recovery without essential information (on hazards, impacts, consequences, stakeholder needs and resources) is like 'driving blind'. There is an unresolved question of who should be the custodian of data, much of which is lost in the production process. Open data sources on recovery from disaster will be needed in the future. 'Perishable' data need to be collected and made available.

Developmental recovery

Considerations of data aside, there is a massive challenge regarding how to go beyond the status quo ante and achieve developmental reconstruction, moreover in a reasonable length of time. Sometimes as a result of disaster, sometimes for other reasons, many areas where calamity repeatedly strikes are economically depressed, especially in developing countries. Democratic consultation tends to slow down progress, but it does, of course, allow for stakeholder representation. As we noted in Chapter 7 (Fourth Dilemma), Herbert Morrison once said that it is better to take action promptly than to wait for all the information to arrive, or the political will to act will evaporate. This may be true, but time is socially necessary in disaster. There is thus a need to accept the time it takes to do a task properly, including the process of consultation. Problems can be solved by on-course corrections, but they must be faced before the recovery is finished or they will grow and fester afterwards.

Strengthening national governments

We have both learned much during the five years it has taken to write this book and two particularly important lessons stand out. The first is the recurring theme of the importance of strengthening, rather than weakening, the critical role of national governments in managing disaster recovery, at both the centre and local government levels. We believe that this has been seriously neglected in the literature and in past official attitudes from international agencies.

We recall four examples. In Chapter 8, we discuss the dilemma of whether universal standards, as opposed to national standards, should apply to shelter. Ian records a heated exchange in a training course between government directors and a representative from Sphere concerning the lack of consultation with governments in formulating the Sphere standards.

In Chapter 10, we challenge the wisdom that was coming from a World Bank-organised conference on recovery in 2014. We object to the global centralisation policy being advocated that sees frameworks for recovery formulated in Washington or at the World Conference on Disaster Reduction, when this task must surely be a key function of national governments who have the authority and responsibility to define the frameworks that suit their own unique situations.

In Chapter 11, we analyse the survey of 51 experts and noted that not one of them referred to the need to assist governments in their own recovery from disasters that may have killed their staff and wrecked their facilities. We highlight this need in one aspect of our hexagonal model of recovery sectors (model 2 in Chapter 3).

Finally, in this chapter we include model 21, which was originally developed by Mary Comerio to indicate the need in effective recovery for strong governments as well as strong community participation.

Listening to survivors' voices

We have also learned to seek out key informants and listen carefully to the voices of the survivors of disasters as well as inspired leaders who seek to manage their recovery. Following guidance notes, observing recovery protocols, conducting any number of Google searches will never become valid substitutes for direct communication with those closely involved in recovery actions.

We vividly recall so many voices recorded within various chapters. In the Malkondji case study in rural India (Chapter 1), Ian records his encounters in 1996 and 2011 with Mr Rangappa, a local stonemason seeking work after the completion of reconstruction. In Chapter 7, Olga Popovic Larsen describes growing up in Skopje, Yogoslavia while the city was being reconstructed. In Chapter 8, we discuss the 'voices of survivors' and David includes a poignant diary entry on an elderly survivor of the 1980 Southern Italian earthquake in Tricarico, surrounded by his valued precious belongings. In Chapter 9, Ian records a visit to Ibagué, a town in Colombia, to meet a socially committed shopkeeper who described how she single-handedly organised an evacuation plan for her community to escape from flash floods.

Also in Chapter 9, we describe our own direct and memorable experiences of disasters experienced in Barrow-in-Furness and Southern Italy. In Chapter 11, we, as well as our readers, are privileged to hear the personal views of 51 of some the world's experts on what makes a successful disaster recovery. Finally in Chapter 12, we listen to the voices of village leaders in Andhra Pradesh, proudly reflecting on the dwellings they were provided with in 1978 following cyclone devastation and what had happened to them and their settlement in subsequent years. Also in this chapter we hear from General Nadeem, in Pakistan, as to why he pursued a community-based user-build approach to reconstruction when managing a vast rural reconstruction project after the 2005 earthquake.

Models

Our book has developed, as well as borrowed, a number of models that we believe will be useful to concerned officials, observers and analysts who need to conceptualise various interpretations of the processes of disaster recovery. Inevitably, they tend to oversimplify highly complex dynamics but, in our view, the potential benefit of throwing light on this complexity outweighs such concerns. We are also unapologetic about the frequent references to our first-hand experiences of disaster recovery. Both of us have written or edited texts in which we have not been free to express our own lifetime experiences and convictions. It has been refreshing to write about a highly politicised recovery environment without 'looking over our shoulders' at our paymasters.

Final words

We have discussed the problems of gathering vital qualitative data needed for recovery management, and we recalled the words of Albert Einstein: 'Not everything that counts can be counted, and not everything that can be counted counts.' We recognise the need to exchange information and personnel between those who have experience and those who lack it. Not everything can be learned from experience alone, but it is an invaluable, indeed essential, component if the knowledge and wisdom exist to interpret it profitably. We trust that some knowledge and experience have been conveyed in this book. As Martial wrote in his epigrams, *Sunt bona, sunt quædam mediocria, sunt mala plura: quæ legis hic. Aliter non fit, Avite, liber.*[2]

Notes

1   This model is used with the kind permission of Professor Mary Comerio (13 January 2015).
2   'Of the things that you read here, some are good, some mediocre, and most bad. Thus and not otherwise, Avitus, books are made.'

APPENDIX 1
SUMMARY OF MODELS

Chapters 3 and 4 contain a wide variety of models that are used in the development, disaster risk management and recovery fields. Chapter 13 contains the final model. Twenty-one are described in these chapters, including six developed specifically for this book where gaps were found in the literature and new representations were needed (models 1, 2, 12, 14, 17 and 20). Each model has its distinctive message or function in the process of disaster recovery. The following list summarises the models and their intentions.

Development and recovery models (Chapter 3)

1 **Progress with recovery**
Proposes the need for recovery to go beyond a return to the status quo to the development of quality and safety.
2 **Recovery sectors**
Recognises 'development recovery' in five key sectors.
3 **Development recovery and elapsed time**
Observes the stages of resilient recovery over time.
4 **Relationship between disaster and development**
Notes the positive and negative aspects of disasters and development in recovery.

Phases of recovery models (Chapter 3)

5 **Disaster cycle**
Reviews the phases of disaster risk and recovery management.
6 **The Kates and Pijawka recovery model**
Links the four phases of recovery to varied levels of activity.

7 **Cost-effectiveness (unit cost)**
Demonstrates how the phases of recovery are related to escalating unit costs.

8 **Disaster timeline**
Considers the 'ebb and flow' of the strands of disaster and recovery management.

Safe recovery models (Chapter 4)

9 **Probability/consequence risk assessment**
A tool to relate two of the key determinants of risk probability and frequency and consequence of impact.

10 **Disaster 'crunch' model**
Describes the causal factors that generate vulnerability and hazards.

11 **Resilient communities and settlements**
Describes the elements of resilient communities that live in resilient settlements.

12 **Disaster risk reduction measures**
Outlines the range of disaster risk reduction measures.

13 **Development of a safety culture**
Indicates the stages of the progressive development of a safety culture.

Organisation of recovery models (Chapter 4: models 14 to 20; Chapter 13: model 21)

14 **Project planning and implementation**
Suggests a logical sequence for implementing recovery.

15 **The pendulum (after Charles Handy's trust-control dilemma)**
Explores the levels of trust and control exercised by authorities.

16 **Two- or three-stage shelter and housing recovery**
Compares alternative approaches to shelter and the reconstruction strategies.

17 **Modes of shelter and housing**
Compares the diverse range of approaches to shelter and the reconstruction of permanent dwellings.

18 **Strengths, weaknesses, opportunities and threats (SWOT)**
Provides a tool to design projects and continually monitor the progress of recovery.

19 **Edward de Bono's 'six hats' model for problem analysis**
A creative analytical tool designed to examine a problem from six different standpoints.

20 **Organisational frameworks of government for recovery management**
Contrasts alternative approaches for governments organising recovery.

21 **The Mary Comerio comparison of recovery management approaches**
Compares approaches of community participation and the role of governments.

APPENDIX 2

Survey answers to the question: what in your view are the most important aspects of a successful recovery operation following a natural disaster?

1 **Yasemin Aysan**
Humanitarian worker, International Consultant, Turkey

Effective recovery factors are:
- *an independent, qualified, staffed national institution for recovery coordination*
- *long-term funding*
- *rapid economic revitalization*
- *safe land and land tenure for reconstruction*
- *beneficiary choices*
- *the protection of the marginalized, landless, renters, etc.*

2 **Sultan Barakat**
Architect, Senior Fellow at the Brookings Institution, Doha and Chairman of the Post-War Reconstruction and Development Unit, York University, UK

Genuine local ownership and capacity is needed to facilitate a transformative transition in which affected societies can build back better. Temptations to privatise industries and public institutions prematurely must be resisted in the interest of facilitating a just and potentially more equitable recovery.

3 **Stephen Bender**
Urban Planner, International Consultant, USA

Recovery is always in the context of a nation's development as the society created the past exposure presently manifests the vulnerability and assigns/negotiates future risk.

4   **Mihir Bhatt**
Architect, Director, All India Disaster Mitigation Institute (AIDMI), India

The most important aspect of successful recovery operations following natural disasters are the operations that increase the income and replenish assets, extend financial, water, sanitation and transport services; and upgrade of education and health standards of the victims. In addition, successful recovery deepens democracy, widens risk reduction, and pushes poverty back.

5   **Camillo Boano**
Architect, urbanist and educator, Senior Lecturer at The Bartlett Development Planning Unit, University College London, UK

Governments and donors need to be realistic about what is achievable in the short and longer term. Ideally, a comprehensive reconstruction strategy should combine with development planning and address the long-term issues related to disaster reduction:

- *Interventions are enhanced and far more responsive to the needs of the affected population through the adoption of a rights-based approach.*
- *A balanced position on recovery, combining a community-based, enabler approach plus a technology-based provider approach is better able to take.*
- *Account of contextual variations in societies and events at different scales.*
- *Appropriate integration of hardware (physical and material interventions) and software (regulation, local capacities, contacts) can improve interventions in the complex phases of relief and reconstruction and then facilitate the transition.*
- *While reconstruction of damaged infrastructure is critical, it is inadequate unless local vulnerabilities are identified and effective ways of reducing them are available.*
- *Recovery does need to be imagined, practised and institutionalised through a rights-based approach. Ideally, this guarantees access to opportunities, compensation and choices as an individual and as a community.*
- *Democracy and equity are fundamental preconditions for the creation of a recovery space.*
- *Post-disaster reconstruction is far more complex than has been generally acknowledged to date, despite many decades of experience. It involves both the planning of material things and the resolution of competing social interests in a complex process of mediation and negotiation.*
- *The drivers of recovery and reconstruction must be located at the individual and community level, balancing centralised decisions that ensure equity in reconstruction and decentralised adaptation.*

6   **Teddy Boen**
Structural and Earthquake Engineer and International Consultant, Indonesia

Governments must:
- *develop a solid strategy*
- *appoint a professional with strong leadership*

- *identify and utilize resources intelligently*
- *implement transparency and accountability.*

7   Ian Burton

Geographer, International Consultant, Scientist Emeritus, Adaptation and Impacts Research Group, Meteorological Service of Canada

Identification and holding to account of persons, organizations, and systemic drivers that allowed or promoted growth in vulnerability and exposure.

8   Terry Cannon

Geographer, specialist in vulnerability analysis for disasters and climate change; Research Fellow, Institute of Development Studies, University of Sussex, UK

That the organisations involved in reconstruction stop competing and acting to promote themselves and collaborate much more. And that the priority is restoration or substitution of people's damaged livelihoods and income earning opportunities. Reconstruction must ensure that it does not undermine this, and is based on a full understanding of the local political economy and culture.

9   Omar Dario Cardona

Engineer, International Consultant, Professor at the University of Manizales, Colombia

Avoiding reconstructing pre-existing vulnerability conditions with communities and the public and private sector having opportunities for transformation, sustainability and well-being.

10   Eric Cesal

Architect and Executive Director at Architecture for Humanity, San Francisco, USA

Successful recovery not only rebuilds physically, but addresses the embedded spatial and socio-economic inequities that create the conditions for disaster.

11   Esther Charlesworth

Architect, Founding Director: Architecture without Frontiers (AWF); Associate Professor, School of Architecture and Design, RMIT University, Australia

Planning for both the emergency phase of restoring basic shelter, water and infrastructure, while also having a clear and practical strategy for the longer-term reconstruction of the affected city and community.

12 **Mary Comerio**
Architect, International Consultant, Department of Architecture, University of California, Berkeley, USA

There are three critical components of disaster recovery that are necessary for equity, mitigation and sustainable development:
1   adaptable government programs,
2   locally targeted agendas, and
3   community participation and capacity building.

13 **Tom Corsellis**
Architect, Director: Shelter Centre, Geneva, Switzerland

Respectful, responsive, resilient, informed, impactful, impartial, inclusive, collaborative, coordinated, consistent, targeted, timely, transparent, appropriate, accountable, legal, prioritised, efficient, developmental and humanitarian.

14 **Jennifer Duyne-Barenstein**
Social Anthropologist, Head, World Habitat Research Centre, University of Applied Sciences of Southern Switzerland, Canobbio, Switzerland

A successful recovery operation following a natural disaster recognises all people's social, economic and cultural rights. It is inclusive and empowering, promotes peace and social justice, mitigates pre-disaster vulnerabilities and inequities and is environmentally, socially, economically and culturally sustainable.

15 **Russell Dynes**
Sociologist, Joint Founder and Emeritus Professor, Disaster Research Centre, University of Delaware, Newark, USA

Disaster recovery utilizing victims as key players moves beyond replace and restore towards creativity and change.

16 **David Etkin**
Geographer, Associate Professor of Disasters and Emergency Management, York University, Toronto, Canada

Successful recovery should not reconstruct past vulnerability, but it should:
* *involve local stakeholders, and be . . .*
* *contextual: reflecting local narratives, not cookie-cutter solutions, and be . . .*
* *ethical: reflecting the normative values of the culture affected.*

17 **Maureen Fordham**
Geographer, Professor of Gender and Disaster Resilience, Department of Geography, University of Northumbria, UK

Social justice, human rights and equality of opportunity in all aspects. This must include participation from grassroots women on up.

18 Jean-Christophe Gaillard
Associate Professor at the School of Environment, University of Auckland, New Zealand

Recovery should balance continuity, to consider survivors' grounding in places and fasten the process, and change, to foster DRR.

19 Tony Gibbs
International Consultant, Consulting Civil Engineer, Consulting Engineering Partnership (CEP) Barbados

Poor communities must remember that 'The most expensive buildings are those that fail' and poor communities cannot afford expensive buildings. Complying with the minimum standards in codes is aiming to construct the worst buildings that the law would allow.

20 Mohammed Hamza
Architect, Professor of Disaster Risk Management, University of Copenhagen, Denmark

Successful recovery is the one that not only manages to reduce existing and known vulnerabilities and risk, but goes further into not creating new ones in the long term.

21 John Handmer
International Consultant, Professor and Director, Centre for Risk and Community Safety and Human Security Program, RMIT, Melbourne, Australia

There are very many important facets to recovery – and the importance of each will often be determined by the circumstances of the event and place. However, after life itself, in both rich places (e.g. Christchurch, New Zealand) and poor places (e.g. Phuket, Thailand), people need their livelihoods. Therefore, these livelihoods and local economies, both formal and informal, are fundamental to recovery as they provide local control and the necessary money and confidence. Unfortunately, in their enthusiasm to help, governments and NGOs often undermine local control, livelihoods and economies.

22 Terry Jeggle
Field-based Educator, International Policy Consultant, Senior Officer, IDNDR and UNISDR, Visiting Scholar, Graduate School of Public and International Affairs, Pittsburgh University, USA

The question is ambiguous. If you mean, aspects, i.e. 'characteristics' that demonstrate the results of a successful recovery . . . : A satisfied, productive population proud of its joint accomplishments creating a safer, sustainable, better social, environmental habitat – on their terms. Or, if you mean the aspects i.e. 'means' through which a successful recovery can be realized . . . : Recognized competent directing authority; predominantly driven by affected population's priority requirements; tempered by strategic systems infrastructure, implemented by the people.

23 Rumana Kabir
Disaster risk reduction specialist, Bangladesh and UK

- *Resourceful government*
- *Active civil society which addresses people's priority needs*
- *Corruption-free aid distribution*
- *Previous disaster management experience (not merely emergency aid but also overall recovery and preparedness overviews from grass roots to donor level)*
- *Insight from the aid industry (we assume that people are willing to wait and live under the plastic sheets or tents, for a better shelter which might take long or might not happen, as we don't commit ourselves long enough, as we can't engage ourselves as aid industry in the recovery process at all because it is assumed to be the insider's – the people's and the government's – job, not ours as we are outsiders).*

24 Ilan Kelman
Reader, Institute for Risk and Disaster Reduction and Institute for Global Health, University College London (UCL), UK

Long-term disaster risk reduction (including preparedness, prevention, and mitigation). Without thinking and acting before, success afterwards is difficult. As well, I was surprised to see the term 'natural disaster' in the question. I thought that we were avoiding the term?

25 Randolph Kent
Former Senior UN official and the Head of the Humanitarian Futures Programme, King's College London, UK

If we view recovery in terms of 'natural disasters', we miss the multidimensional causes of disasters that need to be overcome to ensure durable and sustained recovery.

26 Fred Krimgold
Architect, Director, Disaster Risk Reduction Program, Advanced Research Institute, Virginia Tech, USA

Successful reconstruction should reduce future disaster risk through safe siting and construction for economically and socially viable communities.

27 **Allan Lavell**
Geographer, International Consultant, Coordinator, Social Study Programme on Disaster Risk, Latin American Social Science Faculty, San Jose, Costa Rica

Understanding past and present disaster risk contexts on site and their underlying causes, livelihood-based approaches, sensitivity to persons and their needs, consulted and participatory response from outsiders.

28 **Tony Lloyd-Jones**
Urban Planner, Director of Consultancy and Research, Max Lock Centre, University of Westminster, London, UK

Co-ordinate short-term relief efforts and simultaneously agree a framework for sustainable, long-term recovery and restoring local livelihoods and social infrastructure.

29 **Franklin MacDonald**
Geologist, International Consultant, Visiting Scholar, York University, Toronto, Canada and Jamaica

Effective timely recovery is facilitated when functional, genuine PPP (public, private partnerships and civil society inclusive of the marginalised) mechanisms processes, plans, shared values and good governance exist in the affected areas prior to any 'shock'. In addition all available knowledge (including innovative combinations of traditional, cultural and scientific knowledge) needs to be mobilized before and after shocks, crises and extreme events to contribute to the resilience and sustainability of society including our livelihoods and revenue streams (at all levels – household, local, sub-national and national).

30 **Andrew Maskrey**
Planner, Coordinator of UN Global Assessment Reports on Disaster Risk Reduction UNISDR, Geneva, Switzerland

Effective disaster recovery would focus less on past losses and more on future risks and opportunities to transform development.

31 **Babar Mumtaz**
International Consultant, Urban Planner and Development Economist specialising in urban management and housing finance, London, UK

Improved land, construction and infrastructure systems that enable households and communities to have greater access to and control over their housing.

32 **John Norton**
Architect, President and Founder: Development Workshop France (DWF) France

Survivor consultation, involvement and engagement, encouraging replicable safe skills, techniques and resource use and consideration of local knowledge and values.

33 Paul Oliver

Writer and Lecturer on the Anthropology of Shelter, Emeritus Professor, Oxford Institute for Sustainable Development, Oxford Brooks University, UK

In a disaster recovery situation there are haves and have-nots, the powerful and the powerless, the relief organisations and the victims of disaster. They are thrown together by the unique and peculiar circumstances of a catastrophe, obliged to relate in ways which would never occur in normal times.

These cultural contexts in disasters are complex, and in the ultimate sense incompatible. One can but hope that with informed research and sensitive design the gap between the relief culture, the victim culture and the indigenous culture can be effectively narrowed.

34 Anthony Oliver-Smith

Anthropologist, International Consultant, Professor Emeritus of Anthropology, University of Florida, USA

Successful disaster recovery reduces systemic economic, social and political causes of vulnerability and exposure to improve peoples' security and lives.

35 Robert B. Olshansky

Professor, Department Head, Department of Urban and Regional Planning, University of Illinois at Urbana-Champaign, USA

Transparency, communications systems, information clearing houses, widespread and substantial involvement of affected citizens and groups, and money. And skilled leaders who can facilitate all these things.

36 Marcus Oxley

Executive Director, Global Network of Civil Society Organisations for Disaster Reduction, UK

- *Ensure affected populations are central to the assessment and planning processes.*
- *Build on local capacities and sources of resilience guided by a disaster recovery framework supported by dedicated institutional arrangements.*
- *Adopt a build back safer resilience recovery approach.*

37 David Sanderson

Professor, Department of Urban Design and Planning, Norwegian University of Science and Technology, Trondheim, Norway

Support to allow people to determine their own future.

38 Graham Saunders

Architect; Head, Shelter and Settlements, International Federation of Red Cross and Red Crescent Societies; Coordinator, Interagency Global Shelter Cluster, Geneva, Switzerland

Participatory, appropriate and sustainable – locally owned, locally driven. Beyond projects and programmes to established communities of homes, families and viable livelihoods.

39 Theo Schilderman

Architect, Housing and Reconstruction Advisor, Rugby, UK

A people-centred integrated approach, covering livelihood diversification, safer housing and infrastructure, restoration of social networks, and community empowerment.

40 Anshu Sharma

Architect, Founder of SEEDS, Delhi, India

Safety, sustainability, imbibement of cultural values, meeting of aspirations, and overall happiness of the families, in particular the women and children.

41 Rajib Shaw

Professor, Graduate School of Global Environmental Studies, Kyoto University, Japan

Disaster recovery is a development and innovation opportunity, and needs to meet the demand of community for a better life.

42 Jo da Silva

Civil Engineer, Director Arup International Development, London, UK

Communities that are safer and more resilient, as a result of cumulative action taken by them, their government, and humanitarian agencies.

43 Anil Sinha

Vice Chairman, Bihar State Disaster Management Authority, Government of Bihar, Patna, India

Given the huge opportunity that a recovery operation offers, 'build back better' is widely recognised and accepted as a sound overall approach. To accomplish this, I recommend three mantras:
* *first, never recreate a risk that existed earlier*
* *secondly, never create a new risk in the process of recovery*
* *thirdly, constantly reduce risk in whatever one does as a matter of practice and promoting a culture of prevention.*

44 Maggie Stephenson

Architect, International Consultant, Technical Advisor UN-Habitat, London, UK

1   *Recovery is a moving target and involves multiple processes. Keep watching. Keep listening. Keep trying to understand and be relevant.*

2   *Be ambitious. Be ambitious for others. Enable their ambitions. Build collective purpose. Leave no one out. Leave no one behind.*

   I guess one word I keep thinking about is 'nimble', not intrusive or oppressive, fast and smart, skilful and respectful.

45 Paul Thompson

Architect, Founder: INTERWORKS Madison, USA

- *Do no harm*
- *Support local initiative*
- *Mind the environment*
- *Position the community for a sustainable, healthy and prosperous future.*

46 John Twigg

Researcher, Co-Director, Centre for Urban Sustainability and Resilience, Department of Civil, Environmental and Geomatic Engineering, University College London, UK

Effective disaster recovery: (a) recognises that recovery is complex and multifaceted, with no fixed end point, and (b) is underpinned by a robust and verifiable theory of change.

47 Krishna Vatsa

Disaster Management Practitioner, Regional Disaster Reduction Advisor, Asian Region BCPR-UNDP, Delhi, India

Guided by needs assessment, recovery aims at inclusive social and physical recovery through adequate resource commitment in a well-defined time frame.

1   *Recovery extends beyond relief and humanitarian assistance. It is about restoring services and infrastructure, developing resilience through rebuilding shelter and livelihoods, and reducing risks through disaster risk reduction and social protection.*

2   *Recovery is an important form of assistance to the disaster-affected people. Such assistance prevents them from sliding into long-term poverty and deprivation.*

3   *Recovery is planned and implemented dynamically along a continuum of time, commencing in the first three months of a disaster event to almost a decade of implementation.*

4   *Often, recovery policies and plans are developed after a disaster event. While it is important to recognize specific recovery needs arising from a disaster event and*

*formulate policies accordingly, a recovery programme is very much helped by a
long-term policy.*

5   *Along with the recovery policy, an institutional set up dedicated to the
implementation of recovery programme can implement and deliver recovery
programme with greater efficiency.*

6   *While recovery is organized with government leadership, families and communities are the key drivers.*

7   *A recovery programme needs to be supported with expertise, technical assistance,
and supervision on a continuous basis.*

8   *In a recovery programme, there are competing priorities related to recovery and
restoration of infrastructure, shelter, livelihoods, social services, and markets.*

9   *Recovery needs to be driven through a comprehensive assessment of damages,
losses, and needs.*

10   *Cash transfers have emerged as an innovative arrangement for transferring
recovery assistance.*

11   *Recovery requires allocation and mobilization of resources through multiple
sources.*

12   *A recovery programme offers a window of opportunity to rebuild better.*

48 Christine Warmsler

International Consultant, Associate Professor at Lund University Centre for
Sustainability Studies (LUCSUS), Lund, Sweden

*The systematic mainstreaming of disaster risk reduction and climate change adaptation
at both local and institutional levels are key to successful recovery, leading to operations
where institutional and citizens' localized efforts can complement each other and top-
down and bottom-up approaches are united.*

49 Kenneth Westgate

Geographer, International Consultant, Australia

*Ban the phrase 'back to normal', listen to community priorities, dig deep to address
the root causes of vulnerability. And of course, NEVER refer to a natural disaster.
The wrong starting point!*

50 Gustavo Wilches-Chaux

International consultant, author and photographer

*Most of the time the 'outsiders' intervention is focused in repairing the spider's web,
not in enforcing the spider ('outsiders' comprising all who are not part of the affected
territory). A reconstruction process is successful if the main result is that enforcement:
spiders who will by themselves be able to reinforce the web to avoid new disasters
occurring and to repair it (or build a different one) when a disaster could not be
avoided.*

51 Ben Wisner

Geographer, grandfather, UCL Hazard Research Centre, University College London, UK

Planners embrace evacuees' needs: affordable housing, jobs, services, infrastructure; inclusive planning/implementation includes all income, age, gender, ethnic and health-ability groups.

APPENDIX 3

Key books and websites on disaster recovery

Selected books

The following list is confined to books, not all of which are in print. Websites are accurate at the time of writing (May 2015) but may be subject to change.

Alexander, D. E. 2002. *Principles of Emergency Planning and Management*. Harpenden, UK: Terra Publishing, and New York: Oxford University Press.

Alexander, D., Davidson, C. H., Fox, A., Johnson, C. and Lizzarralde, G. (eds). 2007. *Post-Disaster Reconstruction: Meeting Stakeholder Needs*. Florence: Firenze University Press.

Amaratunga, D. and Haigh, R. (eds). 2011. *Post-Disaster Reconstruction of the Built Environment: Rebuilding for Resilience*. Oxford, UK: Wiley-Blackwell.

Anderson, M. B. and Woodrow, P. J. 1989. *Rising from the Ashes: Development Strategies in Times of Disaster*. Boulder, Colorado: Westview Press, and Paris: UNESCO.

Aquilino, M. (ed.). 2011. *Beyond Shelter: Architecture and Human Dignity*. New York: Metropolis Books.

Ashmore, J. (ed.). 2013. *Shelter Projects 2011–2012*. Geneva: International Federation of Red Cross and Red Crescent Societies, UN-Habitat and UN High Commission for Refugees.

Awotona, A. (ed.). 1997. *Reconstruction after Disaster: Issues and Practices*. Aldershot, UK: Ashgate.

Aysan, Y. and Oliver, P. 1987. *Housing and Culture after Earthquakes: A Guide for Future Policy Making in Seismic Areas*. Oxford, UK: Oxford Polytechnic Press.

Aysan, Y. and Davis, I. (eds). 1992. *Disasters and the Small Dwelling: Perspectives for the UN IDNDR*. London: James & James.

Aysan, Y., Clayton, A., Cory, A., Davis, I. and Sanderson, D. 1995. *Developing Building for Safety Programmes: Guidelines for Organizing Safe Building Improvement Programmes in Disaster-Prone Areas*. Oxford, UK: Oxford Centre for Disaster Studies, and London: Intermediate Technology Publications.

Bacon, P. and Hobson, C. (eds). 2014. *Human Security and Japan's Triple Disaster: Responding to the 2011 Earthquake, Tsunami and Fukushima Crisis.* Abingdon, UK: Routledge.

Barakat, S. 2003. *Housing Reconstruction after Conflict and Disaster.* Humanitarian Practice Network (HPN) Report No. 43. London: Overseas Development Institute.

Charlesworth, E. 2006. *Architects without Frontiers: War, Reconstruction and Design Responsibility.* Oxford, UK: Architectural Press, Elsevier.

Charlesworth, E. 2014. *Humanitarian Architecture: Fifteen Stories of Architects Working after Natural Disasters.* Abingdon, UK: Routledge.

Charlesworth, E. and Ahmed, I. 2015. *Sustainable Housing Reconstruction: Designing Resilient Housing after Natural Disasters.* Abingdon, UK: Routledge.

Comerio, M. C. 1998. *Disaster Hits Home: New Policy for Urban Housing Recovery.* Berkeley, California: University of California Press.

Corsellis, T. and Vitale, A. (eds). 2008. *Transitional Settlement and Reconstruction after Natural Disasters.* Geneva: United Nations.

Da Silva, J. 2010. *Lessons from Aceh: Key Considerations in Post-Disaster Reconstruction.* Rugby, UK: Practical Action Publishing.

Davis, I. 1978. *Shelter after Disaster.* Oxford, UK: Oxford Polytechnic Press.

Davis, I. (ed.). 1981. *Disasters and the Small Dwelling.* Oxford, UK: Pergamon.

Davis, I. 2007. *Learning from Disaster Recovery: Guidance for Decision Makers.* Geneva: International Recovery Platform.

Davis, I. (ed.). 2015. *Shelter after Disaster, Guidelines for Assistance* (second edition). Geneva: UN Office for the Co-ordination of Humanitarian Affairs, UN-Habitat, and International Federation of Red Cross and Red Crescent Societies.

Duyne Barenstein, J. E. and Leemann, E. (eds). 2012. *Post-Disaster Reconstruction and Change: Communities' Perspectives.* Boca Raton, Florida: CRC Press.

Erikson, K. T. 1976. *Everything in its Path: Destruction of Community in the Buffalo Creek Flood.* New York: Simon & Schuster.

Esnard, A-M. and Sapat, A. 2014. *Displaced by Disaster: Recovery and Resilience in a Globalizing World.* Abingdon, UK: Routledge.

FEMA. 2000. *Rebuilding for a More Sustainable Future: An Operational Framework.* Washington, DC: US Federal Emergency Management Agency.

Geipel, R. 1982. *Disaster and Reconstruction: The Friuli (Italy) Earthquakes of 1976* (trans. Wagner, P.). London: George Allen & Unwin.

Geipel, R. 1990. *The Long-Term Consequences of Disasters: The Reconstruction of Friuli, Italy, in its International Context, 1976–1988.* Heidelberg: Springer-Verlag.

Haas, J. E., Kates, R. W. and Bowden, M. J. (eds). 1977. *Reconstruction Following Disaster.* Cambridge, Massachusetts: MIT Press.

Hannigan, J. 2012. *Disasters Without Borders. The International Politics of Natural Disasters.* Cambridge, UK: Polity Press.

Jha, A., Duyne Barenstein, J., Phelps, P. M., Pittet, D. and Sena, S. 2010. *Safer Homes, Stronger Communities: A Handbook for Reconstructing after Natural Disasters.* Washington, DC: Global Facility for Disaster Risk Reduction, World Bank.

Langenbach, R. 2009. *Don't Tear it Down! Preserving the Earthquake Resistant Vernacular Architecture of Kashmir.* New Delhi: UNESCO.

Lizarralde, G., Johnson, C. and Davidson, C. (eds). 2010. *Rebuilding after Disasters: From Emergency to Sustainability.* Abingdon, UK: Spon Press.

Lyons, M., Schilderman, T. and Boano, C. 2010. *Building Back Better: Delivering People-Centred Housing Reconstruction at Scale.* Rugby, UK: Practical Action Publishing.

Oliver, P. 2006. *Built to Meet Needs: Cultural Issues in Vernacular Architecture. Part IV: Cultures, Disasters and Dwellings.* Oxford, UK: Architectural Press, Elsevier.

Oliver-Smith, A. 1986. *The Martyred City: Death and Rebirth in the Andes*. Albuquerque, New Mexico: University of New Mexico Press.

Olshansky, R. B. and Johnson, L. A. 2010. *Clear as Mud: Planning for the Rebuilding of New Orleans*. Washington, DC: American Planning Association.

Phillips, B. 2009. *Disaster Recovery*. Boca Raton, Florida: CRC Press.

Sanderson, D. and Burnell, J. 2013. *Beyond Shelter after Disaster: Practice, Process and Possibilities*. Abingdon, UK: Routledge.

Schilderman, T. and Parker, E. (eds). 2014. *Still Standing? Looking Back at Reconstruction and Disaster Risk Reduction in Housing*. Rugby, UK: Practical Action Publishing.

Schwab, J. C. (ed.). 2014 *Planning for Post-Disaster Recovery: Next Generation*. PAS Report No. 576. Chicago: American Planning Association.

Shaw, R. (ed.). 2014. *Disaster Recovery: Used or Misused Development Opportunity*. Berlin: Springer.

Shelter Centre. 2010. *Shelter after Disaster: Strategies for Transitional Settlement and Reconstruction*. Geneva: UN Shelter Centre.

Spangle, W. and Associates. 1991. *Rebuilding after Earthquakes: Lessons from Planners*. Portola Valley, California: William Spangle and Associates.

UNDRO. 1982. *Shelter after Disaster: Guidelines for Assistance*. Geneva: Office of the United Nations Disaster Relief Co-ordinator.

United Nations. 1970. *Skopje Resurgent: The Story of a United Nations Special Fund Town Planning Project*. New York: United Nations Development Programme.

Vale, L. J. and Campanella, T. J. (eds). 2005. *The Resilient City: How Modern Cities Recover from Disaster*. New York: Oxford University Press.

Walter, J. 2001. *World Disasters Report: Focus on Recovery*. Geneva: International Federation of Red Cross and Red Crescent Societies.

Warmsler, C. 2014. *Cities, Disaster Risk and Adaptation*. Abingdon, UK: Routledge.

Selected websites

Disaster Research Centre, University of Delaware
Disaster recovery, theory, practice and case studies
http://drc.udel.edu/

Global Facility for Disaster Reduction and Recovery, World Bank
Reconstruction planning, economic recovery, housing reconstruction
www.gfdrr.org/

International Federation of Red Cross and Red Crescent Societies
Global shelter cluster; shelters and settlements
www.ifrc.org/what-we-do/disaster-management/responding/services-for-the-disaster-affected/shelter-and-settlement/

International Recovery Platform
Recovery background and strategies
www.recoveryplatform.org/

i-Rec information and research for reconstruction, University of Montreal
International network
www.grif.umontreal.ca/i-rec.htm

Natural Hazards Center, University of Colorado at Boulder
Studies of recovery and reconstruction; case histories
www.colorado.edu/hazards/

Œuvre durable, University of Montreal
Disaster resilience and sustainable reconstruction research alliance
www.grif.umontreal.ca/observatoire/index_EN.html

Shelter Case Studies .org
Initiative funded by the UN High Commission for Refugees, International Federation of Red Cross and Red Crescent Societies, and UN-Habitat
www.sheltercasestudies.org

UN-Habitat
The lead agency within the United Nations system for coordinating activities in the field of human settlements
http://unhabitat.org/urban-themes/reconstruction/

United Nations Environment Programme
Post-crisis environmental recovery methods
www.unep.org/disastersandconflicts/

Collection

Ian Davis has donated his collection of books on all aspects of disasters, (including disaster recovery) to Oxford Brookes University Library as a Special Collection. It is called 'The Ian Davis Disaster Management Collection'. This is a reference rather than loan collection and is available to readers and researchers seven days a week during semesters and weekdays during vacation. Full details of opening hours are given on the library website (www.brookes.ac.uk/library). It is not necessary to make a prior booking before visiting the library. The collection is positioned immediately adjacent to the 'Paul Oliver Vernacular Architecture Library'. The titles of books catalogued in either collection can be determined online at: http://capitadiscovery.co.uk/brookes/

REFERENCES

Abbott, D. and Porter, S. 2013. Environmental hazard and disabled people: from vulnerable to expert to interconnected. *Disability and Society*, 28 (6), 839–52.

ADW. 2012. *World Risk Report 2012*. Berlin: Alliance Development Works.

Agarwal, A. 2007. *Cyclone Resistant Building Architecture*. New Delhi: Government of India and UN Development Programme. Available at: http://nidm.gov.in/PDF/safety/flood/link2.pdf (accessed 5 February 2015).

Albala-Bertrand, J. M. 2007. Globalization and localization: an economic approach. In Rodríguez, H., Quarantelli, E. L. and Dynes, R. R. (eds), *Handbook of Disaster Research*. New York: Springer, pp. 147–67.

Aldrich, D. P. 2012. *Building Resilience: Social Capital in Post-Disaster Recovery*. Chicago: University of Chicago Press.

Alexander, D. E. 1984. Housing crisis after natural disaster: the aftermath of the November 1980 Southern Italian earthquake. *Geoforum*, 15 (4), 489–516.

Alexander, D. E. 1989. Preserving the identity of small settlements during post-disaster reconstruction in Italy. *Disasters*, 13 (3), 228–36.

Alexander, D. E. 2000. *Confronting Catastrophe: New Perspectives on Natural Disasters*. Harpenden, UK: Terra Publishing, and New York: Oxford University Press.

Alexander, D. E. 2002. *Principles of Emergency Planning and Management*. Harpenden, UK: Terra Publishing, and New York: Oxford University Press.

Alexander, D. E. 2007. 'From rubble to monument' revisited: modernised perspectives on recovery from disaster. In Alexander, D., Davidson, C. H., Fox, A., Johnson, C. and Lizzarralde, G. (eds), *Post-Disaster Reconstruction: Meeting Stakeholder Needs*. Florence, Italy: Firenze University Press, pp. xiii–xxii.

Alexander, D. E. 2013. Resilience and disaster risk reduction: an etymological journey. *Natural Hazards and Earth System Sciences*, 13 (11), 2707–16.

Alexander, D. and Davis, I. 2012. Disaster risk reduction: an alternative viewpoint. *International Journal of Disaster Risk Reduction*, 2, 1–5.

Alexander, D. and Sagramola, S. 2014. *Major Hazards and People with Disabilities: Their Involvement in Disaster Preparedness and Response*. Strasbourg: Council of Europe.

Alexander, D., Davidson, C. H., Fox, A., Johnson, C. and Lizzarralde, G. (eds). 2007. *Post-Disaster Reconstruction: Meeting Stakeholder Needs*. Florence, Italy: Firenze University Press.

ALNAP. n.d. Our role [online]. *ALNAP.* Available at: www.alnap.org/who-we-are/our-role (accessed 8 May 2015).

Ambraseys, N. and Bilham, R. 2011. Corruption kills. *Nature,* 469 (7329), 153–5.

Anderson, S. 1999. *The Man Who Tried to Save the World: The Dangerous Life and Mysterious Disappearance of Fred Cuny.* New York: Doubleday.

Angotti, T. 1977. Playing politics with disaster: the earthquakes of Fruili and Belice (Italy). *International Journal of Urban and Regional Research,* 1 (1–4), 327–31.

Arshad, S. and Ather, S. 2013. *Rural Housing Reconstruction Program Post-2005. Earthquake Learning from the Pakistan Experience. A Manual for Post-Disaster Housing Program Managers.* Washington, DC: Global Facility for Disaster Risk Reduction and Recovery (GFDRR). Available at: www.gfdrr.org/sites/gfdrr.org/files/publication/RHRP_PAKISTAN_WEB.pdf (accessed 5 February 2015).

Ashdown, P. 2011. *Humanitarian Emergency Response Review.* London: Department for Overseas Development. Available at: www.gov.uk/government/uploads/system/uploads/attachment_data/file/67579/HERR.pdf (accessed 5 February 2015).

Ashmore, J. (ed.). 2010. *Shelter Projects 2009.* Geneva: UN-Habitat and International Federation of Red Cross and Red Crescent Societies.

Ashmore, J. (ed.). 2013. *Shelter Projects 2011–2012.* Geneva: International Federation of Red Cross and Red Crescent Societies, UN-Habitat and UN High Commission for Refugees.

ATC. 2005. *ATC-20 Building Safety Evaluation Forms and Placards.* Redwood City, California: Applied Technology Council. Available at: www.atcouncil.org/45-downloadable/downloads/107-atc-20-download (accessed 5 February 2015).

Aysan, Y. and Oliver, P. 1987. *Housing and Culture after Earthquakes: A Guide for Future Policy Making in Seismic Areas.* Oxford, UK: Oxford Polytechnic Press.

Aysan, Y. and Davis, I. (eds). 1992. *Disasters and the Small Dwelling: Perspectives for the UN IDNDR.* London: James & James.

Aysan, Y., Clayton, A., Cory, A., Davis, I. and Sanderson, D. 1995. *Developing Building for Safety Programmes: Guidelines for Organizing Safe Building Improvement Programmes in Disaster-Prone Areas.* Oxford, UK: Oxford Centre for Disaster Studies, and London: Intermediate Technology Publications.

Ayyub, B. M., McGill, W. L. and Kaminskiy, M. 2007. Critical asset and portfolio risk analysis: an all-hazards framework. *Risk Analysis,* 27 (4), 789–802.

Barber, R. 2009. The responsibility to protect the survivors of natural disaster: Cyclone Nargis, a case study. *Journal of Conflict and Security Law,* 14 (1), 3–34.

Barrows, H. H. 1923. Geography as human ecology. *Annals of the Association of American Geographers,* 13, 1–14.

Becker, P. and Tehler, H. 2013. Constructing a common holistic description of what is valuable and important to protect: a possible requisite for disaster risk management. *International Journal of Disaster Risk Reduction,* 6, 18–27.

Bendimerad, F. 2004. Disaster risk reduction and sustainable development. Paper presented at the *Thirteenth World Conference on Earthquake Engineering,* Vancouver, Canada, 1–6 August 2004.

Bengtsson, L., Lu, X., Garfield, R., Thorson, A. and von Schreeb, J. 2010. *Internal Population Displacement in Haiti. Preliminary Analyses of Movement Patterns of Digicel Mobile Phones: 1 January to 11 March 2010.* Stockholm: Karolinska Institute, and New York: Colombia University.

Berkes, F. and Ross, H. 2013. Community resilience: toward an integrated approach. *Society and Natural Resources,* 26 (1), 5–20.

Bernal, V. A. and Procee, P. 2012. Four years on: what china got right when rebuilding after the sichuan earthquake [online]. *World Bank.* Available at: http://blogs.worldbank.

org/eastasiapacific/four-years-on-what-china-got-right-when-rebuilding-after-the-sichuan-earthquake (accessed 5 February 2015).

Bhatia, G. 1986. Laurie Baker: architect for the common man – an interview with Gautam Bhatia. In Kagal, C. and Bhatia, G. (eds), *Vistāra – The Architecture of India, Catalogue of the Exhibition*. New Delhi: The Festival of India, pp. 215–21.

Bhatia, G. 1991. *Laurie Baker: Life, Works, Writings*. London: Penguin.

Bilham, R. 2013. Obituary: Nicholas Ambraseys, 1929–2012. *International Journal of Disaster Risk Reduction*, 4, ii–iv.

Birkmann, J., Cardona, O. D., Carreño, M. L., Barbat, A. H., Pelling, M., Schneiderbauer, S., Kienberger, S., Keiler, M., Alexander, D., Zeil, P. and Welle, T. 2013. Framing vulnerability, risk and societal responses: the MOVE framework. *Natural Hazards*, 67 (2), 193–211.

Boano, C. and García, M. 2011. Lost in translation? The challenges of an equitable post-disaster reconstruction process: lessons from Chile. *Environmental Hazards*, 10 (3), 293–309.

Boyce, J. K. 2000. Let them eat risk? Wealth, rights and disaster vulnerability. *Disasters*, 24 (3), 254–61.

Bretherton, D. and Ride, A. 2011. *Community Resilience in Natural Disasters*. New York: Palgrave MacMillan.

Bullock, Nicholas 2002. *Building the Post-War World, Modern Architecture and Reconstruction*. Abingdon, UK: Routledge.

Burton, I. 1998. Adapting to climate change in the context of national economic planning and development. In Veit, P. (ed.), *Africa's Valuable Assets: A Reader in Natural Resource Management*. Washington, DC: World Resources Institute, pp. 198–222.

Cabot Venton, C. and Peters, K. 2014. Dare to prepare: taking disaster risk seriously [online]. *Huffington Post*, 7 March. Available at: www.huffingtonpost.com/...cabot.../dare-to-prepare_b_4859476.html (accessed 5 February 2015).

Cardona, O. D. 2004. The need for rethinking the concepts of vulnerability and risk from a holistic perspective: a necessary review and criticism for effective risk management. In Bankoff, G., Frerks, G. and Hilhorst, D. (eds), *Mapping Vulnerability: Disasters, Development and People*. London: Earthscan, pp. 37–51.

Charlesworth, E. 2014. *Humanitarian Architecture: Fifteen Stories of Architects Working after Natural Disasters*. Abingdon, UK: Routledge.

Chubb, J. 2002. Three earthquakes: political response, reconstruction and the institutions: Belice (1968), Friuli (1976), Irpinia (1980). In Dickie, J., Foot, J. and Snowden, F. M. (eds), *Disastro! Disasters in Italy since 1860: Culture, Politics, Society*. New York: Palgrave MacMillan, pp. 186–233.

Clermont, C., Sanderson, D., Sharma, A. and Spraos, H. 2011. *Urban Disasters: Lessons from Haiti*. London: Disasters Emergency Committee.

Comerio, M. C. 1998. *Disaster Hits Home: New Policy for Urban Housing Recovery*. Berkeley, California: University of California Press.

Comerio, M. 2013. *Housing Recovery in Chile: A Qualitative Mid-Program Review*. PEER Report No. 2013/01, February 2013. Berkeley, California: Pacific Earthquake Engineering Research Center, University of California at Berkeley.

Comerio, M. 2014a. Disaster recovery and community renewal: housing approaches. *Cityscape*, 16 (2), 51–68.

Comerio, M. 2014b. Housing recovery lessons from Chile. *Journal of the American Planning Association*, 80 (4), 340–50.

Covey, S. 1989. *The Seven Habits of Highly Effective People*. New York: Simon & Schuster.

Cuny, F. C. 1977. Refugee camps and camp planning: the state of the art. *Disasters*, 1 (2), 125–44.

Cuny, F. C. 1983. *Disasters and Development*. New York: Oxford University Press.

Da Silva, J. 2010. *Lessons from Aceh. Key Considerations in Post-Disaster Reconstruction*. Rugby, UK: Practical Action Publishing.

Davidson, C. H., Johnson, C., Lizarralde, G., Dikmen, N. and Sliwinski, A. 2007. Truths and myths about community participation in post-disaster housing projects. *Habitat International*, 31 (1), 100–15.

Davis, I. 1975a. Disaster housing: a case study of Managua. *Architectural Design*, 45 (1), 42–7.

Davis, I. 1975b. Skopje rebuilt: reconstruction following the 1963 earthquake. *Architectural Design*, 45 (11), 660–3.

Davis, I. 1977a. The interveners. *New Internationalist*, 53, July, 22–3.

Davis, I. 1977b. Housing and shelter provision following the Guatemala earthquakes of February 4 and 6 1976. *Disasters*, 1 (1), 82–90.

Davis, I. 1978. *Shelter after Disaster*. Oxford, UK: Oxford Polytechnic Press.

Davis, I. 1981. Disasters and settlements: towards an understanding of the key issues. In Davis, I. (ed.), *Disasters and the Small Dwelling*. Oxford, UK: Pergamon, pp. 11–23.

Davis, I. 1983. Disasters as agents of change, or, form follows failure. *Habitat International*, 7 (5–6), 277–310.

Davis, I. 1985. *Shelter after Disaster*. PhD dissertation. London: Development Planning Unit (DPU), University College London.

Davis, I. 1987. Safe shelter within unsafe cities. *Open House International*, 12 (3), 5–15.

Davis, I. 2006. Sheltering from extreme hazards. In Asquith, L. and Vellinga, M. (eds), *Vernacular Architecture in the 21st Century: Theory, Education and Practice*. Abingdon, UK: Taylor and Francis, pp. 145–54.

Davis, I. 2007. *Learning from Disaster Recovery: Guidance for Decision Makers*. Geneva: International Recovery Platform.

Davis, I. 2011a. *What is the Vision for Sheltering and Housing in Haiti? Summary Observations of Reconstruction Progress following the Haiti Earthquake of January 12th 2010*. Port-au-Prince, Haiti: UN-Habitat. Available at: www.onohabitat.org/haiti (accessed 6 February 2015).

Davis, I. 2011b. What have we learned from 40 years' experience of disaster shelter? *Environmental Hazards*, 10 (3), 193–212.

Davis, I. 2012. What is the vision for sheltering and housing in Haiti? Summary observations of reconstruction. Progress following the Haiti earthquake of 12th January 2010. Port-au-Prince, Haiti: UN-Habitat, 19. http://onuhabitat.org/index.php?option=com_docman&task=cat_view&gid=223&Itemid=235 (accessed 5 February 2015).

Davis, I. 2014a. Epilogue: 'architecture as service'. In Charlesworth, E., *Humanitarian Architecture: Fifteen Stories of Architects Working after Natural Disasters*. Abingdon, UK: Routledge, pp. 220–8.

Davis, I. (ed.). 2014b. The vulnerability challenge. In Davis, I. (ed.), *Disaster Risk Management in Asia and the Pacific*. Abingdon, UK: Routledge, pp. 65–108.

Davis, I. (ed.). 2015. *Shelter after Disaster* (second edition). Geneva: UN office for the Co-ordination for Humanitarian Affairs, UN-Habitat, International Federation of Red Cross and Red Crescent Societies.

Davis, I. and Bhatt, M. 1996. *Evaluation of the Reconstruction of Malkondji village, Maharashtra following the Latur Earthquake of 1993*. Oxford, UK: Oxford Centre for Disaster Studies, and Ahmedabad, India: Disaster Mitigation Institute.

de Bono, E. 1985. *Six Thinking Hats: An Essential Approach to Business Management*. London: Penguin.

de Ville de Goyet, C. and Morinière, L. 2006. *The Role of Needs Assessment in the Tsunami Response*. London: Tsunami Evaluation Coalition (TEC), Overseas Development Institute.

Desai, R. and Desai, R. 2008. *Manual on Hazard Resistant Construction In India: For Reducing Vulnerability in Buildings Built Without Engineers*. Ahmedabad, India: National Centre for People's Action in Disaster Preparedness.

Domoto, A., Ohara, M., Hara, H., Aoki, R. and Amano, K. 2011. *Japanese Women's Perspective on 3/11*. Tokyo: Japanese Women's Network for Disaster Risk Reduction.

Drabek, T. E. 1986. *Human System Response to Disaster: An Inventory of Sociological Findings*. New York: Springer-Verlag.

Drabek, T. E. and McEntire, D. A. 2003. Emergent phenomena and the sociology of disaster: lessons, trends and opportunities from the research literature. *Disaster Prevention and Management*, 12 (2), 97–112.

Drucker, P. F. 1974. *Management: Tasks, Responsibilities, Practices*. New York: Harper & Row.

Duyne Barenstein, J. E. and Leemann, E. (eds). 2012. *Post-Disaster Reconstruction and Change: Communities' Perspective*. Boca Raton, Florida: CRC Press.

Duyne Barenstein, J., Modan, A. N., Talha, K., Upadhyay, N. and Khandhadai, C. 2014. Looking back at agency-driven housing reconstruction in India: case studies from Maharashtra, Gujarat and Tamil Nadu. In Schilderman, T. and Parker, E. (eds), *Still Standing? Looking Back at Reconstruction and Disaster Risk Reduction in Housing*. Rugby, UK: Practical Action Publishing, pp. 38–59.

ECA. 2012. *The European Union's Solidarity Fund's Response to the 2009 Abruzzi Earthquake: The Relevance and Cost of Operations*. Special Report No. 24. Luxembourg: European Court of Auditors.

Economist, The. 2009. The Sichuan earthquake: salt in the wounds. Bereaved parents treated like criminals. *The Economist*, 14 May.

Economist, The. 2014. The world this week. *The Economist*, 12 July, p. 6.

EEFIT. 2013. *Recovery Two Years after the 2011 ToHoku Earthquake and Tsunami: A Return Mission Report by EEFIT*. London: Earthquake Engineering Field Investigation Team, Institution of Structural Engineers. Available at: www.istructe.org/webtest/files/23/23e209f3-bb39-42cc-a5e5-844100afb938.pdf (accessed 5 February 2015).

Eisenhower, D. 1957. Remarks at the National Defense Executive Reserve Conference, November 14, 1957 [online]. *University of Michigan Digital Library. The Public Papers of the Presidents of the United States: Dwight D. Eisenhower, 1957*. Available at: http://quod.lib.umich.edu/p/ppotpus?cginame=text-idx;id=navbarbrowselink;page=browse (accessed 5 February 2015).

Elegant, S. 2009. Transcript of interview with Ai Weiwei on the first anniversary of the Sichuan earthquake. *Time* magazine, 12 May. Available at: http://world.time.com/2009/05/12/ai-weiwei-interview-transcript/ (accessed 5 February 2015).

Elhawary, S. and Aheeyar, M. 2008. *Beneficiary Perceptions of Corruption in Humanitarian Assistance: A Sri Lanka Case Study*. London: Humanitarian Policy Group, Overseas Development Institute. Available at: www.odi.org.uk/search/site/sri%20lanka%20corruption (accessed 5 February 2015).

Elliot, L. and Pilkington, E. 2015. Half global wealth held by the 1%, Oxfam warns of widening inequality gap days ahead of Davos economic summit. *The Guardian*, 19 January, p. 1.

Elliot, M. 2015. The age of miracles. *Time* magazine, 26 January, p. 26.

Enarson, E. P. 2012. *Women Confronting Natural Disaster: From Vulnerability to Resilience*. Boulder, Colorado: Lynne Rienner.

Enarson, E. P. and Dhar Chakrabarti, P. B. 2009. *Women, Gender and Disaster: Global Issues and Initiatives.* Thousand Oaks, California: Sage.

Erikson, K. T. 1976. *Everything in its Path: Destruction of Community in the Buffalo Creek Flood.* New York: Simon & Schuster.

ERRA. 2008. *Compliance Catalogue: Guidelines for the Construction of Compliant Rural Houses.* Islamabad, Pakistan: UN Habitat and Earthquake Reconstruction and Rehabilitation Authority.

Escaleras, M., Anbarci, N. and Register, C. A. 2007. Public sector corruption and major earthquakes: a potentially deadly interaction. *Public Choice*, 132 (1–2), 209–30.

Esnard, A-M. and Sapat, A. 2014. *Displaced by Disaster: Recovery and Resilience in a Globalizing World.* Abingdon, UK: Routledge.

Fallahi, A. 2007. Lessons learned from the housing reconstruction following the Bam earthquake in Iran. *Australian Journal of Emergency Management*, 22 (1), 26–35.

Farmer, P. 2011. *Haiti after the Earthquake.* New York: Public Affairs.

Faure Walker, J. and Alexander, D. (eds). 2014. *The Post-Disaster Phase of Transitional Settlement: A Perspective from Typhoon Yolanda (Haiyan) in Eastern Philippines.* IRDR Special Report 2014-01. London: Institute for Risk and Disaster Reduction, University College London.

Feireiss, K. (ed.). 2009. *Architecture in Times of Need: Make it Right Rebuilding New Orleans' Lower Ninth Ward.* New York: Prestel Publishing.

Flinn, B. 2013. Changing approaches to post-disaster shelter. *Humanitarian Exchange Magazine,* 58, July. Available at: www.odihpn.org/humanitarian-exchange-magazine/issue-58/changing-approaches-to-post-disaster-shelter (accessed 4 May 2015).

Foster, H. D. 1976. Assessing disaster magnitude: a social science approach. *Professional Geographer*, 28 (3), 241–7.

Foster, H. D. 1980. *Disaster Planning: The Preservation of Life and Property.* New York: Springer-Verlag.

Franchina, L., Carbonelli, M., Gratta, L., Crisci, M. and Perucchini, D. 2011. An impact-based approach for the analysis of cascading effects in critical infrastructures. *International Journal of Critical Infrastructures*, 7 (1), 73–90.

Gerdin, M., Chataigner, P., Tax, L., Kubai, A. and von Schreeb, J. 2014. Does need matter? Needs assessments and decision-making among major humanitarian health agencies. *Disasters*, 38 (3), 451–64.

GFDRR. 2014a. *Conference Statement: World Reconstruction Conference 2: Changing the Paradigm: Disaster Recovery as a Means to Sustainable Development.* Washington, DC: Global Facility for Disaster Reduction and Recovery, World Bank.

GFDRR. 2014b. *Guide to Developing Disaster Recovery Frameworks.* World Reconstruction Conference Version. Washington, DC: Global Facility for Disaster Reduction and Recovery, World Bank.

Ghafory-Ashtiany, M. 2009. View of Islam on earthquakes, human vitality and disaster. *Disaster Prevention and Management*, 18 (3), 218–32.

Gilbert, C. 2013. Hidden tent camps persist in Haiti four years after the earthquake [online]. *Grassroots Online*. Available at: www.grassrootsonline.org/news/blog/claire-gilbert (accessed 5 February 2015).

Glancy, J. 2001. The folly of Fallingwater. *The Guardian*, 9 September.

Gokhale, S. and Mistry, J. 2012. Participatory post disaster reconstruction. In Lizarralde, G., Jigyasu, R., Vasavada, R., Havelka, S. and Duyne Barenstein, J. (eds), *Proceedings of the 2010 International i-Rec Conference, Participatory Design and Appropriate Technology for Disaster Reconstruction,* Ahmedabad, India. Montreal: Groupe de recherche IF, University of Montreal, pp. 30–40.

Government of China. 2010. *National Human Rights Action Plan of China 2009–2010.* Beijing: People's Republic of China Government. Available at: http://news.xinhuanet. com/english/2009-04/13/content_11177126.htm (accessed 5 February 2015).

Granot, H. 1996. Disaster subcultures. *Disaster Prevention and Management,* 5 (4), 36–40.

Gray, B. and Bayley, S. 2015. *Case Study: Shelter Innovation Ecosystem.* CENTRIM Working Paper for the Humanitarian Innovation Ecosystem Project. Brighton: University of Brighton.

Greenberg, M. R., Lahr, M. and Mantell, N. 2007. Understanding the economic costs and benefits of catastrophes and their aftermath: a review and suggestions for the U.S. Federal Government. *Risk Analysis,* 27 (1), 83–96.

Greene, M., Godavitarne, C., Krimgold, F., Nikolic-Brzev, S. and Pantelic, J. 2000. Overview of the Maharashtra, India emergency earthquake rehabilitation program. Paper presented at *Twelfth World Conference on Earthquake Engineering,* Auckland, New Zealand, 30 January–4 February.

Grube, K. 2009. Ai Weiwei's challenge to China's government over earthquake [online]. *ArtAsiaPacific Magazine,* 64, July–August. Available at: http://artasiapacific.com/ Magazine/64/AiWeiweiChallengesChinasGovernmentOverEarthquake (accessed 4 May 2015).

GSDMA. 2014a. Introduction [online]. *Gujarat State Disaster Management Authority.* Available at: http://gsdma.org/about-us/introduction.aspx (accessed 5 May 2015).

GSDMA. 2014b. Constitution and history [online]. *Gujarat State Disaster Management Authority.* Available at: http://gsdma.org/about-us/constitution-history.aspx (accessed 5 May 2015).

Haas, J. E. and Ayre, R. S. 1969. *The Western Sicily Earthquake of 1968.* Washington, DC: National Academy of Sciences for the National Academy of Engineering.

Haas, J. E., Kates R. W. and Bowden, M. J. (eds). 1977. *Reconstruction following Disaster.* Cambridge, Massachusetts: MIT Press.

Hagerman, J. and Doherty, B. 2009. Keeping the recovery safe. *Federation of American Scientists,* 29 April. Available at: http://fas.org/programs/energy/btech/China%20IAQ. pdf (accessed 5 February 2015).

Han, G., Gu, L., Zhang, D. and Xiang, M. 2014. Reconstruction and sustainable development in Bechuan. Reconstruction model of Beichuan for the world. Presentation at workshop organised by the Asian Development Bank Institute and the Institute for Disaster Management and Reconstruction, Sichuan University, *Building Strong Disaster Risk Management Systems in Asia,* Chengdu, P.R. China, 10–12 June.

Handmer, J. and Dovers, S. 2013. *Handbook of Disaster Policies and Institutions: Improved Emergency Management and Climate Change Adaptation.* London: Earthscan.

Handy, C. 1995. Trust and the virtual organisation. *Harvard Business Review,* May–June. Available at: http://hbr.org/1995/05/trust-and-the-virtual-organization/ (accessed 5 February 2015).

Harris, E. 1998. Struck powerless: January's ice storm robbed millions of heat, light and the conceit that natural disasters don't happen here. *Canadian Geographic,* 118 (2), 38–42.

Hinshaw, R. E. 2006. *Living with Nature's Extremes: The Life of Gilbert Fowler White.* Boulder, Colorado: Johnson Books.

Hoffman, D. 1993. *Frank Lloyd Wright's Fallingwater: The House and its History* (second edition). New York: Dover.

Hogg, S. J. 1980. Reconstruction following seismic disaster in Venzone, Friuli. *Disasters,* 4 (2), 173–85.

Hosseini, M. and Izadkhah, Y. 2004. Lessons learnt from the Bam earthquake of December 2003. Paper presented at the *Asian Seismological Commission Conference*, Yerevan, Armenia, 18–20 October.

Hosseini, M., Izadkhah. Y. and Pir-Ata, P. 2008. Lessons learnt from shelter actions and reconstruction of Bam, after the destructive earthquake of December 26, 2003. Paper presented at the *Fourteenth World Conference on Earthquake Engineering*, Beijing, China, 12–17 October.

Human Rights Center. 2005. *After the Tsunami: Human Rights of Vulnerable Populations*. Berkeley, California: Human Rights Center and East-West Center, University of California, Berkeley.

Humanitarian Practice Network. 1995. Accountability in disaster response: assessing the impact and effectiveness of relief assistance. *Humanitarian Exchange Magazine,* 3 April.

Hyslop, M. P. and Collins, A. E. 2013. Hardened institutions and disaster risk reduction. *Environmental Hazards*, 12 (1), 19–31.

IDMC. 2013. One in ten Filipinos affected by Haiyan, as picture of mass displacement emerges [online]. *Internal Displacement Monitoring Centre*, 13 November. Available at: www.internal-displacement.org/blog/2013/1-in-10-filipinos-affected-by-haiyan-as-picture-of-mass-displacement-emerges (accessed 4 May 2015).

IFRC. n.d. From crisis to recovery [online]. *International Federation of Red Cross and Red Crescent Societies*. Available at: www.ifrc.org/en/what-we-do/recovery/ (accessed 4 May 2015).

IFRC. 1974. *The Code of Conduct for the International Red Cross and Red Crescent Movement and NGOs in Disaster Relief*. Prepared jointly by the International Federation of Red Cross and Red Crescent Societies and the ICRC. Available at: www.ifrc.org/Global/Publications/disasters/code-of-conduct/code-english.pdf (accessed 5 May 2015).

IFRC. 2006. *What is VCA? An introduction to vulnerability and capacity assessment*. Geneva: International Federation of Red Cross and Red Crescent Societies. Available at: www.ifrc.org/Global/Publications/disasters/vca/whats-vca-en.pdf (accessed 5 February 2015).

IFRC. 2012. *Assisting Host Families and Communities after Crises and Natural Disaster: A Step-by-Step Guide*. Geneva: International Federation of Red Cross and Red Crescent Societies.

IFRC. 2013. *World Disasters Report 2010. Focus on Technology and the Future of Humanitarian Action*. Geneva: International Federation of Red Cross and Red Crescent Societies.

IFRC. 2014. *World Disasters Report: Focus on Culture and Risk*. Geneva: International Federation of Red Cross and Red Crescent Societies.

ILO. 2014. Fragile states and disaster response [online]. *International Labour Organization*. Available at: www.ilo.org/employment/areas/crisis-response/lang—en/index.htm (accessed 5 February 2015).

Inaba, Y. 2013. What's wrong with social capital? Critiques from social science. In Kawachi, I., Takao, S. and Subramanian, S.V. (eds), *Global Perspectives on Social Capital and Health*. Berlin: Springer, pp. 323–42.

Ingram, J., Guillermo, F., Rumbaitis-del Rio, C. and Khazai, B. 2006. Post-disaster recovery dilemmas: challenges in balancing short-term and long-term needs for vulnerability reduction. *Environment Science and Policy*, 9 (7–8), 607–13.

IOM. 2014. Four and a half years after devastating quake, 92% of displaced population has left camps in Haiti [online]. Press briefing by International Organization for Migration, Port-au-Prince, Haiti, 4 June, Available at: www.iom.int/cms/en/sites/iom/home/news-and-views/press-briefing-notes/pbn-2014b/pbn-listing/four-and-a-half-years-after-deva.html (accessed 5 February 2015).

IRP. n.d. About IRP [online]. *International Recovery Platform*. Available at: www.recoveryplatform.org/about_irp/ (accessed 5 May 2015).

IRP. 2014. Glossary of terms [online]. *International Recovery Platform*. Available at: www.recoveryplatform.org/resources/glossary/ (accessed 5 February 2015).

Jackson, J. 2006. Declaration of taking twice: the Fazendeville community of the Lower Ninth Ward. *American Anthopologist*, 108 (4), 765–80.

James, W. 1906. On Some Mental Effects of the Earthquake. *William James: Writings 1902–1910*. Washington, DC: The Library of America.

Jha A., Duyne Barenstein, J., Phelps, P. M., Pittet, D. and Sena, S. (eds) 2010. *Safer Homes, Stronger Communities: A Handbook for Reconstructing after Natural Disasters*. Washington, DC: World Bank. Available at: www.preventionweb.net/files/12229_gfdrr.pdf (accessed 5 May 2015).

Jing, L. 2014. '5:12' Sichuan earthquake: disaster relief and recovery. Presentation at workshop organised by the Asian Development Bank Institute and the Institute for Disaster Management and Reconstruction, Sichuan University, *Building Strong Disaster Risk Management Systems in Asia*, Chengdu, P.R. China, 10–12 June.

John of Salisbury. 1955 [1159]. *The Metalogicon of John Salisbury* (translated with an introduction and notes by Daniel D. McGarry). Berkeley and Los Angeles, California: University of California Press.

JRCS. 2013. *Japan: Earthquake and Tsunami Operations Update No. 10*. Tokyo: Japanese Red Cross Society. Available at: http://reliefweb.int/report/japan/earthquake-and-tsunami-operations-update-n°-10 (accessed 5 May 2015).

Kates, R. W. and Pijawka, K. D. 1977. From rubble to monument: the pace of reconstruction. In Haas, J. E., Kates, R. W. and Bowden, M. J. (eds), *Reconstruction Following Disaster*. Cambridge, Massachusetts: MIT Press, pp. 1–23.

Keller, A. Z., Wilson, H. C. and Al-Madhari, A. 1992. Proposed disaster scale and associated model for calculating return periods for disasters of given magnitude. *Disaster Prevention and Management*, 1 (1), 26–33.

Kennedy, J., Ashmore, J., Babister, E. and Kelman, I. 2008. The meaning of 'build back better': evidence from post-tsunami Aceh and Sri Lanka. *Journal of Contingencies and Crisis Management*, 16 (1), 24–36.

Khatam, A. 2006. The destruction of Bam and its reconstruction following the earthquake of December 2003. *Cities*, 23 (6), 462–4.

Kollewe, J. 2014. Lloyd's calls on insurers to take into account climate-change risk. *The Guardian*, 8 May, p. 30.

Kreimer, A., Eriksson, J., Muscat, R., Arnold, M. and Scott, C. 1998. *The World Bank's Experience with Post-Conflict Reconstruction*. Washington, DC: Evaluation Department, World Bank.

Krogerus, M. and Tschäppeler, R. 2011. *The Decision Book: Fifty Models for Strategic Thinking*. London: Profile Books.

Kundu, D. 2011. Elite capture in participatory urban governance. *Economic and Political Weekly*, 66 (10), 23–5.

Kynaston, D. 2015. Banging the drum for British culture. *Sunday Times: Culture*, 25 January, p. 5.

Ladinski, V. 1997. Post-1963 Skopje earthquake reconstruction: long-term effects. In Awotona, A. (ed.), *Reconstruction after Disaster: Issues and Practices*. Aldershot, UK: Ashgate, pp. 73–107.

Lall, R. R. 2014. Port-au-Prince: collision of ideals and aid has yoked progress. *The Guardian*, 27 January. Available at: www.theguardian.com/cities/series/cities-back-from-the-brink (accessed 5 February 2015).

Lane, S., Klauser, F. and Kearnes, M. B. (eds). 2012. *Critical Risk Research: Practices, Politics and Ethics*. Chichester, UK: Wiley-Blackwell.

Leake, J. and Trump, S. 2014. Bankers to be kept dry as villagers flood. *Sunday Times*, 2 March.

Leighton, F. B. 1976. Urban landslides: targets for land-use planning in California. In Coates, D. R. (ed.), *Urban Geomorphology*. Special Paper No. 174. Boulder, Colorado: Geological Society of America, pp. 37–60.

Leon, E., Kelman, I., Kennedy, J. and Ashmore, J. 2009. Capacity building lessons from a decade of transitional settlement and shelter. *International Journal of Strategic Property Management*, 13 (3), 247–65.

Lewis, J. 1999. *Development in Disaster-Prone Places: Studies of Vulnerability*. London: Intermediate Technology Group.

Lindell, M. K. and Prater, C. S. 2002. Risk area residents' perceptions and adoptions of seismic hazard adjustments. *Journal of Applied Social Psychology*, 32 (11), 2377–92.

Lloyd-Jones, T. 2007. Building back better: how action research and professional networking can make a difference to disaster reconstruction and risk reduction. Paper presented at *Reflections on Practice: Capturing Innovation and Creativity*. RIBA Research Symposium 2007, London, 19 September.

Loper, J. 2002. Fallingwater restoration uncovers more damage. *Civil Engineering*, 72 (2), 24.

McEntire, D. A. 2005. Why vulnerability matters: exploring the merit of an inclusive disaster reduction concept. *Disaster Prevention and Management*, 14 (2), 206–22.

McEntire, D. A. 2011. Understanding and reducing vulnerability: from the approach of liabilities and capabilities. *Disaster Prevention and Management*, 20 (3), 294–313.

McEntire, D. A., Crocker, C. G. and Peters, E. 2010. Addressing vulnerability through an integrated approach. *International Journal of Disaster Resilience in the Built Environment*, 1 (1), 50–64.

McMahon, J. P. 2009. Ai Weiwei: 'Remembering' and the politics of dissent [online]. *Khan Academy*. Available at: www.khanacademy.org/humanities/global-culture/global-art-architecture/a/ai-weiwei-remembering-and-the-politics-of-dissent (accessed 6 February 2015).

Madge, J. 1945. *The Rehousing of Britain*. London: The Pilot Press.

Make it Right Foundation. 2015. New Orleans [online]. *Make it Right*. Available at: http://makeitright.org/where-we-work/new-orleans/ (accessed 5 February 2015).

Mallet, R. 1862. *Great Neapolitan Earthquake of 1857: The First Principles of Observational Seismology*. London: Chapman & Hall for the Royal Society, 2 vols.

Manyena, S. B., O'Brien, G., O'Keefe, P. and Rose, J. 2011. Disaster resilience: a bounce back or bounce forward ability? *Local Environment*, 16 (5), 417–24.

Margottini, C. (ed.). 2014. *After the Destruction of Giant Buddha Statues in Bamiyan (Afghanistan) in 2001: A UNESCO's Emergency Activity for the Recovering and Rehabilitation of Cliff and Niches*. Berlin: Springer.

Mason, W. 2014. 'The danger artist': photograph of brain scans of Ai Weiwei showing celebral haemorrhage from police brutality in Chengdu, China, 12 August 2009. In Ai, W. and Pins, A. (eds). *Spatial Matters: Art Architecture and Activism*. London: Tate Publishing, p. 432.

Mayunga, J. S. 2012. Assessment of public shelter users' satisfaction: lessons learned from south-central Texas flood. *Natural Hazards Review*, 13 (1), 82–7.

Meyer, V. Priest S. and Kuhlicke, C. 2012. Economic evaluation of structural and non-structural flood risk management measures: examples from the Mulde River. *Natural Hazards*, 62 (2), 301–24.

Miller, J. B. 1974. *Aberfan: A Disaster and its Aftermath*. London: Constable.

Mobasser, M. 2006. *Factors impeding success in post-disaster housing with specific reference to Bam, Iran*. Master's thesis. Oxford, UK: Oxford Brookes University.

Montz, B. E. and Tobin, G. A. 2008. Livin' large with levees: lessons learned and lost. *Natural Hazards Review*, 9 (3), 150–7.

Mooney, M. A. 2010. Haiti's resilient faith. *America: The National Catholic Review*, 202 (6), 21–3.

Moss, S. 2008. *Local Voices, Global Choices for Successful Disaster Risk Reduction: A Collection of Case Studies about Community-Centre Partnerships for Disaster Risk Reduction*. London: British Overseas NGOs for Development. Available at: www.preventionweb.net/english/professional/publications/v.php?id=10883 (accessed 5 February 2015).

Mumtaz, B. Mughal, H. Stephenson, M. and Bothara, J. 2008. The challenges of reconstruction after the October 2005 Kashmir earthquake. Paper presented at *Engineering an Earthquake Resilient New Zealand*, 2008 New Zealand Society of Earthquake Engineers Conference, Wairakei, New Zealand, 11–13 April.

Munich Re. 2000. *Topics. Natural Catastrophes: The Current Position*. Munich: Munich Reinsurance Company.

Myers, M. F. and White, G. F. 1993. The challenge of the Mississippi flood. *Environment*, 35 (10), 6–9, 25–35.

National Diet of Japan. 2012. *The Official Report of the Fukushima Nuclear Accident Independent Investigation Commission: Executive Summary*. Tokyo: National Diet of Japan.

Neal, D. M. 1997. Reconsidering the phases of disasters. *International Journal of Mass Emergencies and Disasters*, 15 (2), 239–64.

New York Times, The. 2009. Sichuan earthquake. *New York Times*, 6 May. Available at: http://topics.nytimes.com/top/news/science/topics/earthquakes/sichuan_province_china/index.html (accessed 5 February 2015).

Norberg-Schultz, C. 1980. *Genius Loci: Towards a Phenomenology of Architecture*. New York: Rizzoli International.

O'Brien, G. and O'Keefe, P. 2013. *Managing Adaptation to Climate Risk: Beyond Fragmented Responses*. London: Routledge.

Ogasawara, T., Matsubayashi, Y., Sakai, S. and Yasuda, T. 2012. Characteristics of the 2011 Tohoku earthquake and tsunami and its impact on the northern Iwate coast. *Coastal Engineering Journal*, 54 (1), 1–16.

Oi, H., Sato, Y. and Koirala, G. 1998. *Sufferings of People and Problems of Communities in the Aftermath of a Disaster. Interviews with Victims of 1993 Flood Disaster in Nepal*. Tokyo: Nepal-Japan Friendship Association for Water Induced Disaster Prevention.

Oliver, P. 2006. *Built to Meet Needs: Cultural Issues in Vernacular Architecture*. Oxford, UK: Architectural Press, Elsevier.

Oliver-Smith, A. 1986. *The Martyred City: Death and Rebirth in the Andes*. Albuquerque, New Mexico: University of New Mexico Press.

Olson, R. A. 2003. Legislative politics and seismic safety: California's early years and the 'Field Act', 1925–1933. *Earthquake Spectra*, 19 (1), 111–31.

Olson, R. S. and Olson, R. A. 1987. Urban heavy rescue. *Earthquake Spectra*, 3 (4), 645–58.

Özerdem, A. and Bowd, R. (eds). 2010. *Participatory Research Methodologies: Development and Post-Disaster/Conflict Reconstruction*. Farnham, UK: Ashgate.

Parrinello, G. 2013. The city-territory: large-scale planning and development policies in the aftermath of the Belice valley earthquake (Sicily, 1968). *Planning Perspectives*, 28 (4), 571–93.

PBS. 2013. Intertect: the international disaster specialists [online]. *PBS*. Available at: www. pbs.org/wgbh/pages/frontline/shows/cuny/bio/intertect.html (accessed 5 May 2015).

Peppiatt, D., Mitchell, J. and Holzmann, P. 2001. *Cash Transfers in Emergencies: Evaluating Benefits and Assessing Risks*. Humanitarian Practice Network Paper No. 35. London: Overseas Development Institute.

Perry, R. W. and Quarantelli, E. L. (eds). 2005. *What is a Disaster? New Answers to Old Questions*. Philadelphia: Xlibris Press.

Pham, D. 2011. Shigeru Ban develops modular shelter for Japanese displaced by earthquake [online]. *Inhabit*, 17 March. Available at: http://inhabitat.com/shigeru-ban-develops-modular-shelter-for-japanese-displaced-by-earthquake/ (accessed 5 February 2014).

Pitt, M. 2008. *Learning Lessons from the 2007 Floods: An Independent Review by Sir Michael Pitt*. London: Cabinet Office.

Platt, S. 2012. *Reconstruction in Chile. Post 2010 Earthquake*. ReBuilDD Field Trip, September 2011. Cambridge: Cambridge Architectural Research (CAR).

Platteau, J. P. 2004. Monitoring elite capture in community-driven development. *Development and Change*, 35 (2), 223–46.

Plyer, A. 2014. Facts for features: Katrina impact [online]. *The Data Center*. Available at: www.datacenterresearch.org/data-resources/katrina/facts-for-impact/ (accessed 5 May 2015).

Polack, E. 2008. A right to adaptation: securing the participation of marginalised groups. *IDS Bulletin*, 39 (4), 16–23.

Prince, S. 1920. *Catastrophe and Social Change: Based upon a Sociological Study of the Halifax Disaster*. Studies in History, Economics and Public Law No. 94. New York: Colombia University Press.

Quarantelli, E. L. 1982. General and particular observations on sheltering and housing in American Disasters. *Disasters*, 6 (4), 277–81.

Quarantelli, E. L. 1995. Editor's introduction: what is a disaster? *International Journal of Mass Emergencies and Disasters*, 13 (3), 221–9.

Quarantelli, E. L. (ed.). 1998a. *What is a Disaster? A Dozen Perspectives on the Question*. London: Routledge.

Quarantelli, E. L. 1998b. *Disaster Recovery: Research Based Observations on What it Means, Success and Failure, Those Assisted and Those Assisting* [online]. Preliminary Paper No. 263. Delaware: Disaster Research Center, University of Delaware. Available at: http://udspace.udel.edu/bitstream/handle/19716/282/PP+263.pdf.txt?sequence=3 (accessed 5 May 2015).

Reniers, G. 2012. Integrating risk and sustainability: a holistic and integrated framework for optimizing the risk decision and expertise rad (ORDER). *Disaster Advances*, 5 (2), 25–32.

Rice-Oxley, M. 2014. Disaster cities: after tragedy strikes, what next? *The Guardian*, 27 January. Available at: www.theguardian.com/cities/2014/jan/27/disaster-cities-tragedy-strikes-homes-communities (accessed 5 February 2015).

Richardson, B. 2005. The phases of disaster as a relationship between structure and meaning. *International Journal of Mass Emergencies and Disasters*, 23 (3), 27–54.

Richardson, S. 2010. How not to respond to an earthquake [online]. *The Daily Beast*, 14 April. Available at: www.thedailybeast.com/articles/2010/04/14/how-not-to-respond-to-an-earthquake.html?cid=hp:beastoriginalsL1 (accessed 5 February 2015).

Richter, R. and Flowers, T. 2008. Gendered dimensions of disaster care: critical distinctions in female psychosocial needs, triage, pain assessment, and care. *American Journal of Disaster Medicine*, 3 (1), 31–7.

Rolandi, G. 2010. Volcanic hazard at Vesuvius: an analysis for the revision of the current emergency plan. *Journal of Volcanology and Geothermal Research*, 189 (3–4), 347–62.

Ruskin, J. 1853. *The Stones of Venice Volume II: The Sea-Stories*. London: Smith, Elder, and Co.

Rybczynski, W. 2005. There's no place like home: the historical problems with emergency housing [online]. *Slate*, 7 September. Available at: www.slate.com/articles/arts/architecture/2005/09/theres_no_place_like_home.html (accessed 5 February 2015).

Saviano, R. 2008. *Gomorrah: Italy's Other Mafia*. London: Pan Macmillan.

Schanze, J. 2006. Flood risk management: a basic framework. In Schanze, J., Zeman, E. and Marsalek, J. (eds), *Flood Risk Management: Hazards, Vulnerability and Mitigation Measures*. NATO Science Series Vol. 67. Dordrecht, The Netherlands: Springer, pp. 1–20.

Schilderman, T. 2010. Putting people at the centre of reconstruction. In Lyons, M., Schilderman, T. and Boano, C. (eds), *Building Back Better: Delivering People-Centred Housing Reconstruction at Scale*. Rugby, UK: Practical Action Publishing, pp. 7–37.

Schwab, J. C. (ed.). 2014. *Planning for Post-Disaster Recovery: Next Generation*. PAS Report No. 576. Chicago: American Planning Association.

Setchell, C. A. 2012. Hosting support: an overlooked humanitarian shelter solution. *Monthly Developments*, 30, January–February, 17–18.

Shabnam, N. 2014. Natural disasters and economic growth: a review. *International Journal of Disaster Risk Science*, 5 (2), 157–63.

Sharma, A. and Davis, I. 2013. Seven golden rules of recovery management [online]. *Active Learning Network for Accountability and Performance in Humanitarian Action*, 14 November. Available at: www.alnap.org/blog/94 (accessed 5 February 2015).

Shi, P. 2012. On the role of government in integrated disaster risk governance: based on practices in China. *International Journal of Disaster Risk Science*, 3 (3), 139–46.

Shughart II, W. F. 2006. Katrinanomics: the politics and economics of disaster relief. *Public Choice*, 127 (1–2), 31–53.

Sibley, C. G. and Bulbulia, J. 2012. Faith after an earthquake: a longitudinal study of religion and perceived health before and after the 2011 Christchurch New Zealand earthquake [online]. *PLOS ONE*. http://journals.plos.org/plosone/article?id=10.1371/journal.pone.0049648 (accessed 5 February 2015).

Smith Wiltshire, D. 2004. *Voices of Victims and their Families Five Years after Hurricane Mitch*. Panama City, Panama: Centro de Coordinación para la Prevención de los Desastres Naturales en América Central (CEPREDENAC) and Japan International Cooperation Agency (JICA).

Soliman, H. and Cable, S. 2011. Sinking under the weight of corruption: neoliberal reform, political accountability and justice. *Current Sociology*, 59 (6), 735–53.

Spaliviero, M. (ed.). 2007. *National Baseline Report for Mozambique: Legal, Policy and Institutional Framework for Sustainable Land-Use Planning, Land-Use Management and Disaster Management*. Maputo, Mozambique: UN-Habitat.

Sphere Project. 2011. Minimum standards in shelter, settlement and non-food items. In *Humanitarian Charter and Minimum Standards in Humanitarian Response*. Geneva: Sphere Project, pp. 239–86.

Starr, C. 1969. Societal benefit versus technological risk. *Science*, 165 (3899), 1232–8.

Stephenson, R. S. 1991. *Disasters and Development*. Washington, DC: Pan American Health Organisation.

Stough, L. M. and Mayhorn, C. B. 2013. Population segments with disabilities. *International Journal of Mass Emergencies and Disasters*, 31 (3), 384–402.

Strachnyi, K. 2012. *Risk Management Quotes*. Available at: http://riskarticles.com/wp-content/uploads/2013/12/Risk-Management-Quotes-eBook.pdf (accessed 5 February 2015).

Szakats, G. A. J. 2006. *Improving the Earthquake Resistance of Small Buildings, Houses and Community Infrastructure.* Wellington, New Zealand: AC Consulting Group.

Tatham, P. 2012. Some reflections on the breadth and depth of the field of humanitarian logistics and supply chain management. *Journal of Humanitarian Logistics and Supply Chain Management,* 2 (2), 108–11.

Thawnghmung, A. M. 2013. *The 'Other' Karen in Myanmar: Ethnic Minorities and the Struggle Without Arms.* New York: Asiaworld, Lexington Books.

Thompson, C. and Thompson, P. M. 1976. *Reconstruction of Housing in Guatemala: A Survey of Programs Proposed after the Earthquake of February 1976.* Geneva: UNDRO-Intertect.

Tierney, K. 2008. Hurricane Katrina: catastrophic impacts and alarming lessons. In Quigley, J. M. and Rosenthal, L. M. (eds), *Risking House and Home: Disasters, Cities, Public Policy.* Berkeley, California: Institute of Governmental Studies, Berkeley Public Policy Press, pp. 119–36.

Time. 1976. Moss, the tentmaker. *Time* magazine, 26 July, 108 (4): 60.

Tosatti, G. (ed.). 2003. *A Review of the Scientific Contributions on the Stava Valley Disaster (Eastern Italian Alps), 19th July 1985.* Bologna: Consiglio Nazionale delle Ricerche, Pitagora Editrice.

Transparency International. 2013. *Corruption Perceptions Index 2013.* Berlin: Transparency International.

Trescatherick, B. 1989. *Barrow's Home Front 1939–1945.* Kendal, UK: Titus Wilson.

Turner, J. F. C. 1976. *Housing by People: Towards Autonomy in Building Environments.* London: Marion Boyars.

Turner, J. F. C. and Fitcher, R. (eds). 1972. *Freedom to Build: Dweller Control of the Housing Process.* New York: Macmillan.

Twain, M. 1924 [1887]. *Mark Twain's Autobiography.* Berkeley, California: University of California Press (reprinted).

UNDP. 1970. *Skopje Resurgent: The Story of a United Nations Special Fund Town Planning Project.* New York: United Nations Development Programme.

UNDP. 2010. *Human Development Report 2010. The Real Wealth of Nations: Pathways to Human Development.* Nairobi: United Nations Development Programme.

UNDRO. 1982. *Shelter after Disaster: Guidelines for Assistance.* Geneva: Office of the United Nations Disaster Relief Co-ordinator.

UNISDR. 2005. *Hyogo Framework for Action 2005–2015: Building the Resilience of Nations and Communities to Disasters.* Geneva: United Nations International Strategy for Disaster Reduction.

UNISDR. 2009. *UNISDR Terminology on Disaster Risk Reduction.* Geneva: United Nations International Strategy for Disaster Reduction. Available at: www.unisdr.org/files/7817_UNISDRTerminologyEnglish.pdf (accessed 5 February 2015).

UNISDR. 2013. *Making Cities Resilient: Summary for Policymakers.* Geneva: United Nations International Strategy for Disaster Reduction.

UNOCHA. 2006. *Exploring Key Changes and Developments in Post-Disaster Settlement, Shelter and Housing, 1982–2006.* Geneva: UN Office for the Coordination of Humanitarian Affairs.

US Census Bureau. 2015. State and county quick facts [online]. US Census Bureau. Available at: http://quickfacts.census.gov/qfd/states/22/2255000.html (accessed 5 February 2015).

US Chamber of Commerce. 2012. *The Role of Business in Disaster Response.* Washington, DC: US Chamber of Commerce.

US Government. 2009. Chengdu courts held trials of earthquake activists [online]. *Congressional Executive Commission on China,* 6 November 2009. Available at: www.cecc. gov/publications/commission-analysis/chengdu-courts-hold-trials-of-earthquake-activists (accessed 5 February 2015).

Vale, L. J. and Campanella, T. J. (eds). 2005. *The Resilient City: How Modern Cities Recover from Disaster.* New York: Oxford University Press.

van der Gaag, N. and collaborators. 2013. *Because I am a Girl. The State of the World's Girls, 2013. In Double Jeopardy: Adolescent Girls and Disasters.* Woking, UK: Plan International.

Van Essen, M., Van den Berg, G. and Pietersma, P. 2003. *Key Management Models: The 60+ Models Every Manager Needs to Know.* Harlow, UK: Pearson Education.

Versluis, A. 2014. Formal and informal material aid following the 2010 Haiti earthquake as reported by camp dwellers. *Disasters,* 38 (S1), S94–S109.

Walker, B. 1998. Resilience, instability and disturbance in ecosystem dynamics. *Environment and Development Economics,* 3 (2), 221–62.

Walker, P. 1996. Whose disaster is it anyway? Rights, responsibilities and standards in crisis. *Journal of Humanitarian Assistance,* 13 August. Available at: https://sites.tufts.edu/jha/archives/100 (accessed 8 May 2015).

Weber, L. and Peek, L. 2012. Documenting displacement: an introduction. In Weber, L. and Peek, L. (eds), *Displaced: Life in the Katrina Diaspora.* Austin, Texas: University of Texas Press, pp. 1–20.

Webster, B. 2014. We can't protect everyone from flooding, experts say. *The Times,* 7 March 2013.

Weichselgartner, J. 2001. Disaster mitigation: the concept of vulnerability revisited. *Disaster Prevention and Management,* 10 (2), 85–94.

Wiles. P., Selvester, K. and Fidalgo, L. 2005. *Learning Lessons from Disaster Recovery: The Case of Mozambique.* Disaster Risk Management Working Paper Series, No. 12. Washington, DC: World Bank. Available at: www.recoveryplatform.org/assets/publication/mozambique.pdf (accessed 5 February 2015).

Wisner, B., Blaikie, P., Cannon, T. and Davis, I. 2004. *At Risk: Natural Hazards, People's Vulnerability and Disasters* (second edition). Abingdon, UK: Routledge.

World Bank. 2013. Cyclone devastation averted: India weathers Phailin [online]. *World Bank,* 17 October. Available at: www.worldbank.org/en/news/feature/2013/10/17/india-cyclone-phailin-destruction-preparation (accessed 5 February 2015).

Yap, D. J. 2013. Days after Yolanda's wrath, looting erupts in Tacloban City. *Philippine Daily Enquirer,* 10 November. Available at: http://newsinfo.inquirer.net/524731/days-after-yolandas-wrath-looting-erupts-in-tacloban-city (accessed 5 February 2015).

Young, S., Balluz, L. and Malilay, J. 2004. Natural and technologic hazardous material releases during and after natural disasters: a review. *Science of the Total Environment,* 322 (1–3), 3–20.

EPILOGUE

.כַּאֲהָרֵים וַאֲפֵלָתְךָ ,אוֹרֶךָ בַּחֹשֶׁךְ וְזָרַח ;תַשְׁבִּיעַ נַעֲנָה וְנֶפֶשׁ ,נַפְשְׁךָ לָרָעֵב וְתָפֵק.

.מֵימָיו יְכַזְבוּ-לֹא אֲשֶׁר ,מַיִם וּכְמוֹצָא ,רָוָה כְגַן ,וְהָיִיתָ ;יַחֲלִיץ וְעַצְמֹתֶיךָ ,נַפְשְׁךָ בְּצַחְצָחוֹת וְהִשְׂבִּיעַ ,תָמִיד ,יְהוָה וְ

.לָשֶׁבֶת נְתִיבוֹת מְשֹׁבֵב ,פֶּרֶץ גֹּדֵר לְךָ וְקֹרָא ;תְּקוֹמֵם וָדוֹר-דוֹר מוֹסְדֵי ,עוֹלָם חָרְבוֹת מִמְּךָ וּבָנוּ יב

If you do away with the yoke of oppression ... and if you spend
yourselves in behalf of the hungry and satisfy the needs of the oppressed,
then your light will rise in the darkness, and your night will become
like the noonday. The Lord will guide you always; he will satisfy your
needs in a sun-scorched land and will strengthen your frame.

You will be like a well-watered garden, like a spring whose waters
never fail. Your people will rebuild the ancient ruins and will raise
up the age-old foundations; you will be called 'repairer of broken walls',
'restorer of streets with dwellings'.

Isaiah, *Chapter 58, Verses 10–12*
(c. 748 BC)

گر هيچ نشانه نيست اندر وادى
بسيار اميدهاست در نوميدى
اى دل مبر اميد كه در روضهء جان
خرما دهدت شاخ در خت بيدى

When there's no sign of hope in the desert,
So much hope still lives inside despair.
Heart, don't kill that hope: Even willows bear
Sweet fruit in the garden of the soul.

Kolliyaat-e Shams-e Tabrizi
Maulana Jalalu-'d-din Muhammad i Rumi
(c. AD 1000)

INDEX